INSTRUCTOR'S MANUAL AND PRINTED TEST BANK
Laurie Hurley

Calculus and Its Applications

Marvin L. Bittinger
Indiana University Purdue University Indianapolis

Boston   San Francisco   New York
London   Toronto   Sydney   Tokyo   Singapore   Madrid
Mexico City   Munich   Paris   Cape Town   Hong Kong   Montreal

Reproduced by Pearson Addison-Wesley from electronic files supplied by the author.

Copyright © 2004 Pearson Education, Inc.
Publishing as Pearson Addison-Wesley, 75 Arlington Street, Boston, MA 02116

All rights reserved. This publication may be reproduced for classroom use only. Printed in the United States of America.

ISBN   0-321-17310-4

2 3 4 5 6 VHG 06 05 04 03

CONTENTS

I. CHAPTER TESTS, FORMS A-F, and FINAL EXAMINATION 1

There are six **NEW** test forms for each chapter and the final examination. All six test forms are modeled after the Chapter Tests in the text.

Included at the end of each chapter test and the final exams are both synthesis and technology-based exercises. The synthesis exercises are meant to be more challenging and they generally require students to go beyond the immediate objectives of the chapter. Synthesis exercises are separated from the main body of the test by a solid line. The technology-based exercises follow the synthesis exercises and are separated by a bold line. Both the synthesis and the technology-based exercises have been placed at the end of the tests to make it easy to omit them if the instructor wishes to do so.

Chapter 1 1

Chapter 2 37

Chapter 3 61

Chapter 4 85

Chapter 5 109

Chapter 6 133

Chapter 7 157

Final Examination 169

II. ANSWER KEYS FOR TESTS, FORMS A-F, AND FINAL EXAMINATION 217

III. ANSWERS TO THE EVEN-NUMBERED EXERCISE SETS IN THE TEXT 283

If the instructor wants the students to have all the answers for the exercises in the exercise sets in the text, these answers can be easily duplicated.

IV. ANSWERS TO THE EXTENDED TECHNOLOGY APPLICATIONS 341

These answers are provided for the Extended Technology Applications that occur at the end of each chapter in the text. The Extended Technology Applications are designed to consider interesting and topical situations in greater depth; make use of curve fitting, regression, and modeling using a grapher; and allow for possible group or collaborative learning.

Thanks go to Julie Stephansen for accuracy checking the manuscript. Thank you to Mike Penna for producing the numerous graphs and figures required for this manual and test bank. Thanks to Sheri Minkner for her prompt and accurate typing of the manuscript and keen eye for detail.

CALCULUS AND ITS APPLICATIONS

Name:

Chapter 1, Form A

1. *Heart Attacks and Cholesterol.* The following graph relates the annual heart attack rate per 10,000 men and their blood cholesterol level*.

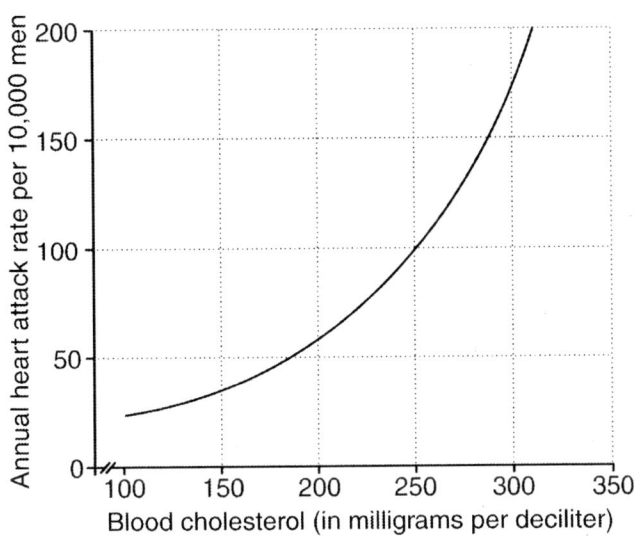

Use the graph to answer the following.

(a) What is the annual heart attack rate per 10,000 men for those whose blood cholesterol level is 275 mg/dl?

1. (a) _____

(b) What is the blood cholesterol level for men with a heart attack rate of 100 attacks per 10,000 individuals?

(b) _____

2. *Business: Compound Interest.* A person makes an investment at 4%, compounded annually. It has grown to $1310.40 at the end of 1 yr. How much was originally invested?

2. _____

3. A function is given by $f(x) = x^3 - 4$. Find (a) $f(-2)$ and (b) $f(x+h)$.

3. (a) _____

(b) _____

4. What are the slope and the y-intercept of $y = -\frac{1}{2}x - 5$?

4. _____

5. Find an equation of the line with slope $\frac{2}{3}$, containing the point $(3, -6)$.

5. _____

---

*Copyright 1989, CSPI. Adapted from *Nutrition Action Healthletter*

6. Find the slope of the line containing the points $(3, -2)$ and $(-6, -8)$.

6. _____

Find the average rate of change.

7.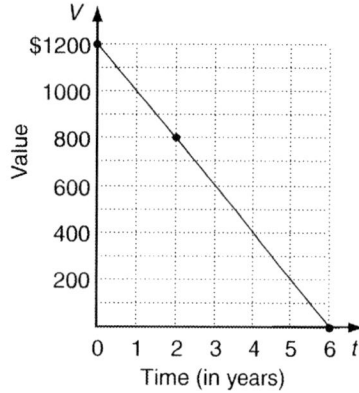

8.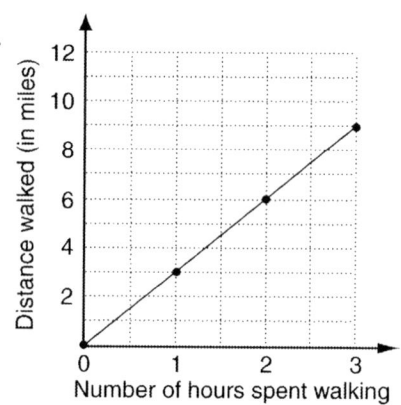

7. _____

8. _____

9. *Ohm's Law.* The electrical current I, in amperes, in a circuit is directly proportional to the voltage V. When 45 volts are applied, the current is 15 amperes. Find an equation of variation expressing I as a function of V.

9. _____

10. *Business: Profit-and-Loss Analysis.* A manufacturing company has fixed costs of \$15,000 for producing a specialty soap mold. Thereafter, the variable costs are \$0.95 for each bar of soap produced in the mold. The revenue from each specialty soap is expected to be \$3.50.

   (a) Formulate a function $C(x)$ for the total cost of producing x specialty soaps.

   (b) Formulate a function $R(x)$ for the total revenue from the sale of x specialty soaps.

   (c) Formulate a function $P(x)$ for the total profit from the production and sale of x specialty soaps.

   (d) How many specialty soaps must the company sell in order to break even?

10. (a) _____

    (b) _____

    (c) _____

    (d) _____

11. *Economics: Equilibrium Point.* Find the equilibrium point for the demand and supply functions

$$D(p) = (p-4)^2, \quad 0 \le p \le 4,$$

and

$$S(p) = p^2 + 2p + 6.$$

11. _____

CALCULUS AND ITS APPLICATIONS Chapter 1, Form A

Use the vertical-line-test to determine whether each of the following is the graph of a function.

12. 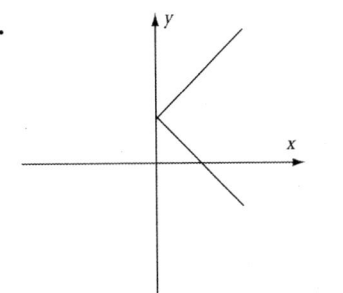   13.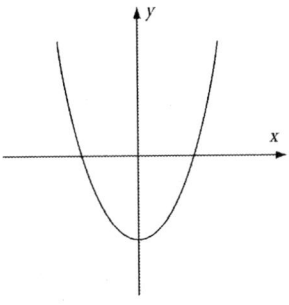

12. _____

13. _____

14. For the following graph of function f, determine (a) $f(1)$; (b) the domain; (c) all x-values such that $f(x) = 3$; and (d) the range.

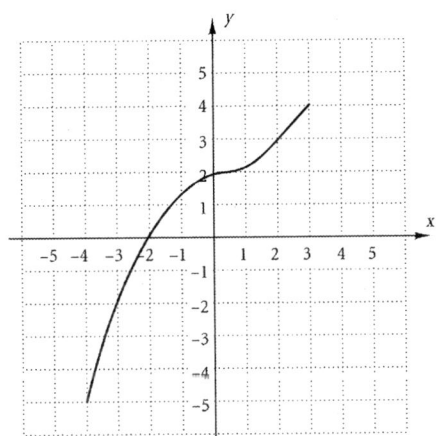

14. (a) _____

(b) _____

(c) _____

(d) _____

15. Graph: $f(x) = \dfrac{2}{x}$.

15.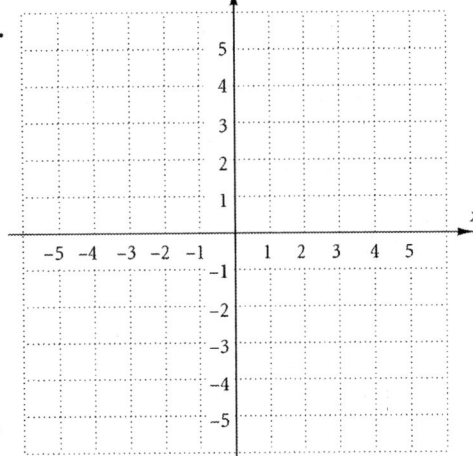

16. Convert to rational exponents: $\dfrac{1}{\sqrt[5]{x^4}}$.

16. _____

17. Convert to radical notation: $y^{-2/3}$.

17. _____

18. Graph: $f(x) = \dfrac{x^2 - 16}{x - 4}$.

18.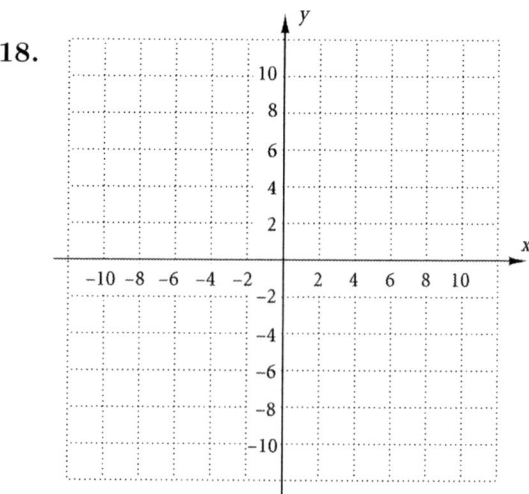

Determine the domain of the function.

19. $f(x) = \dfrac{x^2 + 4}{(x - 3)(x + 2)}$

19. _____

20. $f(x) = \sqrt{2x - 6}$

20. _____

21. Write interval notation for the following graph.

21. _____

22. Graph: $f(x) = \begin{cases} x^2 + 1, & \text{for } x \geq -1 \\ x - 2, & \text{for } x < -1 \end{cases}$

22.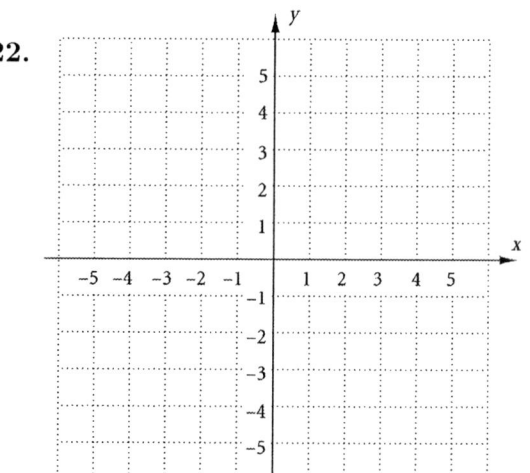

CALCULUS AND ITS APPLICATIONS   Chapter 1, Form A

23. *Milk Consumption.* The number of gallons of lowfat milk consumed by the average American in a year for several years is recorded in the following table.*

Year, x	Number of Gallons, g
1970	4.5
1975	7
1980	9
1985	11
1990	12.5
1995	12

(a) Using the data points $(1975, 7)$ and $(1995, 12)$, find a linear function that fits the data.

(b) Use the linear function to predict the number of gallons of lowfat milk the average American drank in 1988.

23. (a) _____

(b) _____

24. *Wind Friction.* Wind friction, or air resistance, increases with speed. The table below shows some measurements made in a wind tunnel.

Velocity, v (in kilometers per hour)	Force of Resistance, f (in newtons)
10	3
21	4.2
34	6.2
40	7.1
45	15.1
52	29.0

(a) Make a scatterplot of the data.

(b) Decide whether the data seem to fit a quadratic function.

(c) Using the data points $(10, 3)$, $(40, 7.1)$, and $(52, 29.0)$, find a quadratic function that fits the data.

(d) Use the function to estimate force of resistance when velocity is 35 km/h.

24. (a)
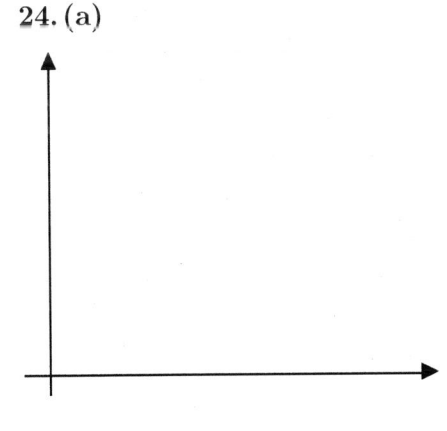

(b) _____

(c) _____

(d) _____

---

*Copyright 1990, CSPI. Adapted from *Nutrition Action Healthletter*

25. *Economics: Demand.* The demand function for a product is given by

$$x = D(p) = 1450 - 3p^3, \quad 0 \le p \le 7.85.$$

(a) Find the number of units sold when the price per unit is \$5.75.

(b) Find the price per unit when 1000 units are sold.

25. (a) _____

(b) _____

26. Use your grapher to graph: $f(x) = x^3 - 4x^2 + 2x - 3$. Then sketch the graph below.

26.
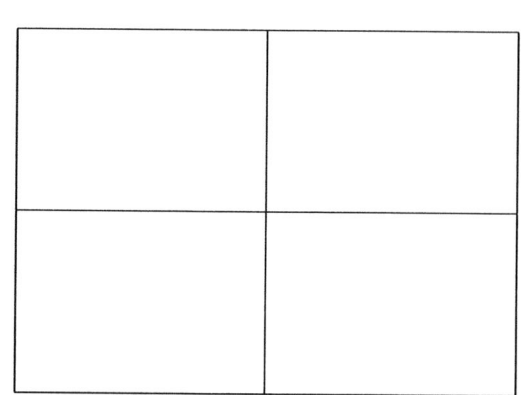

27. Find the zeros of the function: $f(x) = \sqrt[3]{|x^2 - 9|} - 3$.

27. _____

28. *Milk Consumption.* Use the data in Question 23.

(a) Use the REGRESSION feature to fit a linear function to the data.

(b) Use the linear function to predict the lowfat milk consumption of the average American for 1988.

28. (a) _____

(b) _____

CALCULUS AND ITS APPLICATIONS

Name:

Chapter 1, Form B

1. *Medicine.* The following graph relates the number of milligrams of ibuprofen in the bloodstream to the number of hours that have elapsed since 400 mg of the medication was swallowed.

   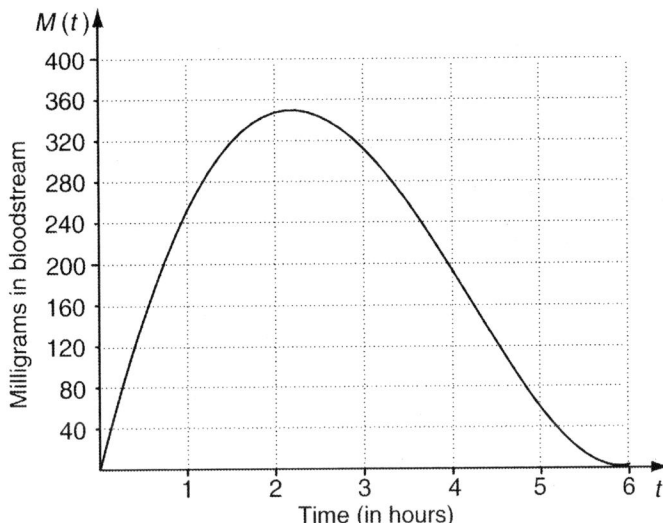

   Use the graph to answer the following.

   (a) How much ibuprofen is in the bloodstream 3 hr after swallowing 400 mg?

   1. (a) _____

   (b) For what lengths of time is there approximately 140 mg of ibuprofen in the bloodstream?

   (b) _____

2. *Business: Compound Interest.* A person makes an investment at 3%, compounded annually. It has grown to $824 at the end of 1 yr. How much was originally invested?

   2. _____

3. A function is given by $f(x) = 2x^3 + 4$. Find (a) $f(-2)$ and (b) $f(x+a)$.

   3. (a) _____

   (b) _____

4. What are the slope and the y-intercept of $y = -4x + 5$?

   4. _____

5. Find an equation of the line with slope $\frac{2}{5}$, containing the point $(5, -2)$.

   5. _____

6. Find the slope of the line containing the points $(4, -2)$ and $(8, 5)$.

6. _____

Find the average rate of change.

7.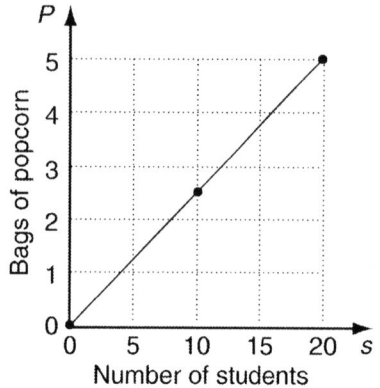

8.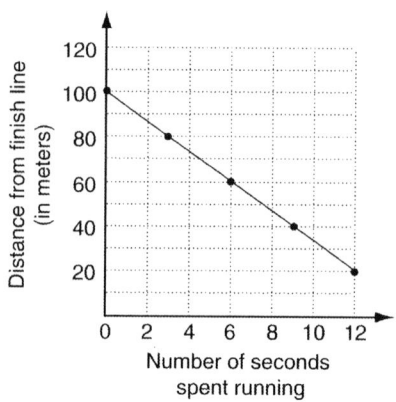

7. _____

8. _____

9. *Use of Aluminum Cans.* The number N of aluminum cans used each year is directly proportional to the number of people P using the cans. It is known that 250 people use 60,000 cans in one year. Find an equation of variation expressing N as a function of P.

9. _____

10. *Business: Profit-and-Loss Analysis.* A cookie company has fixed costs of $8000 for purchasing equipment for producing a new type of cookie. Variable costs are $0.76 for producing each cookie with the new equipment. The revenue from each cookie is expected to be $1.29.

   (a) Formulate a function $C(x)$ for the total cost of producing x cookies.

   (b) Formulate a function $R(x)$ for the total revenue from the sale of x cookies.

   (c) Formulate a function $P(x)$ for the total profit from the production and sale of x cookies.

   (d) How many cookies must the company sell in order to break even?

10. (a) _____

    (b) _____

    (c) _____

    (d) _____

11. *Economics: Equilibrium Point.* Find the equilibrium point for the demand and supply functions

$$D(p) = 5 - p, \quad 0 \leq p \leq 5,$$

and

$$S(p) = 2\sqrt{p+1}.$$

11. _____

CALCULUS AND ITS APPLICATIONS   Chapter 1, Form B

Use the vertical-line-test to determine whether each of the following is the graph of a function.

12. 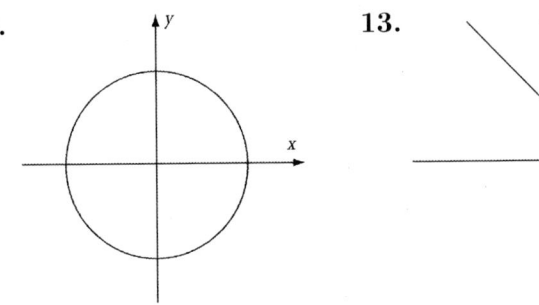   13.

12. _____

13. _____

14. For the following graph of function f, determine (a) $f(-1)$; (b) the domain; (c) all x-values such that $f(x) = 0$; and (d) the range.

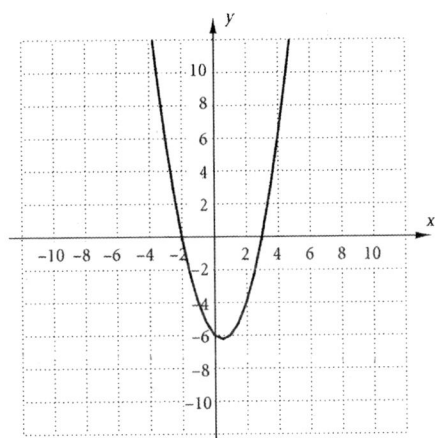

14. (a) _____

(b) _____

(c) _____

(d) _____

15. Graph: $f(x) = \dfrac{3}{x}$.

15.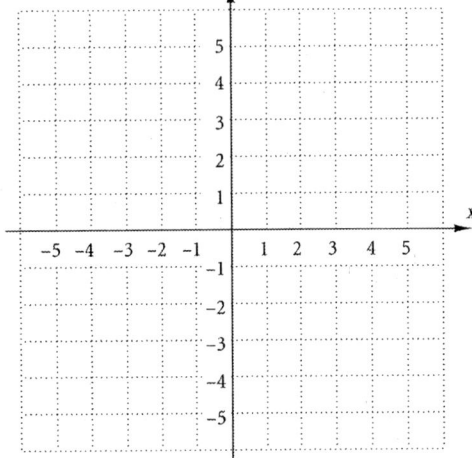

16. Convert to rational exponents: $\dfrac{1}{\sqrt[3]{y^2}}$.

16. _____

17. Convert to radical notation: $p^{-3/4}$.

17. _____

18. Graph: $f(x) = \dfrac{x^2 - x - 6}{x + 2}$.

18.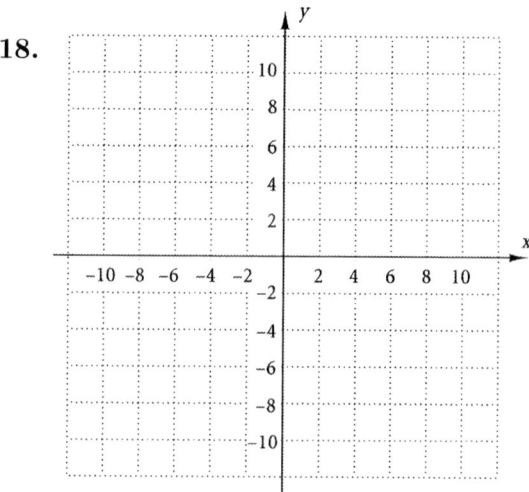

Determine the domain of the function.

19. $f(x) = \dfrac{x^2 - 1}{(x + 3)(x - 4)}$

19. _____

20. $f(x) = \dfrac{1}{\sqrt{2x + 3}}$

20. _____

21. Write interval notation for the following graph.

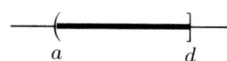

21. _____

22. Graph: $f(x) = \begin{cases} x^2 - 4, & \text{for } x \geq 0 \\ x + 1, & \text{for } x < 0 \end{cases}$

22.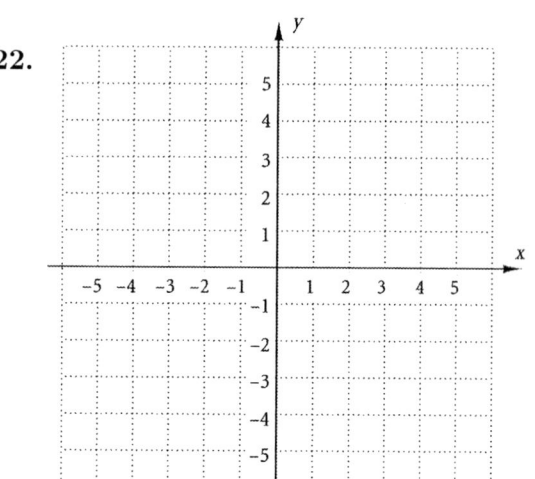

CALCULUS AND ITS APPLICATIONS Chapter 1, Form B

23. *Computer Prices.* The following table lists the average price, in dollars, of a personal computer at Jaytech in several recent years.

Year, x	Average Price, p
1988	1860
1991	1820
1994	1530
1997	1250
2000	800

(a) Using the data points $(1988, 1860)$ and $(1994, 1530)$ find a linear function that fits the data.

23. (a) _____

(b) Use the linear function to estimate the average price of a personal computer in 1992.

(b) _____

24. *Pizza Prices.* Pizza Unlimited has the following prices, in dollars, for pizzas of the given diameter, in inches.

24. (a)

Diameter, d	Price, p
6	5.00
8	6.00
12	8.50
16	11.50
24	18.50

(a) Make a scatterplot of the data.

(b) Decide whether the data seem to fit a quadratic function.

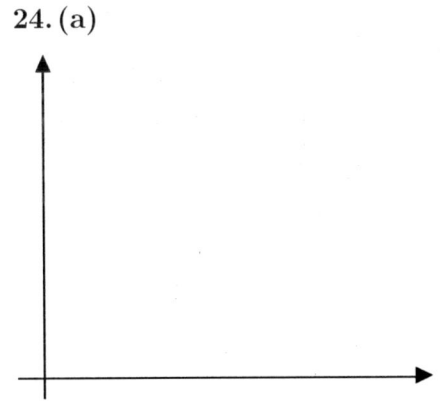

(b) _____

(c) Using the data points $(6, 5)$, $(16, 11.50)$, and $(24, 18.50)$, find a quadratic function that fits the data.

(c) _____

(d) Use the function to estimate the price of a pizza with a 10-in. diameter.

(d) _____

25. *Economics: Demand.* The demand function for a product is given by

$$x = D(p) = 1500 - p^3, \quad 0 \le p \le 11.45.$$

(a) Find the number of units sold when the price per unit is $9.00.

(b) Find the price per unit when 1000 units are sold.

25. (a) _____

(b) _____

26. Use your grapher to graph: $f(x) = 5x^3 - 6x^2 + x - 2$. Then sketch the graph below.

26.

27. Find the zeros of the function:

$$f(x) = \left|\sqrt{x^2 - 4} - 5\right| - 6.$$

27. _____

28. *Computer Prices.* Use the data in Question 23.

(a) Use the REGRESSION feature to fit a linear function to the data.

(b) Use the linear function to estimate the average price of a personal computer in 1992.

28. (a) _____

(b) _____

CALCULUS AND ITS APPLICATIONS

Name:

Chapter 1, Form C

1. *Memory.* The following graph relates the average number of words participants in a psychology experiment were able to memorize in a given amount of time.

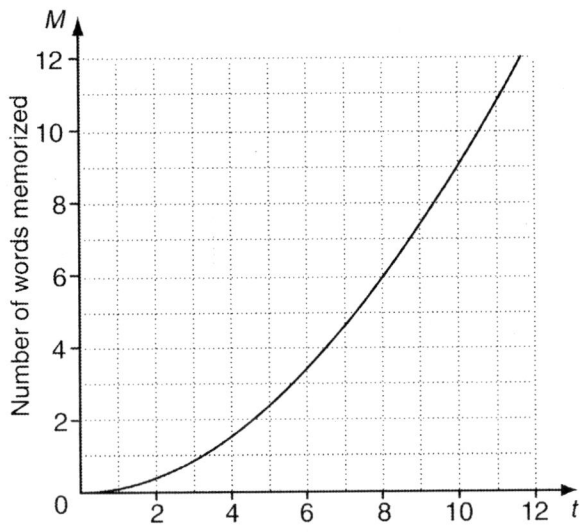

Use the graph to answer the following.

(a) Estimate the number of words memorized after 8 min.

1. (a) _____

(b) After how many minutes is the number of words memorized approximately 10?

(b) _____

2. *Business: Compound Interest.* A person makes an investment at 4%, compounded annually. It has grown to $1406.08 at the end of 1 yr. How much was originally invested?

2. _____

3. A function is given by $f(x) = 3x^2 - 5$. Find (a) $f(-7)$ and (b) $f(x+1)$.

3. (a) _____

(b) _____

4. What are the slope and the y-intercept of $y = -\frac{2}{3}x + 8$?

4. _____

5. Find an equation of the line with slope $-\frac{5}{8}$, containing the point $(4, 0)$.

5. _____

6. Find the slope of the line containing the points $(2, 9)$ and $(-5, -6)$.

6. _____

Find the average rate of change.

7.
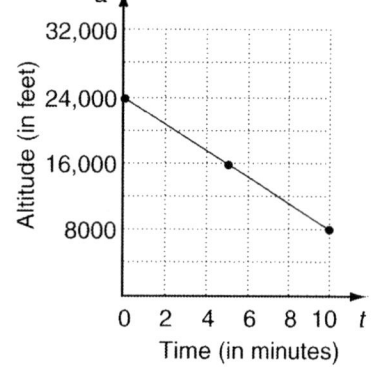

8.
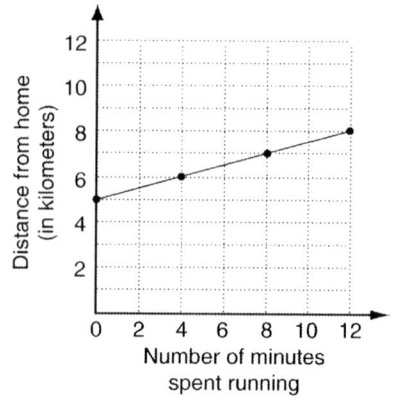

7. _____

8. _____

9. *Weight on the Moon.* The weight M of an object on the moon is directly proportion to its weight E on earth. A person who weighs 192 lb on Earth weighs 32 lb on the moon. Find an equation of variation expressing E as a function of M.

9. _____

10. *Business: Profit-and-Loss Analysis.* Workshop of Westfield is planning on producing a new model hammer. For the first year, the fixed costs are $24,000. The variable costs for producing each hammer are $12. The revenue from each hammer is expected to be $18.

  (a) Formulate a function $C(x)$ for the total cost of producing x hammers.

  (b) Formulate a function $R(x)$ for the total revenue from the sale of x hammers.

  (c) Formulate a function $P(x)$ for the total profit from the production and sale of x hammers.

  (d) How many hammers must the company sell in order to break even?

10. (a) _____

(b) _____

(c) _____

(d) _____

11. *Economics: Equilibrium Point.* Find the equilibrium point for the demand and supply functions

$$D(p) = (p - 5)^2, \quad 0 \le p \le 5,$$

and

$$S(p) = p^2 + p.$$

11. _____

CALCULUS AND ITS APPLICATIONS Chapter 1, Form C

Use the vertical-line-test to determine whether each of the following is the graph of a function.

12.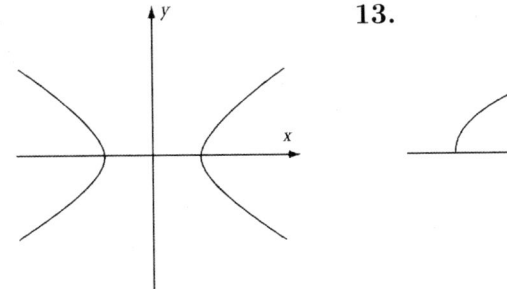

13.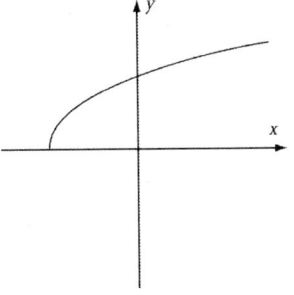

12. _____

13. _____

14. For the following graph of function f, determine (a) $f(-2)$; (b) the domain; (c) all x-values such that $f(x) = 1$; and (d) the range.

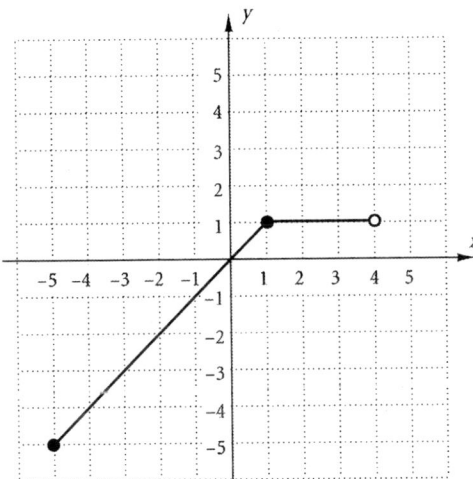

14. (a) _____

(b) _____

(c) _____

(d) _____

15. Graph: $f(x) = \dfrac{6}{(x-3)}$.

15.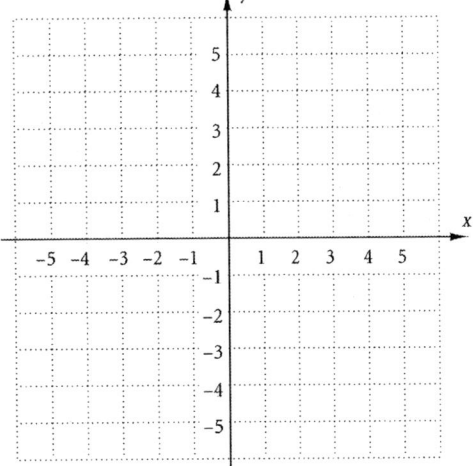

16. Convert to rational exponents: $\dfrac{5}{\sqrt[3]{x}}$.

16. _____

17. Convert to radical notation: $a^{-6/7}$.

17. _____

18. Graph: $f(x) = \dfrac{x^2 - 3x - 10}{x + 2}$.

18.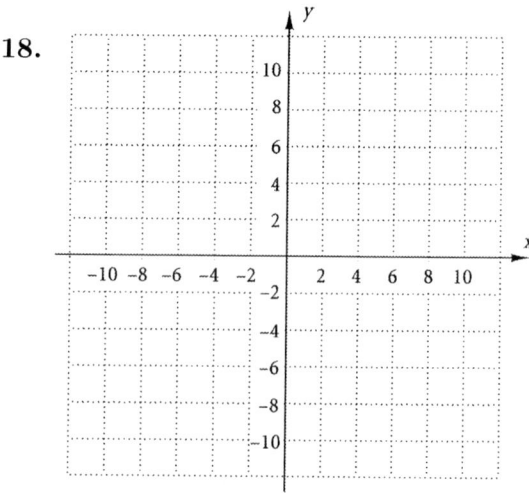

Determine the domain of the function.

19. $f(x) = \dfrac{x^2 - 3}{(x - 6)(x + 2)}$

19. _____

20. $f(x) = \sqrt{4 - 3x}$

20. _____

21. Write interval notation for the following graph.

21. _____

22. Graph: $f(x) = \begin{cases} -x^2 - 3, & \text{for } x \geq -2 \\ x - 1, & \text{for } x < -2 \end{cases}$

22.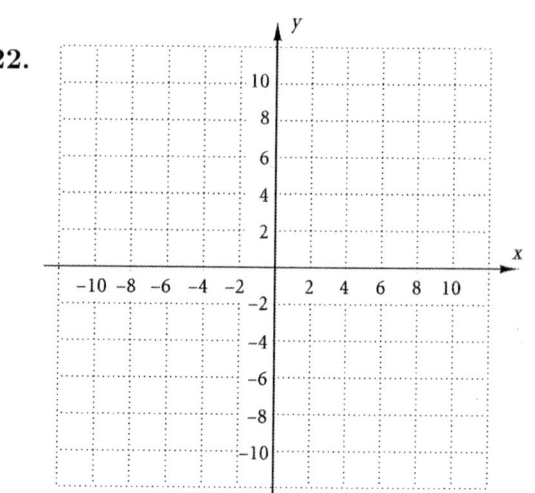

CALCULUS AND ITS APPLICATIONS Chapter 1, Form C

23. *Total Revenue.* The following table lists the revenue in thousands of dollars, of a small bike shop for various years.

Year, x	Total Revenue, R (in thousands)
1994	85
1995	94
1996	97
1998	100
2000	105

(a) Using the data points $(1994, 85)$ and $(1996, 97)$ find a linear function that fits the data.

(b) Use the linear function to predict the revenue of the bike shop in 2002.

23. (a) _____

(b) _____

24. *World Wide Web Sites.* The following table shows the growth of world wide web sites from 1995 through 2000.

Year, x	Number of World Wide Web Sites, W (in millions)
1995	5
1996	8
1997	30
1998	50
1999	80
2000	140

(*Source*: International Data Corporation, 1996)

(a) Make a scatterplot of the data.

(b) Decide whether the data seem to fit a quadratic function.

(c) Using the data points $(1996, 8)$, $(1998, 50)$, and $(2000, 140)$, find a quadratic function that fits the data.

(d) Use the function to estimate the number of web sites in 2005.

24. (a)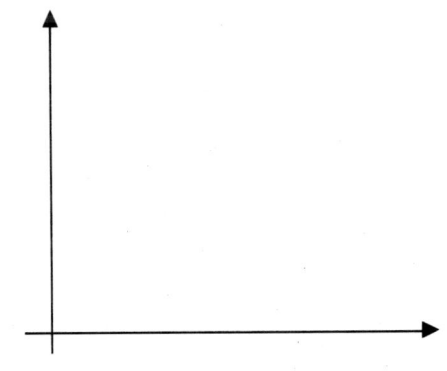

(b) _____

(c) _____

(d) _____

25. *Economics: Demand.* The demand function for a product is given by

$$x = D(p) = 1400 - 2p^3, \ 0 \leq p \leq 8.88.$$

(a) Find the number of units sold when the price per unit is $8.00.

25. (a) _____

(b) Find the price per unit when 900 units are sold.

(b) _____

26. Use your grapher to graph: $f(x) = 2x^3 - 5x^2 + x - 4$. Then sketch the graph below.

26.

27. Use a grapher to find the zeros of the function:

$$f(x) = \left|\sqrt{x^2 - 1} - 3\right| - 5.$$

27. _____

28. *Total Revenue.* Use the data in Question 23.

(a) Use the REGRESSION feature to fit a linear function to the data.

28. (a) _____

(b) Use the linear function to predict the revenue of the bike shop in 2002.

(b) _____

CALCULUS AND ITS APPLICATIONS

Name:

Chapter 1, Form D

1. *Height of a Projectile.* The following graph relates the height H, in feet, of a projectile with an initial velocity of 64 ft/sec to the number of seconds t after a launch.

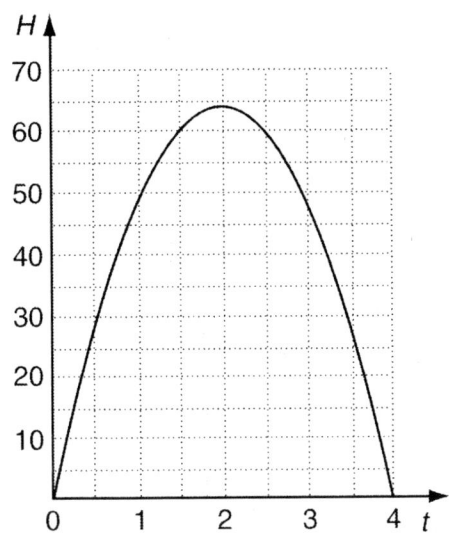

   Use the graph to answer the following.

   (a) What is the height of the projectile 3 sec after launch?

   1. (a) _____

   (b) How many seconds after launch is the projectile 28 ft above ground?

   (b) _____

2. *Business: Compound Interest.* A person makes an investment at 1.5%, compounded annually. It has grown to $761.25 at the end of 1 yr. How much was originally invested?

   2. _____

3. A function is given by $f(x) = 2x^2 + 3$. Find (a) $f(-1)$ and (b) $f(a-3)$.

   3. (a) _____

   (b) _____

4. What are the slope and the y-intercept of $y = 5x - 4$?

   4. _____

5. Find an equation of the line with slope $\frac{5}{8}$, containing the point $(6, -2)$.

   5. _____

6. Find the slope of the line containing the points $(-3, 9)$ and $(5, -3)$.

6. _____

Find the average rate of change.

7.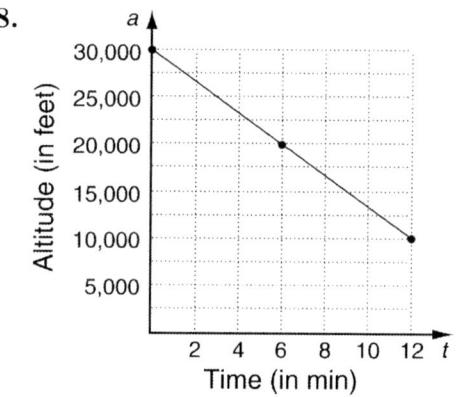

8.

7. _____

8. _____

9. *Weekly Allowance.* According to Fidelity Investments *Investment Vision Magazine*, the average weekly allowance A of children is directly proportional to their grade level G. Recently, the average allowance of a 9th-grade student was $9.66 per week. Find an equation of variation expressing A as a function of G.

9. _____

10. *Business: Profit-and-Loss Analysis.* Aldonna's is planning on producing a ladies' shoe. For the first year, the fixed costs are $135,000. The variable costs for producing each pair of shoes are $35. The revenue from the sale of each pair of shoes is expected to be $70.

   (a) Formulate a function $C(x)$ for the total cost of producing x pairs of shoes.

   (b) Formulate a function $R(x)$ for the total revenue from the sale of x pairs of shoes.

   (c) Formulate a function $P(x)$ for the total profit from the production and sale of x pairs of shoes.

   (d) How many pairs of shoes must Aldonna's sell in order to break even?

10. (a) _____

    (b) _____

    (c) _____

    (d) _____

11. *Economics: Equilibrium Point.* Find the equilibrium point for the demand and supply functions

$$D(p) = (p-5)^2, \quad 0 \leq p \leq 5,$$

and

$$S(p) = p^2 + 3p.$$

11. _____

CALCULUS AND ITS APPLICATIONS Chapter 1, Form D   21

Use the vertical-line-test to determine whether each
of the following is the graph of a function.

12.   13.   12. _____

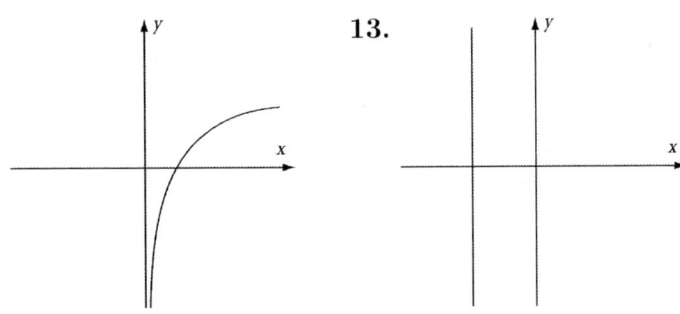

13. _____

14. For the following graph of function f, determine
(a) $f(1)$; (b) the domain; (c) all x-values such that
$f(x) = 2$; and (d) the range.

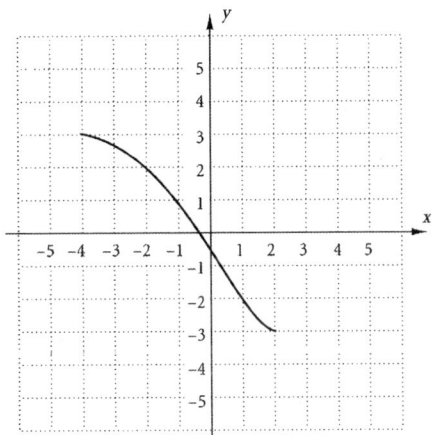

14. (a) _____

(b) _____

(c) _____

(d) _____

15. Graph: $f(x) = \dfrac{-3}{x-4}$.

15.

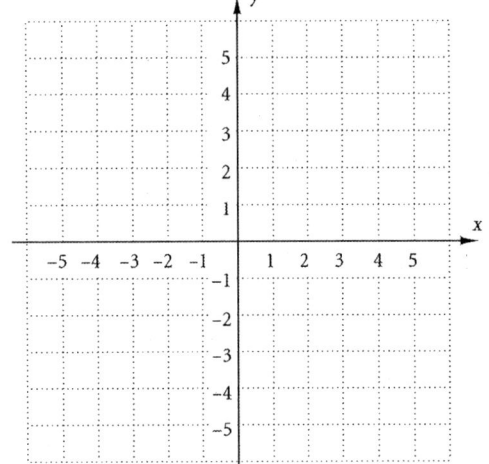

16. Convert to rational exponents: $\dfrac{5}{\sqrt[3]{n}}$.

16. _____

17. Convert to radical notation: $y^{-4/7}$.

17. _____

18. Graph: $f(x) = \dfrac{x^2 - 9}{x + 3}$.

18.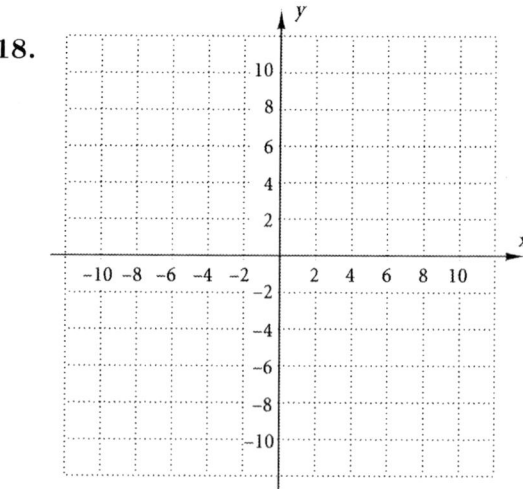

Determine the domain of the function.

19. $f(x) = \dfrac{x^2 + 4x}{(x-4)(x+8)}$

19. _____

20. $f(x) = \dfrac{1}{\sqrt{3-x}}$

20. _____

21. Write interval notation for the following graph.

21. _____

22. Graph: $f(x) = \begin{cases} x^2 - 4, & \text{for } x > 1 \\ x + 3, & \text{for } x \leq 1 \end{cases}$

22.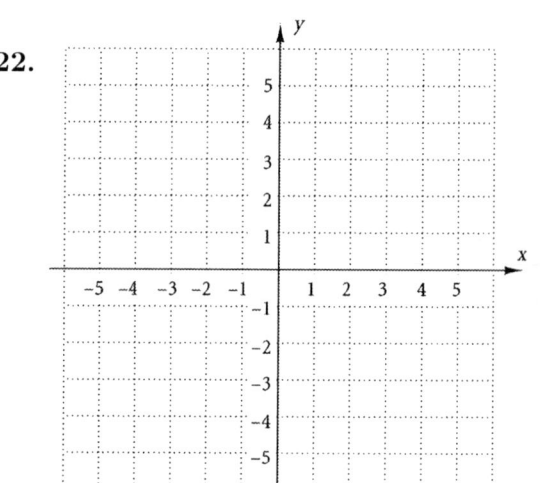

CALCULUS AND ITS APPLICATIONS   Chapter 1, Form D

23. *Average Income.* The following table shows the average yearly income, in dollars, of individuals based on years of schooling.

Years of Schooling, x	Average Income, I
8	16,000
10	19,000
12	25,000
14	28,000

(a) Using the data points $(8, 16{,}000)$ and $(14, 28{,}000)$ find a linear function that fits the data.

23. (a) _____

(b) Use the linear function to predict average income for an individual with 16 years of schooling.

(b) _____

24. *United States Farms.* The following table shows the number of farms, in millions, in the U.S. for various years.

Number of Years since 1850, x	Number of Farms (in millions), f
0	2.1
20	3.2
40	5.0
60	6.7
80	6.6
100	5.8
120	3.5
140	2.1

(*Source:* The Macmillan Visual Almanac, 1996 Statistical Abstract of the United States, 1997)

(a) Make a scatterplot of the data.

24. (a)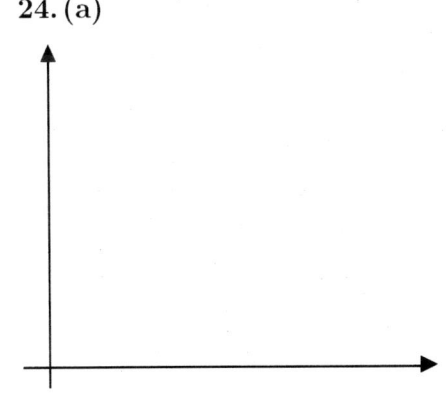

(b) Decide whether the data seem to fit a quadratic function.

(b) _____

(c) Using the data points $(0, 2.1)$, $(60, 6.7)$, and $(120, 3.5)$, find a quadratic function that fits the data.

(c) _____

(d) Use the function to estimate the number of farms in the U.S. in 1980, 130 years after 1850.

(d) _____

25. *Economics: Demand.* The demand function for a product is given by

$$x = D(p) = 1100 - 2p^3, \quad 0 \le p \le 8.19.$$

(a) Find the number of units sold when the price per unit is $7.25.

(b) Find the price per unit when 800 units are sold.

25. (a) _____

(b) _____

26. Use your grapher to graph: $f(x) = x^3 - 2x^2 - 2x + 3$. Then sketch the graph below.

26.

27. Find the zeros of the function:

$$f(x) = \left| \sqrt[3]{x^2 + 1} \right| - 4.$$

27. _____

28. *United States Farms.* Use the data in Question 24.

(a) Use the REGRESSION feature to fit a quadratic function to the data.

(b) Use the function to estimate the number of farms in the U.S. in 1980, 130 years after 1850.

28. (a) _____

(b) _____

CALCULUS AND ITS APPLICATIONS

Name:

Chapter 1, Form E

1. *Medical Concentration.* The concentration C, in parts per million, of a certain antibiotic in the bloodstream after t hours is illustrated in the following graph.

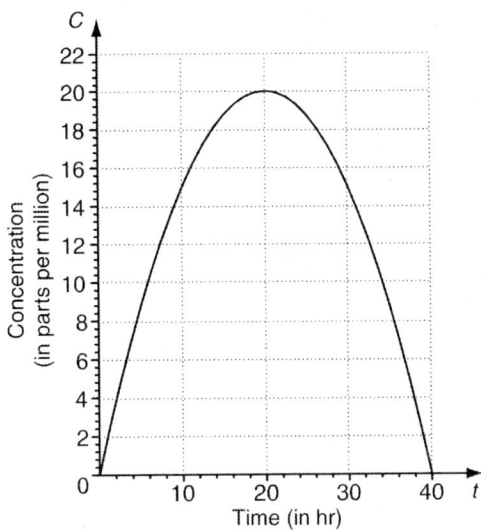

Use the graph to answer the following.

(a) Find the approximate concentration after 10 hr.

(b) For what lengths of time is the concentration approximately 8 parts per million?

1. (a) _____

(b) _____

2. *Business: Compound Interest.* A person makes an investment at 9%, compounded annually. It has grown to $1144.50 at the end of 1 yr. How much was originally invested?

2. _____

3. A function is given by $f(x) = 4x^2 + x$. Find (a) $f(-4)$ and (b) $f(x-2)$.

3. (a) _____

(b) _____

4. What are the slope and the y-intercept of $y = -2x - 6$?

4. _____

5. Find an equation of the line with slope $-\frac{1}{4}$, containing the point $(2, -8)$.

5. _____

6. Find the slope of the line containing the points $(-1, 1)$ and $(2, -8)$.

6. _____

Find the average rate of change.

7.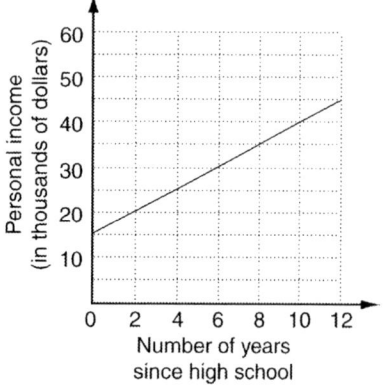

8.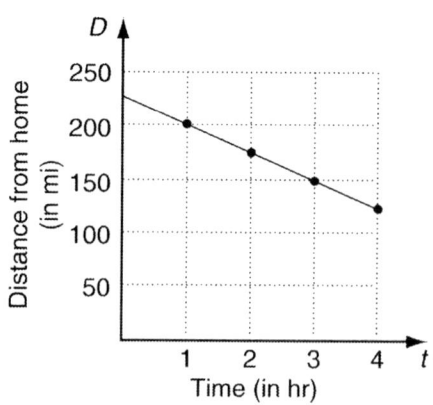

7. _____

8. _____

9. *Relative Aperture.* The relative aperture, or f-stop, A, of a 23.5-mm lens is directly proportional to the focal length F of the lens. A 150-mm focal length has an f-stop of 6.3. Find an equation of variation expressing A as a function of F.

9. _____

10. *Business: Profit-and-Loss Analysis.* Sweet Stuff is planning to introduce lollipops to their line of candies. For the first year, the fixed costs are $10,000. The variable costs for producing each hundred lollipops are estimated to be $10. The revenue from each hundred lollipops is expected to be $30.

    (a) Formulate a function $C(x)$ for the total cost of producing x hundred lollipops.

    10. (a) _____

    (b) Formulate a function $R(x)$ for the total revenue from the sale of x hundred lollipops.

    (b) _____

    (c) Formulate a function $P(x)$ for the total profit from the production and sale of x hundred lollipops.

    (c) _____

    (d) How many hundred lollipops must Sweet Stuff sell in order to break even?

    (d) _____

11. *Economics: Equilibrium Point.* Find the equilibrium point for the demand and supply functions

    $$D(p) = (p-4)^2, \quad 0 \le p \le 4,$$
    and
    $$S(p) = p^2 + p.$$

11. _____

CALCULUS AND ITS APPLICATIONS   Chapter 1, Form E

Use the vertical-line-test to determine whether each of the following is the graph of a function.

12.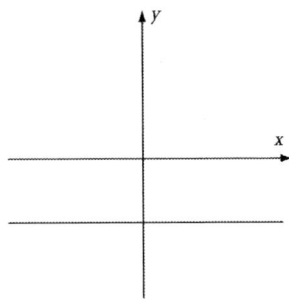

13.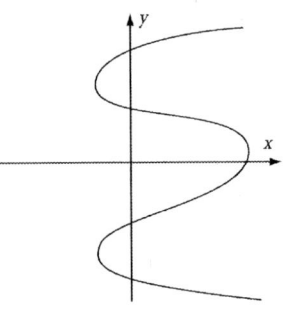

12. _____

13. _____

14. For the following graph of function f, determine (a) $f(3)$; (b) the domain; (c) all x-values such that $f(x) = 2$; and (d) the range.

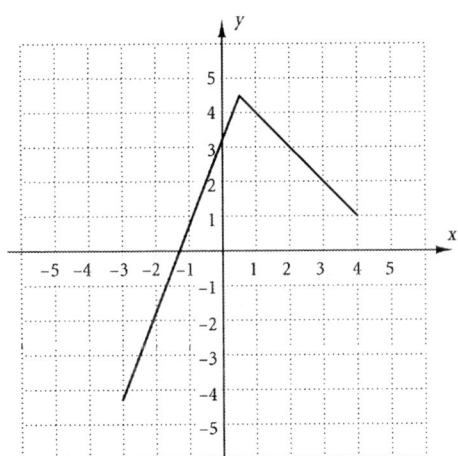

14. (a) _____

(b) _____

(c) _____

(d) _____

15. Graph: $f(x) = \dfrac{4}{x-4}$.

15.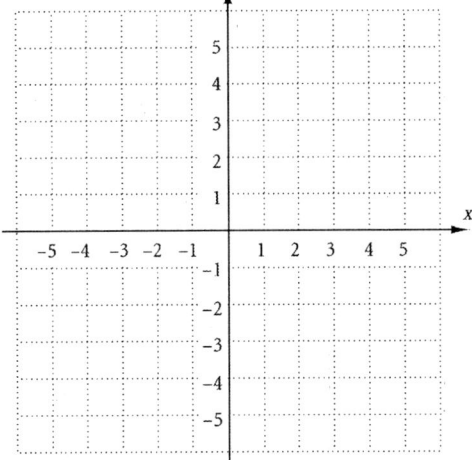

16. Convert to rational exponents: $\dfrac{2}{\sqrt[3]{p^2}}$.

16. _____

17. Convert to radical notation: $y^{-1/2}$.

17. _____

18. Graph: $f(x) = \dfrac{x^2 + 5x + 6}{x + 2}$.

18.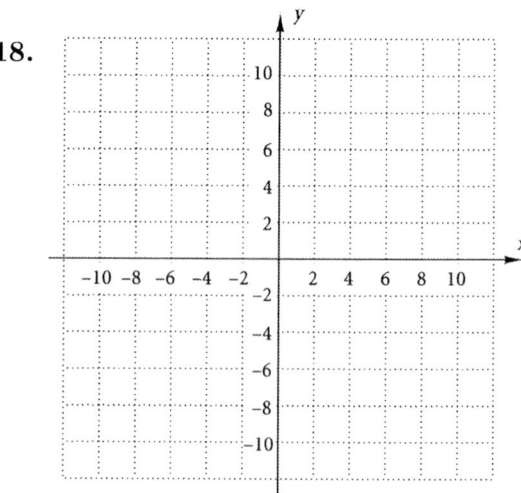

Determine the domain of the function.

19. $f(x) = \dfrac{x^2 + 4x}{(x + 2)(x - 5)}$

19. _____

20. $f(x) = \sqrt{3 - 2x}$

20. _____

21. Write interval notation for the following graph.

21. _____

22. Graph: $f(x) = \begin{cases} x^2 - 3, & \text{for } x \geq 0 \\ -2x, & \text{for } x < 0 \end{cases}$

22.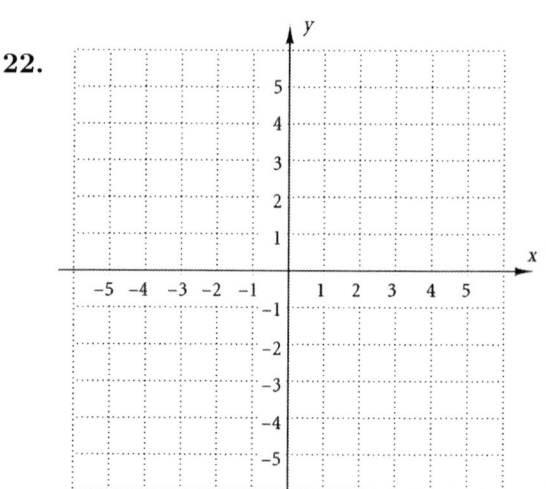

CALCULUS AND ITS APPLICATIONS   Chapter 1, Form E

23. *Test Grades.* The following table shows the study time for a particular chapter in a math class and the corresponding test grade for that chapter.

Study Time (in min), t	Test Grade (in percent), g
40	77
60	83
120	85
200	91
300	95

(a) Using the data points $(60, 83)$ and $(300, 95)$ find a linear function that fits the data.

23. (a)

(b) Use the linear function to predict the test grade for a student who studies 240 min.

(b) _____

24. *Household Income.* The following table shows the median U.S. household income for people of various ages.

Age, a	Median Income in 1996, i
19.5	21,438
29.5	35,888
39.5	44,420
49.5	50,472
59.5	39,815
65	19,448

(*Source:* U.S. Bureau of the Census; The Conference Board: Simmons Bureau of Labor Statistics)

(a) Make a scatterplot of the data.

(b) Decide whether the data seem to fit a quadratic function.

(c) Using the data points $(29.5, 35{,}888)$, $(39.5, 44{,}420)$, and $(65, 19{,}448)$, find a quadratic function that fits the data.

(d) Use the function to estimate the income of a U.S. household with a median age of 45.

24. (a)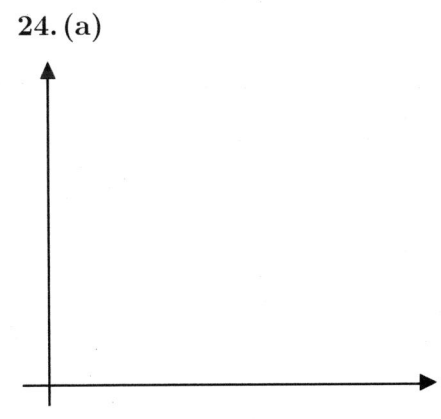

(b) _____

(c)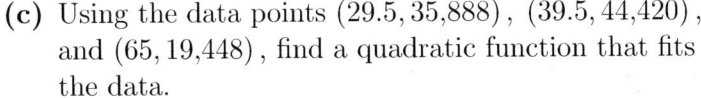

(d)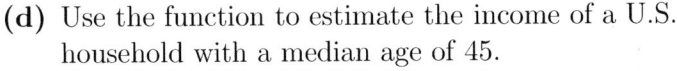

25. *Economics: Demand.* The demand function for a product is given by

$$x = D(p) = 1500 - 3p^3, \quad 0 \le p \le 7.94.$$

(a) Find the number of units sold when the price per unit is $5.00.

(b) Find the price per unit when 1000 units are sold.

25. (a) _____

(b) _____

26. Use your grapher to graph: $f(x) = 2x^3 - 3x^2 + x + 2$. Then sketch the graph below.

26.

27. Find the zeros of the function:

$$f(x) = \left| \sqrt[3]{2 - x^2} \right| - 2.$$

27. _____

28. *Test Grades.* Use the data in Question 23.

(a) Use the REGRESSION feature to fit a linear function to the data.

(b) Use the function to estimate the test grade of a student who studies 30 min.

28. (a) _____

(b) _____

CALCULUS AND ITS APPLICATIONS

Name:

Chapter 1, Form F

1. *Ozone Layer.* The following graph relates the ozone level in parts per billion to the year.

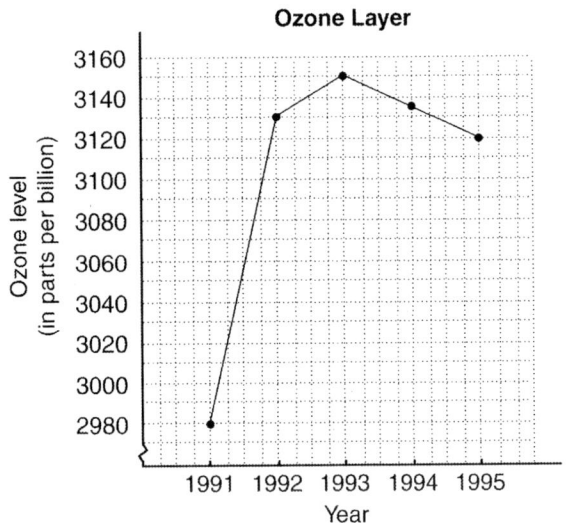

   Use the graph to answer the following.

   (a) What was the ozone level in 1993?

   (b) For what year is the ozone level 2980 parts per billion?

   1. (a) _____

   (b) _____

2. *Business: Compound Interest.* A person makes an investment at 2%, compounded annually. It has grown to $1785 at the end of 1 yr. How much was originally invested?

   2. _____

3. A function is given by $f(x) = 3x^2 - x$. Find (a) $f(-4)$ and (b) $f(x+a)$.

   3. (a) _____

   (b) _____

4. What are the slope and the y-intercept of $y = -2x + 3$?

   4. _____

5. Find an equation of the line with slope $-\frac{2}{3}$, containing the point $(5, -3)$.

   5. _____

6. Find the slope of the line containing the points $(-4, -2)$ and $(3, 10)$.

   6. _____

Find the average rate of change.

7.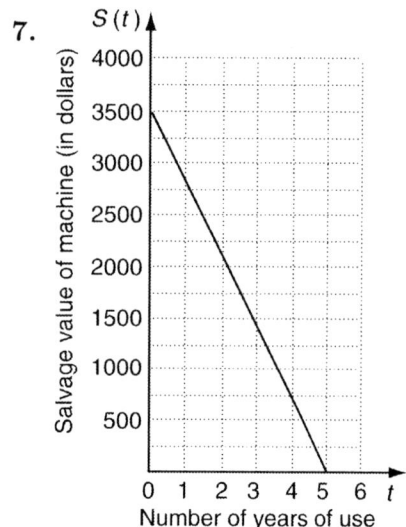

8.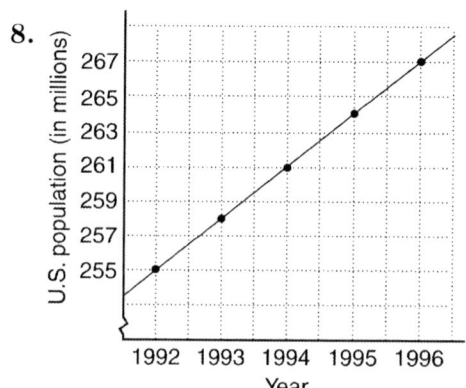

7. _____

8. _____

9. *Hooke's Law.* The distance d that a spring is stretched by a hanging object is directly proportional to the mass m of the object. A 5-kg object stretches a particular spring 28 cm. Find an equation of variation expressing d as a function of m.

9. _____

10. *Business: Profit-and-Loss Analysis.* Office Supplier, Inc. is planning on producing erasers. For the first year, the fixed costs are $5760. The variables costs are estimated to be $6 per dozen erasers. The revenue from each dozen erasers is expected to be $10.80.

   (a) Formulate a function $C(x)$ for the total cost of producing x dozen erasers.

   10. (a) _____

   (b) Formulate a function $R(x)$ for the total revenue from the sale of x dozen erasers.

   (b) _____

   (c) Formulate a function $P(x)$ for the total profit from the production and sale of x dozen erasers.

   (c) _____

   (d) How many dozen erasers must Office Supplier sell in order to break even?

   (d) _____

11. *Economics: Equilibrium Point.* Find the equilibrium point for the demand and supply functions

$$D(p) = \frac{6}{p}, \quad p > 0,$$

and

$$S(p) = 3p.$$

11. _____

CALCULUS AND ITS APPLICATIONS Chapter 1, Form F

Use the vertical-line-test to determine whether each of the following is the graph of a function.

12. 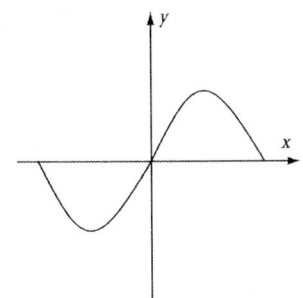   13.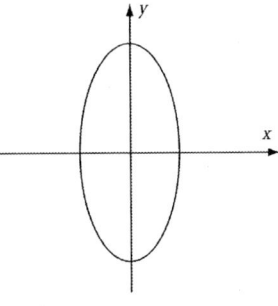

12. _____

13. _____

14. For the following graph of function f, determine
(a) $f(-2)$; (b) the domain; (c) all x-values such that $f(x) = 4$; and (d) the range.

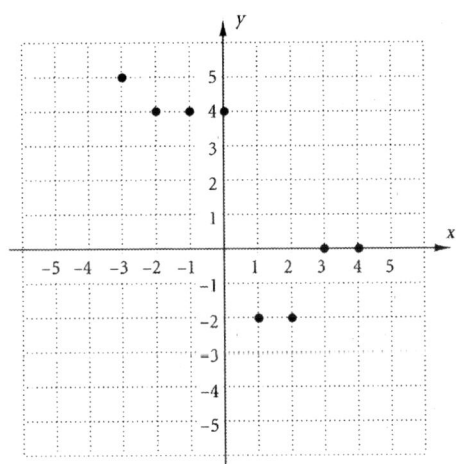

14. (a) _____

(b) _____

(c) _____

(d) _____

15. Graph: $f(x) = -\dfrac{2}{x}$.

15.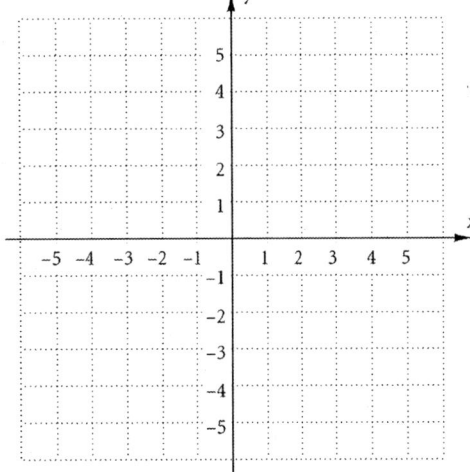

16. Convert to rational exponents: $\dfrac{5}{\sqrt{x}}$.

16. _____

17. Convert to radical notation: $m^{-5/4}$.

17. _____

18. Graph: $f(x) = \dfrac{x^2 + 2x - 8}{x + 4}$.

18.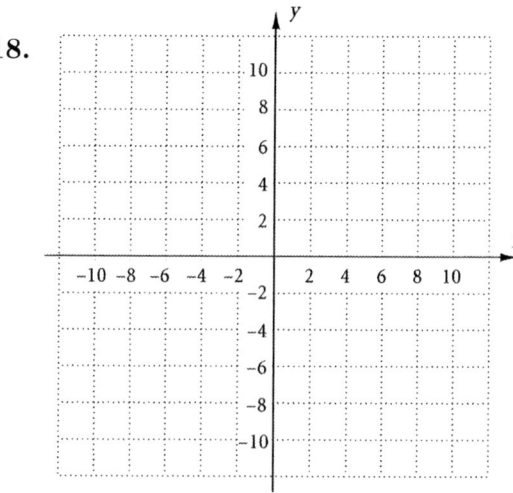

Determine the domain of the function.

19. $f(x) = \dfrac{x^2 - x}{(x + 1)(x - 2)}$

19. _____

20. $f(x) = \dfrac{1}{\sqrt{5 - x}}$

20. _____

21. Write interval notation for the following graph.

21. _____

22. Graph: $f(x) = \begin{cases} x^2 + 1, & \text{for } x > 0 \\ x - 2, & \text{for } x \leq 0 \end{cases}$

22.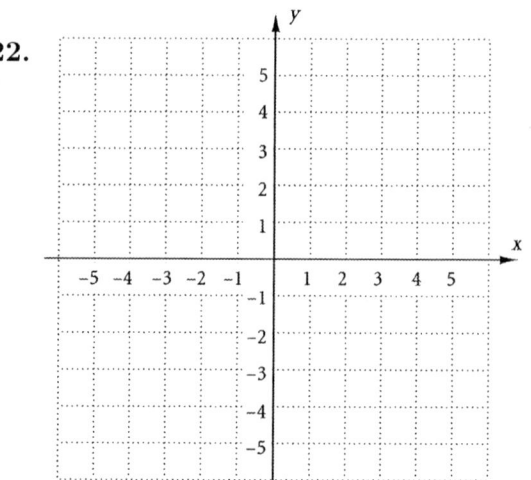

CALCULUS AND ITS APPLICATIONS Chapter 1, Form F

23. *Insurance Rates.* The following table shows the comparison of the cost, in dollars, of a $100,000 life insurance policy for female nonsmokers at certain ages.

Age, a	Cost, c
31	170
32	172
33	176
34	178
35	182

(a) Using the data points $(32, 172)$ and $(35, 182)$ find a linear function that fits the data.

23. (a) _____

(b) Use the linear function to predict the life insurance cost for a female nonsmoker of age 38.

(b) _____

24. *Small Business.* The following table shows the number of new small-business incorporations for various years.

Number of years since 1980, x	Number of New Incorporations, C (in thousands)
0	520
2	560
4	640
6	700
8	695
10	645
11	615

(*Source:* U.S. Small Business Administration)

(a) Make a scatterplot of the data.

24. (a)
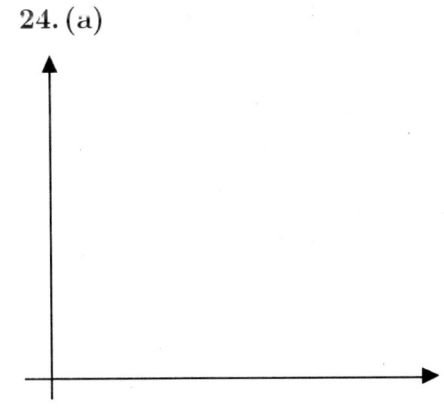

(b) Decide whether the data seem to fit a quadratic function.

(b) _____

(c) Using the data points $(2, 560)$, $(8, 695)$, and $(11, 615)$, find a quadratic function that fits the data.

(c) _____

(d) Use the function to estimate the number of new small-business incorporations 15 years after 1980.

(d) _____

25. *Economics: Demand.* The demand function for a product is given by

$$x = D(p) = 600 - \frac{1}{2}p^3, \quad 0 \le p \le 10.63.$$

(a) Find the number of units sold when the price per unit is $8.00.

25. (a) _____

(b) Find the price per unit when 500 units are sold.

(b) _____

26. Use your grapher to graph: $f(x) = -x^3 + 6x - 3$. Then sketch the graph below.

26.

27. Find the zeros of the function:

$$f(x) = \sqrt[3]{|10 - x^2|} - 4.$$

27. _____

28. *Small Business.* Use the data in Question 24.

(a) Use the REGRESSION feature to fit a quadratic function to the data.

28. (a) _____

(b) Use the function to estimate the number of new small business incorporations 15 years after 1980.

(b) _____

CALCULUS AND ITS APPLICATIONS

Name:

Chapter 2, Form A

Limits Graphically. Consider the following graph of function f for Questions 1-6.

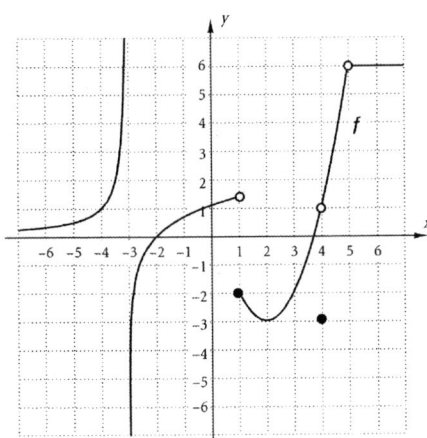

Find the limit, if it exists.

1. $\lim_{x \to -3} f(x)$

2. $\lim_{x \to -2} f(x)$

3. $\lim_{x \to 1} f(x)$

4. $\lim_{x \to 2} f(x)$

5. $\lim_{x \to 4} f(x)$

6. $\lim_{x \to 5} f(x)$

1. _____

2. _____

3. _____

4. _____

5. _____

6. _____

Determine whether the function is continuous.

7.
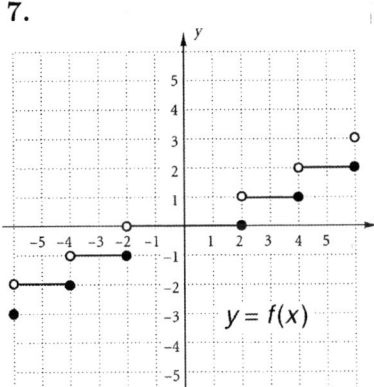

8.
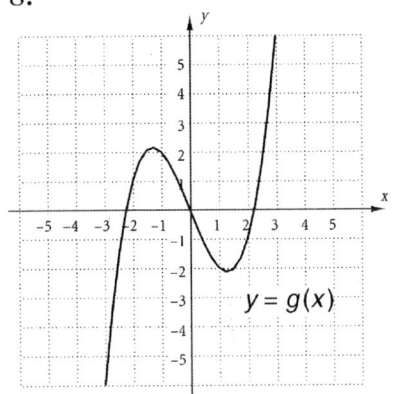

7. _____

8. _____

For the function in Question 7, answer the following.

9. Find $\lim_{x \to 1} f(x)$. 9. _____

10. Find $f(1)$. 10. _____

11. Is f continuous at 1? 11. _____

12. Find $\lim_{x \to 4} f(x)$. 12. _____

13. Find $f(4)$. 13. _____

14. Is f continuous at 4? 14. _____

Find the limit if it exists.

15. $\lim_{x \to 2} (2x^3 - x + 4)$ 15. _____

16. $\lim_{x \to -2^+} \dfrac{x + 2}{x^2 - 4}$ 16. _____

17. $\lim_{x \to 0} \dfrac{8}{x}$ 17. _____

18. Find a simplified difference quotient for
$$f(x) = 4x^2 - 3.$$ 18. _____

19. Find an equation of the tangent line to the graph of $y = 2x + \left(\dfrac{3}{x}\right)$ at the point $(1, 5)$. 19. _____

CALCULUS AND ITS APPLICATIONS Chapter 2, Form A

20. Find the points on the graph of $y = x^3 - 3x^2$ at which the tangent line is horizontal.

20. _____

Find dy/dx.

21. $y = x^{40}$

21. _____

22. $y = 3\sqrt{x}$

22. _____

23. $y = \dfrac{-8}{x^3}$

23. _____

24. $y = x^{4/5}$

24. _____

25. $y = -5x^2 + 0.4x + 16$

25. _____

Differentiate.

26. $y = \dfrac{3}{4}x^4 - 5x^2 + 4x + 1$

26. _____

27. $y = \dfrac{4x + 2}{x^2}$

27. _____

28. $f(x) = (x + 5)^5 (4 - x)^2$

28. _____

29. $y = (3x^4 - 4x^2 + 6)^{-5}$

29. _____

30. $f(x) = x^2 \sqrt{x^3 + 1}$

30. _____

31. For $y = 4x^6 - 9x^3$, find $\dfrac{d^3y}{dx^3}$.

31. _____

32. *Recreation.* The number of games N in a season in which each team plays every other team once is given by $N = \frac{1}{2}(n^2 - n)$, and n is the number of teams.

(a) Find the rate of change of number of games with respect to number of teams.

32. (a) _____

(b) How many games are played if there are 15 teams?

(b) _____

(c) Find the rate of change at $n = 15$ teams.

(c) _____

33. Find $(f \circ g)(x)$ and $(g \circ f)(x)$, given that $f(x) = x - 3x^2$ and $g(x) = x^3$.

33. _____

34. Differentiate $y = (5 - 4x)^{4/3}(5 + 4x)^{2/3}$.

34. _____

35. Find $\lim\limits_{x \to 2} \dfrac{x^3 - 8}{x - 2}$.

35. _____

36. Using a grapher, graph f and f' over the given interval. Then estimate points at which the tangent line to f is horizontal. Sketch the graphs.

$$f(x) = 4x^3 - 25x^2 + 32x + 4\sqrt{x};\ [0, 5]$$

36. _____

37. Find the following limit using tables on a grapher. Start with Step = 0.1 and then go to 0.01, 0.001, and 0.0001. When you think you know the limit, graph and use the TRACE feature to further verify your assertion.

$$\lim_{x \to 0} \dfrac{\sqrt{16 - 4x} - 4}{x}$$

37. _____

CALCULUS AND ITS APPLICATIONS

Name:

Chapter 2, Form B

Limits Graphically. Consider the following graph of function f for Questions 1-6.

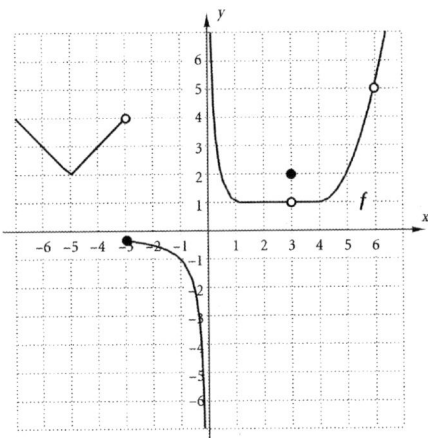

Find the limit, if it exists.

1. $\lim_{x \to -5} f(x)$

2. $\lim_{x \to -3} f(x)$

3. $\lim_{x \to 0} f(x)$

4. $\lim_{x \to 3} f(x)$

5. $\lim_{x \to 5} f(x)$

6. $\lim_{x \to 6} f(x)$

1. _____

2. _____

3. _____

4. _____

5. _____

6. _____

Determine whether the function is continuous.

7.

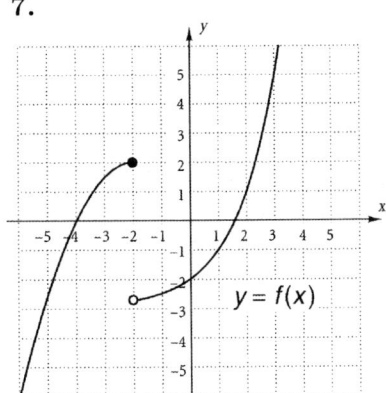

8.

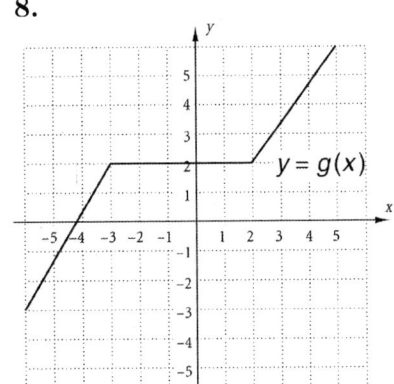

7. _____

8. _____

41

For the function in Question 7, answer the following.

9. Find $\lim_{x \to -2} f(x)$.

9. _____

10. Find $f(-2)$.

10. _____

11. Is f continuous at -2?

11. _____

12. Find $\lim_{x \to 2} f(x)$.

12. _____

13. Find $f(2)$.

13. _____

14. Is f continuous at 2?

14. _____

Find the limit if it exists.

15. $\lim_{x \to 3} (x^4 - 2x^2 + 3)$

15. _____

16. $\lim_{x \to -7^-} \dfrac{7 + x}{49 - x^2}$

16. _____

17. $\lim_{x \to 0} \dfrac{15}{x}$

17. _____

18. Find a simplified difference quotient for
$$f(x) = 3x^2 - 6.$$

18. _____

19. Find an equation of the tangent line to the graph of $y = 3x + \left(\dfrac{8}{x}\right)$ at the point $(2, 10)$.

19. _____

CALCULUS AND ITS APPLICATIONS   Chapter 2, Form B

20. Find the points on the graph of $y = x^3 - 2x^2$ at which the tangent line is horizontal.

20. _____

Find dy/dx.

21. $y = x^{16}$

21. _____

22. $y = 4\sqrt[5]{x}$

22. _____

23. $y = -\dfrac{11}{x^2}$

23. _____

24. $y = x^{3/5}$

24. _____

25. $y = -0.6x^2 + 4.1x - 100$

25. _____

Differentiate.

26. $y = \dfrac{2}{3}x^3 - 4x^2 + 10x + 6$

26. _____

27. $y = \dfrac{-4x}{5-x}$

27. _____

28. $f(x) = (x+1)^3 (6-x)^2$

28. _____

29. $y = \left(6x^2 + 2x^5 + x^6\right)^{-4}$

29. _____

30. $f(x) = x^2 \sqrt{x^3 - 2}$

30. _____

31. For $y = 5x^4 - x^2 + 6$, find $\dfrac{d^3y}{dx^3}$.

31. _____

32. *Surface Area.* The surface area S of a 8-in. high can is given by $S = 2\pi r^2 + 16\pi r$, where r is the length of the radius.

(a) Find the rate of change of surface area with respect to the length of the radius.

32. (a) _____

(b) What is the surface area of a can that is 8 in. high and has a radius of 2 in.?

(b) _____

(c) Find the rate of change at $r = 2$ in.

(c) _____

33. Find $(f \circ g)(x)$ and $(g \circ f)(x)$, given that $f(x) = x^2 + 4x$ and $g(x) = \sqrt{x}$.

33. _____

34. Differentiate $y = (2 - 3x)^{2/3}(5 + x)^{1/2}$.

34. _____

35. Find $\lim\limits_{x \to 3} \dfrac{27 - x^3}{3 - x}$.

35. _____

36. Using a grapher, graph f and f' over the given interval. Then estimate points at which the tangent line to f is horizontal. Sketch the graphs

$$f(x) = 6x^3 - 20x^2 + 10x + 3\sqrt{x}; \ [0, 5]$$

36. _____

37. Find the following limit using tables on a grapher. Start with Step $= 0.1$ and then go to 0.01, 0.001, and 0.0001. When you think you know the limit, graph and use the TRACE feature to further verify your assertion.

$$\lim_{x \to 0} \dfrac{\sqrt{x^2 + 1} - 1}{x}$$

37. _____

CALCULUS AND ITS APPLICATIONS

Name:

Chapter 2, Form C

Limits Graphically. Consider the following graph of function f for Questions 1-6.

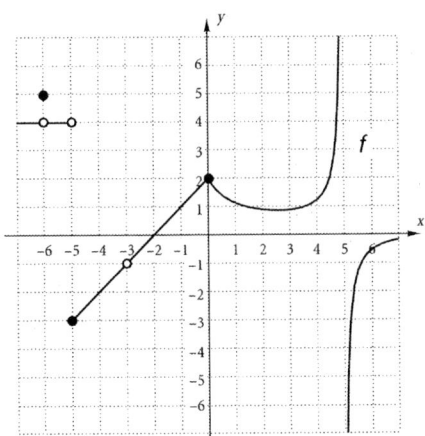

Find the limit, if it exists.

1. $\lim\limits_{x \to -5} f(x)$

2. $\lim\limits_{x \to -3} f(x)$

3. $\lim\limits_{x \to 0} f(x)$

4. $\lim\limits_{x \to -2} f(x)$

5. $\lim\limits_{x \to 5} f(x)$

6. $\lim\limits_{x \to -6} f(x)$

1. _____

2. _____

3. _____

4. _____

5. _____

6. _____

Determine whether the function is continuous.

7.
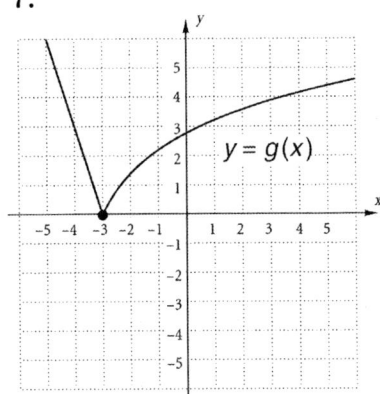

8.
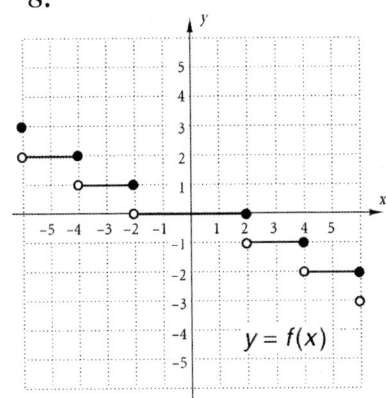

7. _____

8. _____

For the function in Question 8, answer the following.

9. Find $\lim_{x \to 2} f(x)$.

9. _____

10. Find $f(2)$.

10. _____

11. Is f continuous at 2?

11. _____

12. Find $\lim_{x \to 3} f(x)$.

12. _____

13. Find $f(3)$.

13. _____

14. Is f continuous at 3?

14. _____

Find the limit if it exists.

15. $\lim_{x \to -1^+} (3x^2 - x + 10)$

15. _____

16. $\lim_{x \to -6} \dfrac{x^2 + 3x - 18}{x + 6}$

16. _____

17. $\lim_{x \to 0} -\dfrac{3}{x}$

17. _____

18. Find a simplified difference quotient for
$$f(x) = 2x^2 + 5x.$$

18. _____

19. Find an equation of the tangent line to the graph of $y = 4x + \left(\dfrac{-10}{x}\right)$ at the point $(5, 18)$.

19. _____

20. Find the points on the graph of $y = 2x^3 - 3x^2$ at which the tangent line is horizontal.

20. _____

Find dy/dx.

21. $y = x^{15}$

21. _____

22. $y = 6\sqrt{x}$

22. _____

23. $y = -\dfrac{7}{x^7}$

23. _____

24. $y = x^{2/5}$

24. _____

25. $y = -3x^2 + 10.1x - 6$

25. _____

Differentiate.

26. $y = \dfrac{1}{10}x^5 + 3x^4 - 6x - 6$

26. _____

27. $y = \dfrac{3x + 1}{x^4}$

27. _____

28. $f(x) = (x+1)^3 (3-x)^4$

28. _____

29. $y = (6x^2 - 10x + 1)^{-4}$

29. _____

30. $f(x) = x^3 \sqrt{x^4 - 3}$

30. _____

31. For $y = 3x^6 - 4x^3$, find $\dfrac{d^3y}{dx^3}$.

31. _____

32. *Electing Officers.* For a club consisting of n people, the number of ways in which a president, vice-president, and treasurer can be elected is given by $P = n^3 - 3n^2 + 2n$.

(a) Find the rate of change in the number of ways to elect the 3 officers with respect to the number of club members.

(b) How many ways are there to elect the 3 officers when there are 20 club members?

(c) Find the rate of change at $n = 20$.

32. (a) _____

(b) _____

(c) _____

33. Find $(f \circ g)(x)$ and $(g \circ f)(x)$, given that $f(x) = 3 - x^2$ and $g(x) = x^2 + 5$.

33. _____

34. Differentiate $y = (6 - 3x)^{1/3} (10 + x)^{1/6}$.

34. _____

35. Find $\displaystyle\lim_{x \to -2} \dfrac{x^3 + 8}{x + 2}$.

35. _____

36. Using a grapher, graph f and f' over the given interval. Then estimate points at which the tangent line to f is horizontal. Sketch the graphs.

$$f(x) = 3x^5 - 15x^2 + 5x;\ [-3, 3]$$

36. _____

37. Find the following limit using tables on a grapher. Start with Step $= 0.1$ and then go to 0.01, 0.001, and 0.0001. When you think you know the limit, graph and use the TRACE feature to further verify your assertion.

$$\lim_{x \to 6} \dfrac{\sqrt{2x + 4} - 4}{x - 6}$$

37. _____

CALCULUS AND ITS APPLICATIONS

Name:

Chapter 2, Form D

Limits Graphically. Consider the following graph of function f for Questions 1-6.

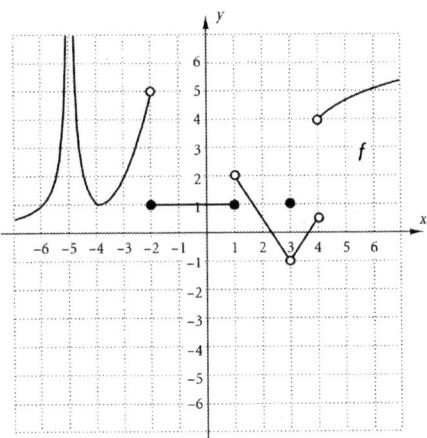

Find the limit, if it exists.

1. $\lim_{x \to -5} f(x)$

2. $\lim_{x \to -2} f(x)$

3. $\lim_{x \to 0} f(x)$

4. $\lim_{x \to 3} f(x)$

5. $\lim_{x \to 4} f(x)$

6. $\lim_{x \to -4} f(x)$

1. _____

2. _____

3. _____

4. _____

5. _____

6. _____

Determine whether the function is continuous.

7.
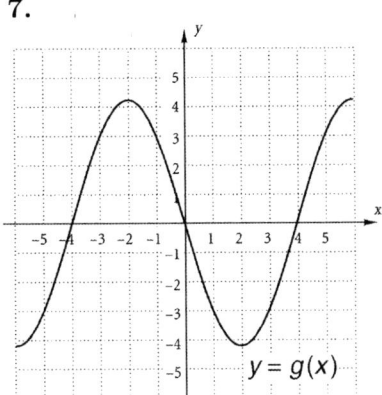

8.
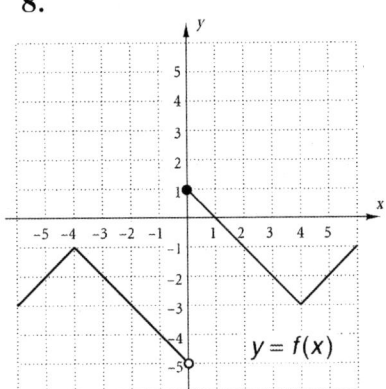

7. _____

8. _____

49

For the function in Question 8, answer the following.

9. Find $\lim\limits_{x \to -4} f(x)$.

9. _____

10. Find $f(-4)$.

10. _____

11. Is f continuous at -4?

11. _____

12. Find $\lim\limits_{x \to 0} f(x)$.

12. _____

13. Find $f(0)$.

13. _____

14. Is f continuous at 0?

14. _____

Find the limit if it exists.

15. $\lim\limits_{x \to -1} \left(2x^3 - 2x^2 + 5x\right)$

15. _____

16. $\lim\limits_{x \to -9^+} \dfrac{9 + x}{81 - x^2}$

16. _____

17. $\lim\limits_{x \to -6} \dfrac{6}{x + 6}$

17. _____

18. Find a simplified difference quotient for
$$f(x) = -2x^2 + 5.$$

18. _____

19. Find an equation of the tangent line to the graph of $y = -x + \left(\dfrac{4}{x}\right)$ at the point $(2, 0)$.

19. _____

CALCULUS AND ITS APPLICATIONS Chapter 2, Form D

20. Find the points on the graph of $y = 6x^3 + 9x^2$ at which the tangent line is horizontal.

20. _____

Find dy/dx.

21. $y = x^{16}$

21. _____

22. $y = 13\sqrt[3]{x}$

22. _____

23. $y = \dfrac{4}{x^8}$

23. _____

24. $y = x^{3/8}$

24. _____

25. $y = -6x^2 - 5x + 4.9$

25. _____

Differentiate.

26. $y = \dfrac{5}{8}x^8 - 4x^6 + 5x + 10$

26. _____

27. $y = \dfrac{x-4}{2-x}$

27. _____

28. $f(x) = (x+2)^4 (3-x)^2$

28. _____

29. $y = (4x^3 - 2x^2 + 5)^{-4}$

29. _____

30. $f(x) = x^3 \sqrt{x^2 - 1}$

30. _____

31. For $y = 2x^6 - 10x^2$, find $\dfrac{d^3y}{dx^3}$.

31. _____

32. *Volume of an icecream cone.* The volume of an icecream cone with a height of 4 in. is given by $V = \dfrac{4}{3}\pi r^2$.

(a) Find the rate of change of volume with respect to length of radius.

32. (a) _____

(b) What is the volume when the radius is 0.5 in.?

(b) _____

(c) Find the rate of change when $r = 0.5$ in.

(c) _____

33. Find $(f \circ g)(x)$ and $(g \circ f)(x)$, given that $f(x) = x^2 + 2x$ and $g(x) = \sqrt{x}$.

33. _____

34. Differentiate $y = (4 - 3x)^{6/5} (1 + x)^{2/5}$.

34. _____

35. Find $\displaystyle\lim_{x \to 5} \dfrac{x^3 - 125}{x - 5}$.

35. _____

36. Using a grapher, graph f and f' over the given interval. Then estimate points at which the tangent line to f is horizontal. Sketch the graphs.

$$f(x) = 2x^5 - 5x^2 - x + 2;\ [-3, 3]$$

36. _____

37. Find the following limit using tables on a grapher. Start with Step $= 0.1$ and then go to 0.01, 0.001, and 0.0001. When you think you know the limit, graph and use the TRACE feature to further verify your assertion.

$$\lim_{x \to 12} \dfrac{\sqrt{x - 3} - 3}{x - 12}$$

37. _____

CALCULUS AND ITS APPLICATIONS

Name:

Chapter 2, Form E

Limits Graphically. Consider the following graph of function f for Questions 1-6.

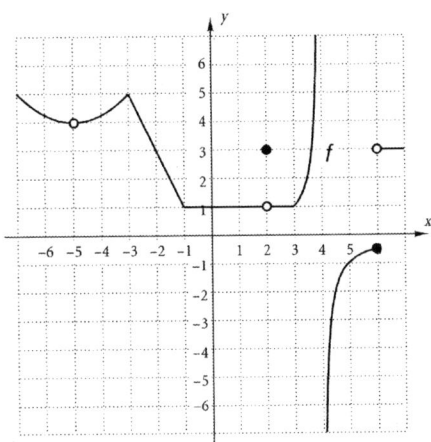

Find the limit, if it exists.

1. $\lim_{x \to -3} f(x)$

2. $\lim_{x \to 4} f(x)$

3. $\lim_{x \to 2} f(x)$

4. $\lim_{x \to 6} f(x)$

5. $\lim_{x \to -2} f(x)$

6. $\lim_{x \to -5} f(x)$

1. _____

2. _____

3. _____

4. _____

5. _____

6. _____

Determine whether the function is continuous.

7.

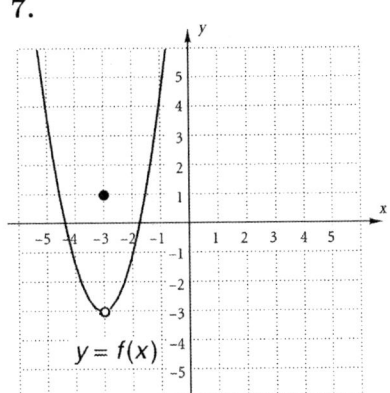

8.

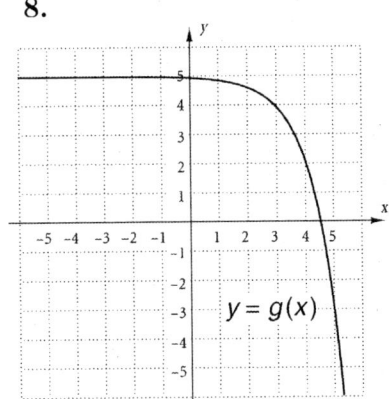

7. _____

8. _____

For the function in Question 7, answer the following.

9. Find $\lim\limits_{x \to -3} f(x)$.

9. _____

10. Find $f(-3)$.

10. _____

11. Is f continuous at -3?

11. _____

12. Find $\lim\limits_{x \to -5} f(x)$.

12. _____

13. Find $f(-5)$.

13. _____

14. Is f continuous at -5?

14. _____

Find the limit if it exists.

15. $\lim\limits_{x \to -3} \left(-x^3 - 3x^2 + 4\right)$

15. _____

16. $\lim\limits_{x \to 6^-} \dfrac{x - 6}{x^2 - 2x - 24}$

16. _____

17. $\lim\limits_{x \to -3} \dfrac{4}{x + 3}$

17. _____

18. Find a simplified difference quotient for
$$f(x) = -x^2 + 5x.$$

18. _____

19. Find an equation of the tangent line to the graph of $y = -2x + \left(\dfrac{6}{x}\right)$ at the point $(2, -1)$.

19. _____

CALCULUS AND ITS APPLICATIONS Chapter 2, Form E

20. Find the points on the graph of $y = x^3 - 2x^2$ at which the tangent line is horizontal.

20. _____

Find dy/dx.

21. $y = x^{106}$

21. _____

22. $y = 5\sqrt[3]{x}$

22. _____

23. $y = \dfrac{120}{x^5}$

23. _____

24. $y = x^{7/3}$

24. _____

25. $y = -5x^2 + \dfrac{1}{2}x - 16$

25. _____

Differentiate.

26. $y = \dfrac{3}{4}x^4 + 8x^2 - 161x + 25$

26. _____

27. $y = \dfrac{x+4}{x^4}$

27. _____

28. $f(x) = (3-x)^4 (x+5)^3$

28. _____

29. $y = (3x^3 - 5x^2 + 8)^{-3}$

29. _____

30. $f(x) = x^3 \sqrt{x^4 - 3}$

30. _____

31. For $y = 280x - 3x^5$, find $\dfrac{d^3y}{dx^3}$.

31. _____

32. *Daily accidents.* The average number of daily accidents involving drivers of age a is approximated by $P = 0.4a^2 - 40a + 1039$.

(a) Find the rate of change of average number of accidents with respect to age.

32. (a) _____

(b) How many daily accidents involve a 20-year old driver?

(b) _____

(c) Find the rate of change when $a = 20$.

(c) _____

33. Find $(f \circ g)(x)$ and $(g \circ f)(x)$, given that $f(x) = x - x^2$ and $g(x) = 4x^2$.

33. _____

34. Differentiate $y = (8 - 2x)^{3/2} (4 + x)^{1/2}$.

34. _____

35. Find $\lim\limits_{x \to -4} \dfrac{x^3 + 64}{x + 4}$.

35. _____

36. Using a grapher, graph f and f' over the given interval. Then estimate points at which the tangent line to f is horizontal.

$$f(x) = 2x^5 + 4x^2 - 7x;\ [-5, 5]$$

36. _____

37. Find the following limit using tables on a grapher. Start with Step = 0.1 and then go to 0.01, 0.001, and 0.0001. When you think you know the limit, graph and use the TRACE feature to further verify your assertion.

$$\lim_{x \to 5} \dfrac{\sqrt{6x - 5} - 5}{x - 5}$$

37. _____

CALCULUS AND ITS APPLICATIONS

Name:

Chapter 2, Form F

Limits Graphically. Consider the following graph of function f for Questions 1-6.

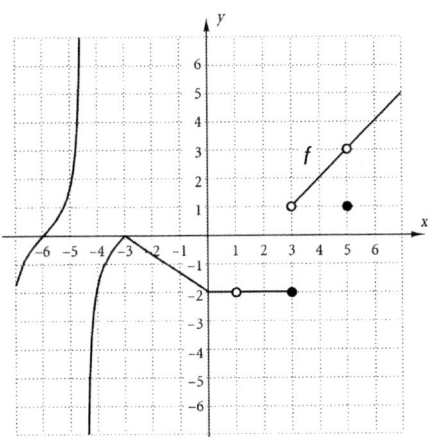

Find the limit, if it exists.

1. $\lim\limits_{x \to -3} f(x)$

2. $\lim\limits_{x \to -4.5} f(x)$

3. $\lim\limits_{x \to 0} f(x)$

4. $\lim\limits_{x \to 3} f(x)$

5. $\lim\limits_{x \to 5} f(x)$

6. $\lim\limits_{x \to 2} f(x)$

1. _____

2. _____

3. _____

4. _____

5. _____

6. _____

Determine whether the function is continuous.

7.

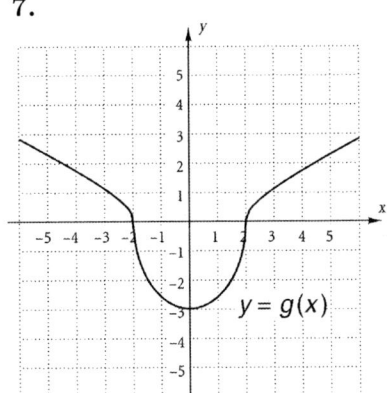

8.

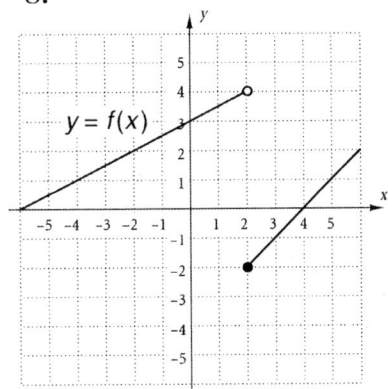

7. _____

8. _____

For the function in Question 8, answer the following.

9. Find $\lim_{x \to -2} f(x)$.

9. _____

10. Find $f(-2)$.

10. _____

11. Is f continuous at -2?

11. _____

12. Find $\lim_{x \to 2} f(x)$.

12. _____

13. Find $f(2)$.

13. _____

14. Is f continuous at 2?

14. _____

Find the limit if it exists.

15. $\lim_{x \to 4} (-x^3 + 6x - 6)$

15. _____

16. $\lim_{x \to -2^-} \dfrac{x+2}{x^2 - 7x - 18}$

16. _____

17. $\lim_{x \to 2} \dfrac{5}{x-2}$

17. _____

18. Find a simplified difference quotient for
$$f(x) = -4x^2 + 1.$$

18. _____

19. Find an equation of the tangent line to the graph of $y = 4x + \left(-\dfrac{6}{x}\right)$ at the point $(3, 10)$.

19. _____

CALCULUS AND ITS APPLICATIONS   Chapter 2, Form F

20. Find the points on the graph of $y = 3x^3 - 9x$ at which the tangent line is horizontal.

20. _____

Find dy/dx.

21. $y = x^{50}$

21. _____

22. $y = 8\sqrt{x}$

22. _____

23. $y = \dfrac{3}{x^4}$

23. _____

24. $y = x^{2/7}$

24. _____

25. $y = -4x^2 - 3.1x + 40$

25. _____

Differentiate.

26. $y = -\dfrac{2}{3}x^3 + 16x^2 + 4x + 11$

26. _____

27. $y = \dfrac{x+6}{3+x}$

27. _____

28. $f(x) = (x+2)^3 (2-x)^2$

28. _____

29. $y = (2x^3 + 16x^2 - 3x)^{-4}$

29. _____

30. $f(x) = x^2 \sqrt{x^6 + 3}$

30. _____

31. For $y = 4x^6 - 3x^2$, find $\dfrac{d^3y}{dx^3}$.

31. _____

32. *Spheres in a pyramid.* The number of spheres in a triangular pyramid with x layers is given by $N = \frac{1}{6}x^3 + \frac{1}{2}x^2 + \frac{1}{3}x$.

(a) Find the rate of change of the number of spheres with respect to the number of layers.

32. (a) _____

(b) How many spheres are there in a pyramid 10 layers deep?

(b) _____

(c) Find the rate of change when $x = 10$.

(c) _____

33. Find $(f \circ g)(x)$ and $(g \circ f)(x)$, given that $f(x) = 2x^2 - 5$ and $g(x) = \sqrt{x}$.

33. _____

34. Differentiate $y = (5 - 2x)^{1/4}(x + 6)^{3/4}$.

34. _____

35. Find $\displaystyle\lim_{x \to -1} \dfrac{1 + x^3}{1 + x}$.

35. _____

36. Using a grapher, graph f and f' over the given interval. Then estimate points at which the tangent line to f is horizontal.

$$f(x) = 2x^3 - 6x^2 + 2x + 2\sqrt{x};\ [0, 5]$$

36. _____

37. Find the following limit using tables on a grapher. Start with Step $= 0.1$ and then go to 0.01, 0.001, and 0.0001. When you think you know the limit, graph and use the TRACE feature to further verify your assertion.

$$\lim_{x \to 2} \dfrac{\sqrt{7x + 2} - 4}{2 - x}$$

37. _____

CALCULUS AND ITS APPLICATIONS   Name:

Chapter 3, Form A

Find the relative extrema of the function. List your answers in terms of ordered pairs. Then sketch a graph of the function.

1. $f(x) = x^2 - 2x - 8$

2. $f(x) = 4x^4 - 12x^2 + 9$

1. _____

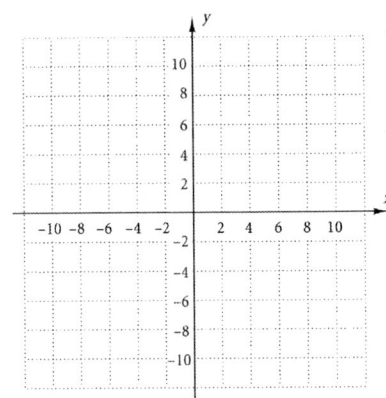 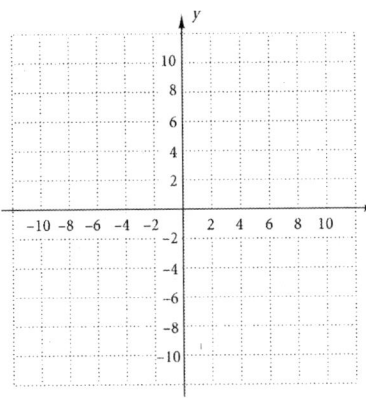

2. _____

3. $f(x) = (x-2)^{2/3} + 1$

4. $f(x) = \dfrac{40}{x^2 + 5}$

3. _____

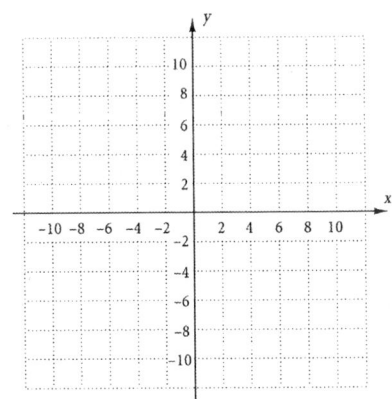 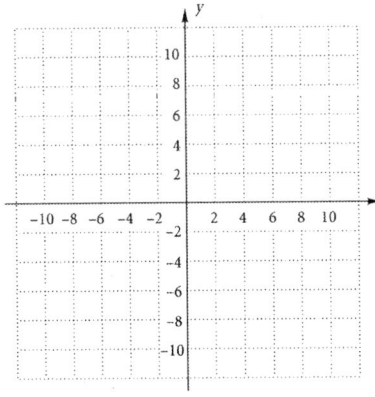

4. _____

5. $f(x) = 4x^3 + 3x^2 - 6x - 5$

6. $f(x) = 1 - 6x + 2x^3$

5. _____

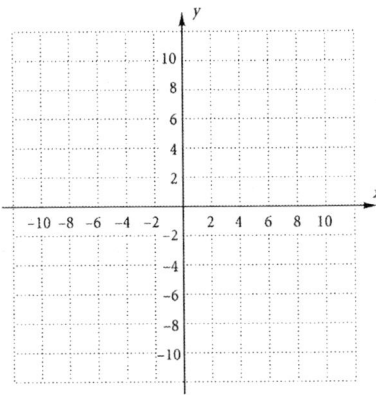 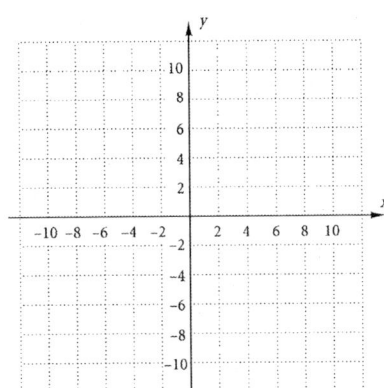

6. _____

61

CALCULUS AND ITS APPLICATIONS Chapter 3, Form A

Find the relative extrema of the function. List your answers in terms of ordered pairs. Then sketch a graph of the function.

7. $f(x) = (x+1)^3 + 3$

8. $f(x) = x\sqrt{16 - x^2}$

7. _____

8. _____

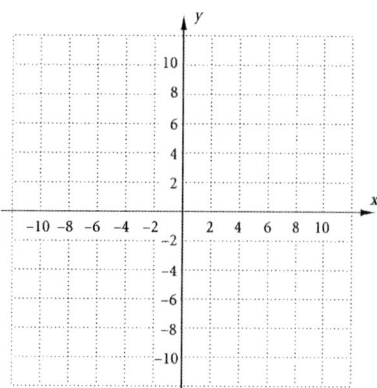
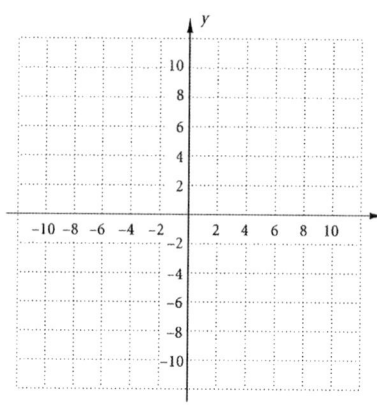

Sketch a graph of the function.

9. $f(x) = \dfrac{3}{x+2}$

10. $f(x) = \dfrac{-4}{x^2 - 1}$

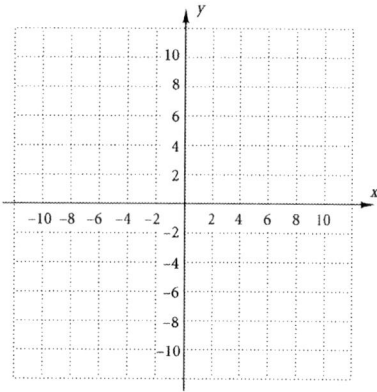
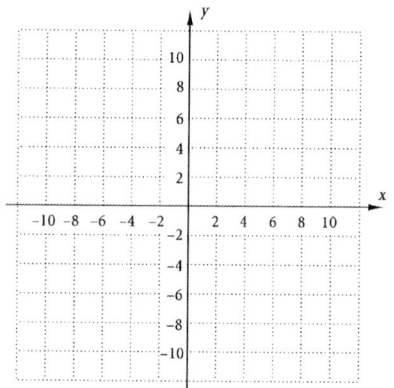

11. $f(x) = \dfrac{x^2 - 9}{x}$

12. $f(x) = \dfrac{x+4}{x-5}$

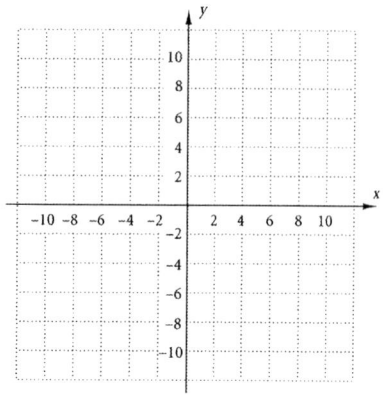
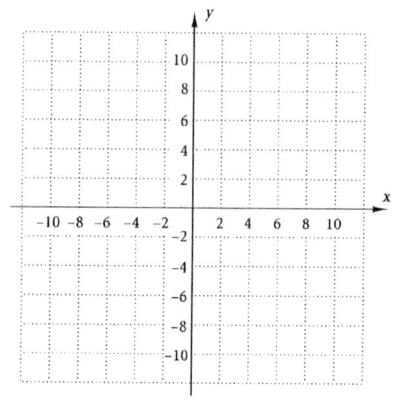

CALCULUS AND ITS APPLICATIONS Chapter 3, Form A

Find the absolute maximum and minimum values of the function, if they exist, over the indicated interval. Where no interval is specified, use the real line.

13. $f(x) = x(3-x)$

13. _____

14. $f(x) = 4x^3 + 3x^2 - 6x - 5;\ [-2, 1]$

14. _____

15. $f(x) = 2x^2 + 5.4x - 6$

15. _____

16. $f(x) = -2x + 3;\ [-2, 2]$

16. _____

17. $f(x) = -2x + 3$

17. _____

18. $f(x) = -5x^2 + 4x - 3$

18. _____

19. $f(x) = x^2 + \dfrac{54}{x};\ (0, \infty)$

19. _____

20. Of all numbers whose difference is 12, find the two that have the minimum product.

20. _____

21. Minimize $Q = x^2 + y^2$, where $x + y = 4$.

21. _____

22. *Business: Maximum profit.* Find the maximum profit and the number of units that must be produced and sold in order to yield the maximum profit.

$$R(x) = x^2 + 100x + 50$$
$$C(x) = 1.2x^2 + 20x + 80$$

22. _____

23. From a thin piece of cardboard 48 in. by 48 in., square corners are cut out so that the sides can be folded up to make a box. What dimensions will yield a box of maximum volume? What is the maximum volume?

23. _____

24. *Business: Minimizing Inventory Costs.* A clothing store sells 2400 pairs of white socks per year. It costs $0.04 to store one pair of white socks for one year. To reorder, there is a fixed cost of $0.75, plus $0.02 for each pair of socks. How many times per year should the store order white socks and in what lot size, in order to minimize inventory costs?

24. _____

25. For $y = f(x) = x^2 - 3$, $x = 2$, and $\triangle x = 0.1$, find $\triangle y$ and $f'(x) \triangle x$.

25. _____

26. Approximate $\sqrt{70}$ using $\triangle y \approx f'(x) \triangle x$.

26. _____

27. For $y = \sqrt{x^2 + 6}$:

(a) Find dy.

(b) Find dy when $x = 2$ and $dx = 0.01$.

27. (a) _____

(b) _____

28. Differentiate the following implicitly to find dy/dx. Then find the slope of the curve at the given point.

$$x^2 + y^3 = 17; \ (4, 1)$$

28. _____

29. A board 10 ft long leans against a vertical wall. If the lower end is being moved away from the wall at a rate of 0.3 ft/sec, how fast is the upper end coming down when the lower end is 8 ft from the wall?

29. _____

30. Find the absolute maximum and minimum values of the function, if they exist, over the indicated interval.

$$f(x) = \frac{x^2}{4 + x^3}; \ [0, \infty)$$

30. _____

31. *Business: Minimizing Average Cost.* The total cost of producing x units of a product is given by

$$C(x) = 50x + 50\sqrt{x} + \frac{\sqrt{x^3}}{50}.$$

(a) Find the average cost $A(x)$.

(b) Find the minimum value of $A(x)$.

31. (a) _____

(b) _____

32. Use a grapher to estimate the relative extrema of the function.

$$f(x) = 4x^3 - 30x^2 + 40x + 3\sqrt{x}$$

32. _____

CALCULUS AND ITS APPLICATIONS

Name:

Chapter 3, Form B

Find the relative extrema of the function. List your answers in terms of ordered pairs. Then sketch a graph of the function.

1. $f(x) = x^2 + 4x - 5$

2. $f(x) = 3x^4 - 12x^2 + 4$

1. _____

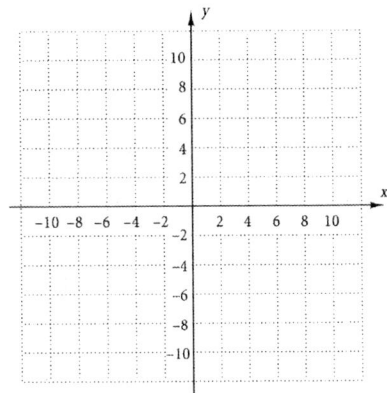 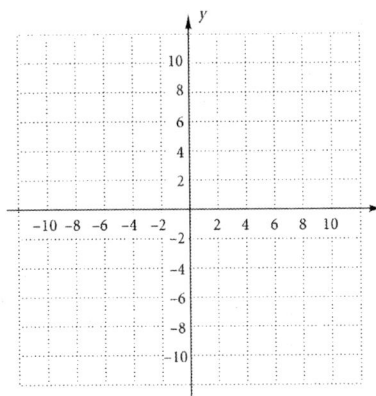

2. _____

3. $f(x) = (x - 2)^{2/3} + 4$

4. $f(x) = \dfrac{-3}{x^2 + 1}$

3. _____

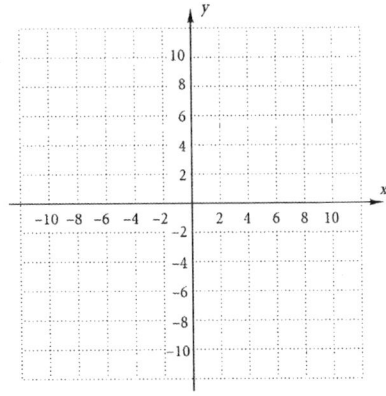 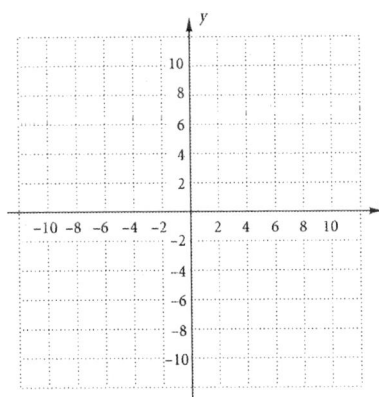

4. _____

5. $f(x) = 2x^3 - 5x^2 - 4x + 3$

6. $f(x) = 6 - 3x + x^3$

5. _____

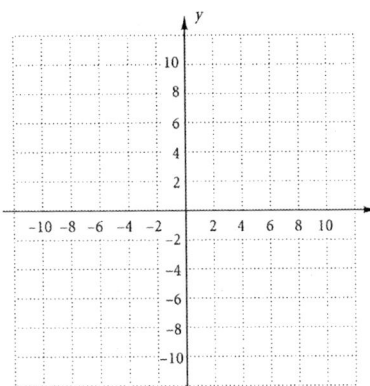 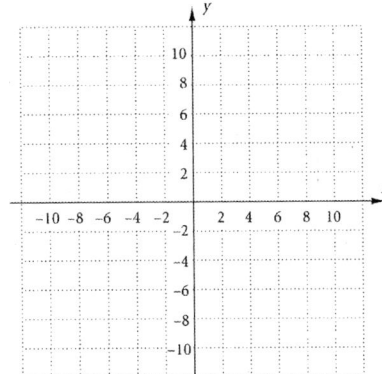

6. _____

Find the relative extrema of the function. List your answers in terms of ordered pairs. Then sketch a graph of the function.

7. $f(x) = (x+3)^3 - 4$

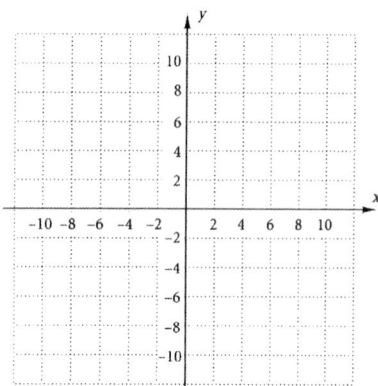

8. $f(x) = x\sqrt{16 - x^2}$

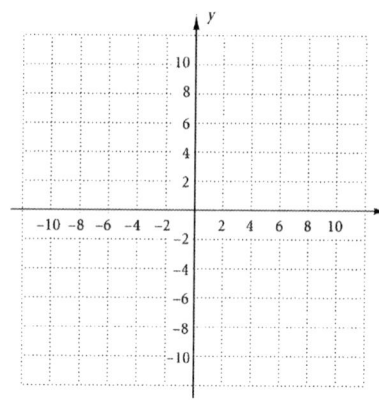

7. _____

8. _____

Sketch a graph of the function.

9. $f(x) = \dfrac{5}{x - 3}$

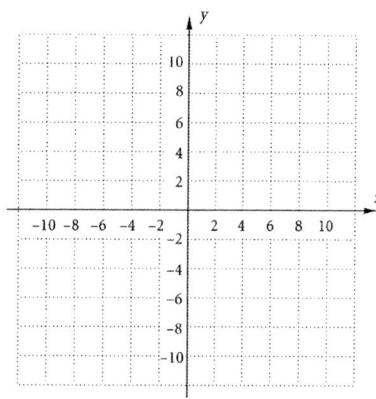

10. $f(x) = \dfrac{-2}{x^2 - 4}$

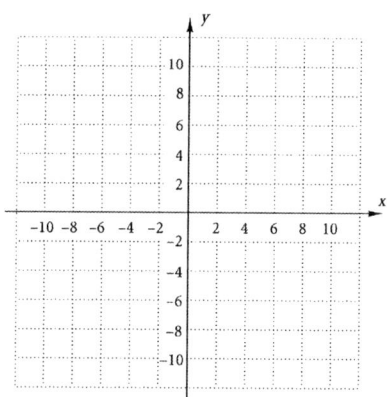

11. $f(x) = \dfrac{x^2 - 1}{x}$

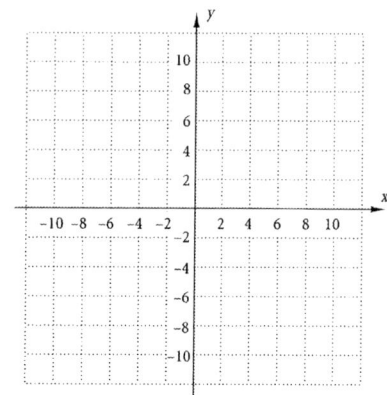

12. $f(x) = \dfrac{x - 1}{x + 4}$

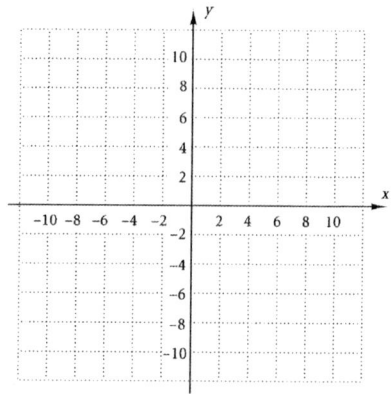

CALCULUS AND ITS APPLICATIONS   Chapter 3, Form B

Find the absolute maximum and minimum values of the function, if they exist, over the indicated interval. Where no interval is specified, use the real line.

13. $f(x) = x(6-x)$

14. $f(x) = 2x^3 - 5x^2 - 4x + 3;\ [-2, 1]$

15. $f(x) = -3x^2 + 6.6x + 4$

16. $f(x) = 4x + 1;\ [-2, 2]$

17. $f(x) = 4x + 1$

18. $f(x) = 5x^2 - 3x + 4$

19. $f(x) = x^2 + \dfrac{2}{x};\ (0, \infty)$

20. Of all numbers whose difference is 9, find the two that have the minimum product.

21. Minimize $Q = x^2 + 2y^2$, where $x + y = 6$.

22. *Business: Maximum profit.* Find the maximum profit and the number of units that must be produced and sold in order to yield the maximum profit.

$$R(x) = x^2 + 90x + 50$$
$$C(x) = 1.1x^2 + 8x + 60$$

23. From a thin piece of cardboard 72 in. by 72 in., square corners are cut out so that the sides can be folded up to make a box. What dimensions will yield a box of maximum volume? What is the maximum volume?

24. *Business: Minimizing Inventory Costs.* A computer store sells 540 printers per year. It costs $10 to store a printer for one year. To reorder, there is a fixed cost of $12, plus $5 for each printer. How many times per year should the store order printers and in what lot size, in order to minimize inventory costs?

13. _____

14. _____

15. _____

16. _____

17. _____

18. _____

19. _____

20. _____

21. _____

22. _____

23. _____

24. _____

25. For $y = f(x) = x^2 - 5$, $x = 8$, and $\triangle x = 0.1$, find $\triangle y$ and $f'(x) \triangle x$.

25. _____

26. Approximate $\sqrt{12}$ using $\triangle y \approx f'(x) \triangle x$.

26. _____

27. For $y = \sqrt{x^2 + 7}$:

(a) Find dy.

(b) Find dy when $x = 2$ and $dx = 0.01$.

27. (a) _____

(b) _____

28. Differentiate the following implicitly to find dy/dx. Then find the slope of the curve at the given point.
$$2x^3 + y^3 = 6; \; (-1, 2)$$

28. _____

29. A board 26 ft long leans against a vertical wall. If the lower end is being moved away from the wall at a rate of 0.1 ft/sec, how fast is the upper end coming down when the lower end is 24 ft from the wall?

29. _____

30. Find the absolute maximum and minimum values of the function, if they exist, over the indicated interval.
$$f(x) = \frac{2x^2}{8 + x^3}; \; [0, \infty)$$

30. _____

31. *Business: Minimizing Average Cost.* The total cost of producing x units of a product is given by
$$C(x) = 60x + 60\sqrt{x} + \frac{\sqrt{x^3}}{60}.$$

(a) Find the average cost $A(x)$.

(b) Find the minimum value of $A(x)$.

31. (a) _____

(b) _____

32. Use a grapher to estimate the relative extrema of the function.
$$f(x) = 4x^3 - 28x^2 + 40x + 2\sqrt{x}$$

32. _____

CALCULUS AND ITS APPLICATIONS

Name:

Chapter 3, Form C

Find the relative extrema of the function. List your answers in terms of ordered pairs. Then sketch a graph of the function.

1. $f(x) = x^2 - 2x - 3$

2. $f(x) = 4x^4 - 4x^2 + 1$

1. _____

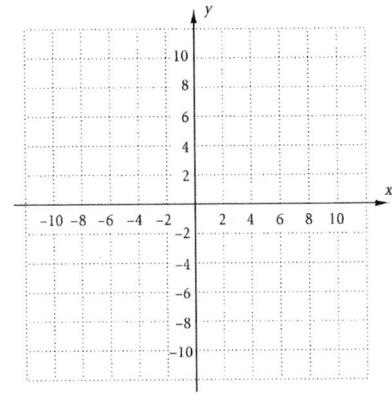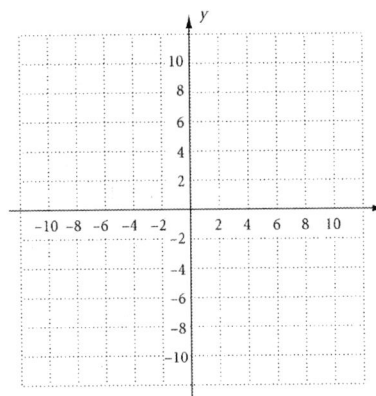

2. _____

3. $f(x) = (x-4)^{2/3} + 1$

4. $f(x) = \dfrac{8}{x^2 + 4}$

3. _____

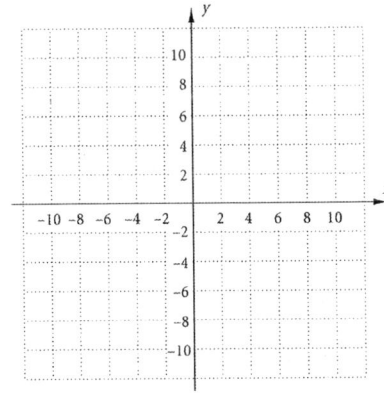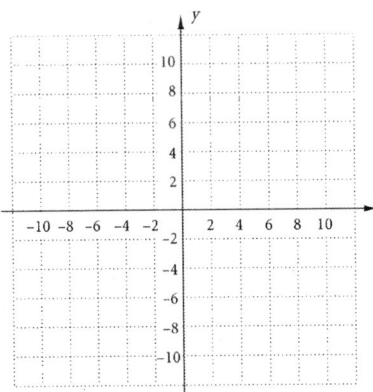

4. _____

5. $f(x) = x^3 + \dfrac{3}{2}x^2 - 6x - 2$

6. $f(x) = 5 - 3x + x^3$

5. _____

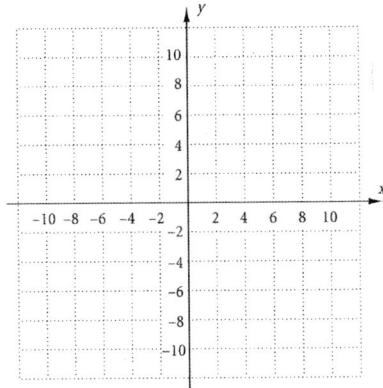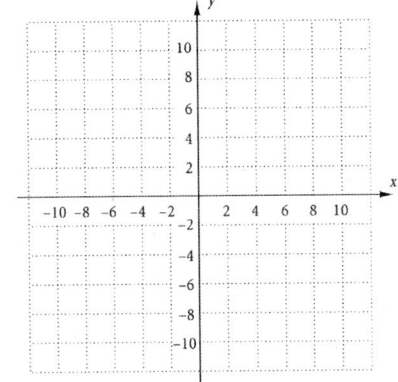

6. _____

Find the relative extrema of the function. List your answers in terms of ordered pairs. Then sketch a graph of the function.

7. $f(x) = (x-1)^3$

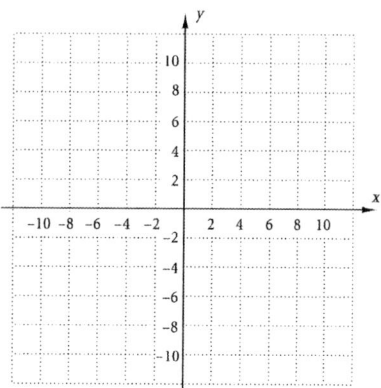

8. $f(x) = x\sqrt{49 - 9x^2}$

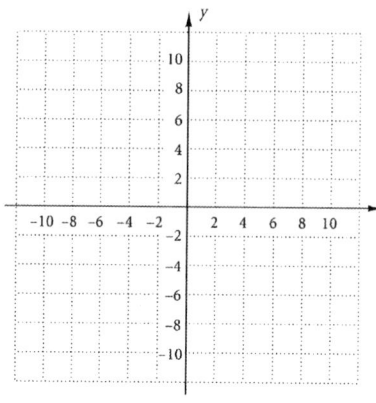

7. _____

8. _____

Sketch a graph of the function.

9. $f(x) = \dfrac{3}{x-6}$

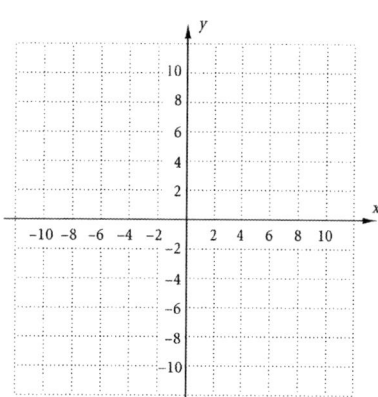

10. $f(x) = \dfrac{-6}{x^2 + 2x - 3}$

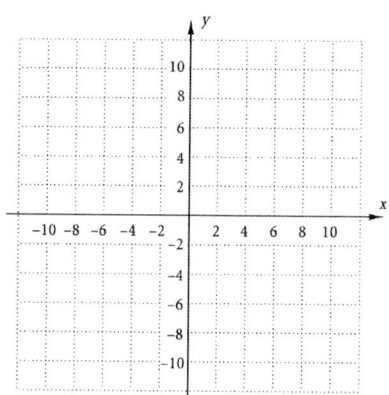

11. $f(x) = \dfrac{2x^2 - 1}{x}$

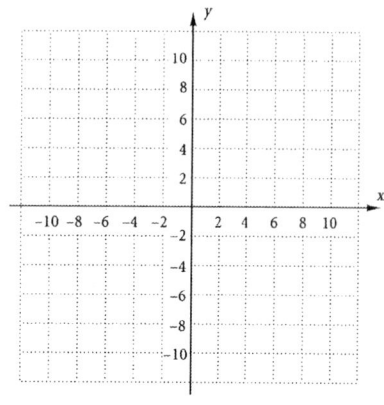

12. $f(x) = \dfrac{x-1}{x+3}$

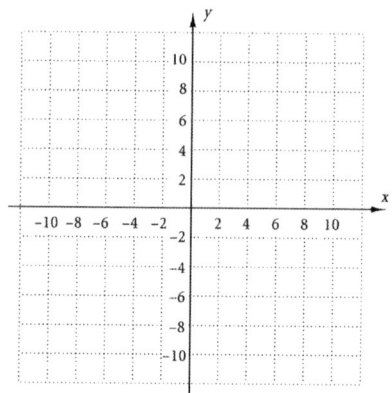

CALCULUS AND ITS APPLICATIONS   Chapter 3, Form C

Find the absolute maximum and minimum values of the function, if they exist, over the indicated interval. Where no interval is specified, use the real line.

13. $f(x) = x(5-x)$

13. _____

14. $f(x) = x^3 + \frac{3}{2}x^2 - 6x - 2;\ [-4, 2]$

14. _____

15. $f(x) = -x^2 + 3.6x + 12$

15. _____

16. $f(x) = -2x + 7;\ [-2, 2]$

16. _____

17. $f(x) = -2x + 7$

17. _____

18. $f(x) = 4x^2 - x + 2$

18. _____

19. $f(x) = x^2 + \dfrac{250}{x};\ (0, \infty)$

19. _____

20. Of all numbers whose difference is 8, find the two that have the minimum product.

20. _____

21. Minimize $Q = 3x^2 + y^2$, where $x + y = 4$.

21. _____

22. *Business: Maximum profit.* Find the maximum profit and the number of units that must be produced and sold in order to yield the maximum profit.

$$R(x) = x^2 + 90x + 80$$
$$C(x) = 1.2x^2 + 6x + 60$$

22. _____

23. From a thin piece of cardboard 84 in. by 84 in., square corners are cut out so that the sides can be folded up to make a box. What dimensions will yield a box of maximum volume? What is the maximum volume?

23. _____

24. *Business: Minimizing Inventory Costs.* A hardware store sells 250 wheelbarrows per year. It costs $3 to store one wheelbarrow for one year. To reorder, there is a fixed cost of $4, plus $0.50 for each wheelbarrow. How many times per year should the store order wheelbarrows and in what lot size, in order to minimize inventory costs?

24. _____

25. For $y = f(x) = x^2 - 4$, $x = 7$, and $\triangle x = 0.1$, find $\triangle y$ and $f'(x) \triangle x$.

25. _____

26. Approximate $\sqrt{60}$ using $\triangle y \approx f'(x) \triangle x$.

26. _____

27. For $y = \sqrt{x^2 + 5}$:

(a) Find dy.

(b) Find dy when $x = 3$ and $dx = 0.01$.

27. (a) _____

(b) _____

28. Differentiate the following implicitly to find dy/dx. Then find the slope of the curve at the given point.

$$2x^3 + y^3 = -15;\ (-2, 1)$$

28. _____

29. A board 13 ft long leans against a vertical wall. If the lower end is being moved away from the wall at a rate of 0.25 ft/sec, how fast is the upper end coming down when the lower end is 12 ft from the wall?

29. _____

30. Find the absolute maximum and minimum values of the function, if they exist, over the indicated interval.

$$f(x) = \frac{-2x^2}{4 + x^3};\ [0, \infty)$$

30. _____

31. *Business: Minimizing Average Cost.* The total cost of producing x units of a product is given by

$$C(x) = 80x + 80\sqrt{x} + \frac{\sqrt{x^3}}{80}.$$

(a) Find the average cost $A(x)$.

(b) Find the minimum value of $A(x)$.

31. (a) _____

(b) _____

32. Use a grapher to estimate the relative extrema of the function.

$$f(x) = 3x^3 - 25x^2 + 30x + \sqrt{x}$$

32. _____

CALCULUS AND ITS APPLICATIONS

Name:

Chapter 3, Form D

Find the relative extrema of the function. List your answers in terms of ordered pairs. Then sketch a graph of the function.

1. $f(x) = x^2 - 7x + 6$

2. $f(x) = x^4 - 4x^2 + 4$

1. _____

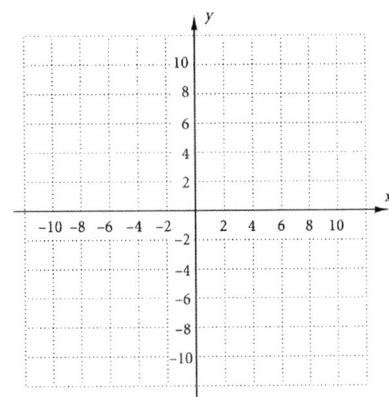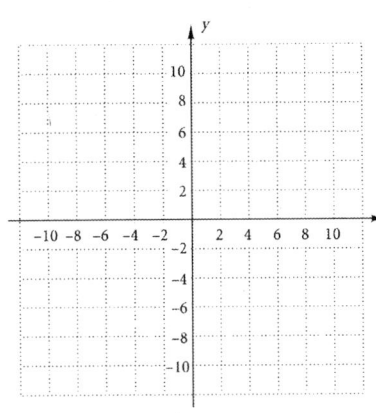

2. _____

3. $f(x) = (x-4)^{2/3} - 1$

4. $f(x) = \dfrac{36}{x^2 + 6}$

3. _____

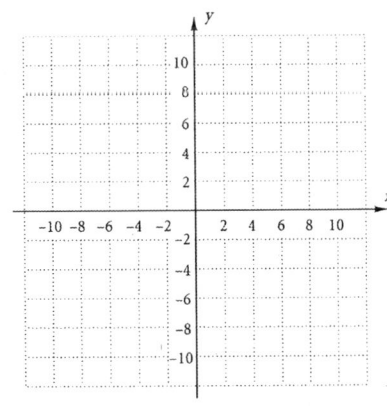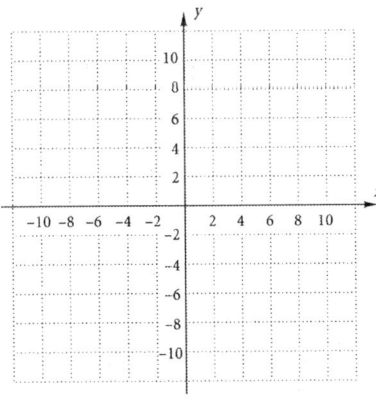

4. _____

5. $f(x) = -8x^3 + 9x^2 + 6x - 2$

6. $f(x) = -5 + 6x - 2x^3$

5. _____

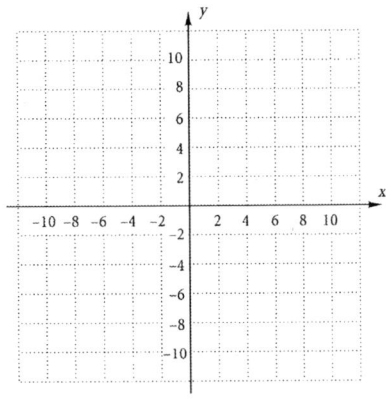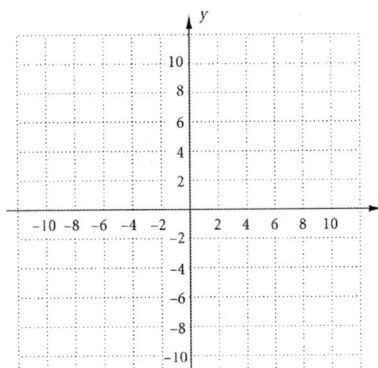

6. _____

73

CALCULUS AND ITS APPLICATIONS Chapter 3, Form D

Find the relative extrema of the function. List your answers in terms of ordered pairs. Then sketch a graph of the function.

7. $f(x) = (x-4)^3$

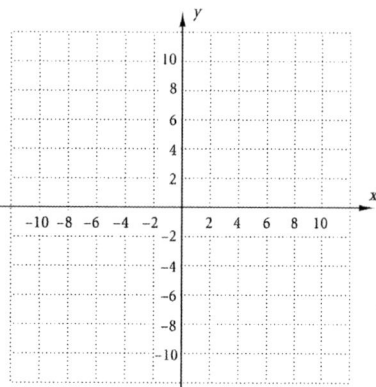

8. $f(x) = x\sqrt{25 - 4x^2}$

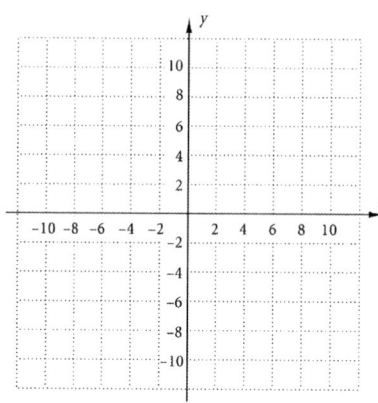

7. _____

8. _____

Sketch a graph of the function.

9. $f(x) = \dfrac{2}{x-3}$

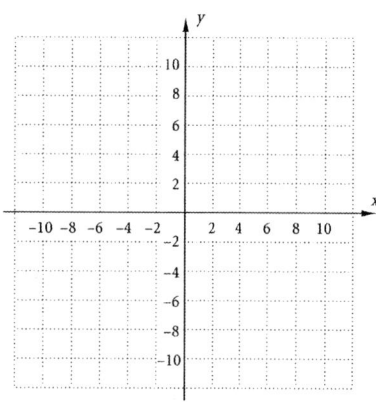

10. $f(x) = \dfrac{-2}{x^2 + 2x + 1}$

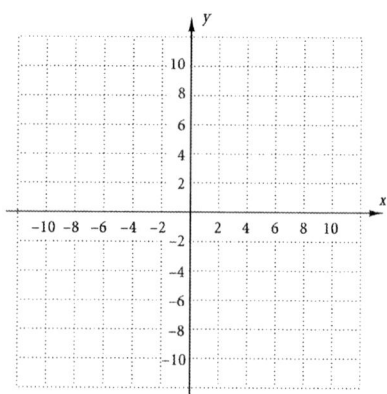

11. $f(x) = \dfrac{x^2 - 9}{x}$

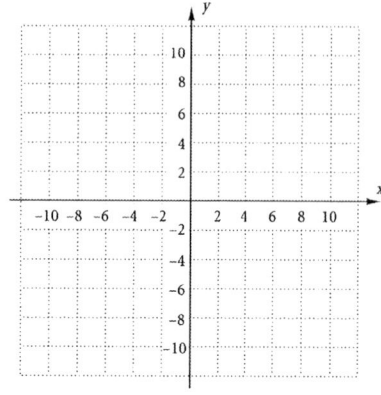

12. $f(x) = \dfrac{x-2}{x+3}$

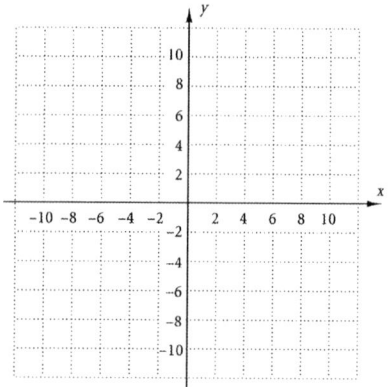

CALCULUS AND ITS APPLICATIONS   Chapter 3, Form D

Find the absolute maximum and minimum values of the function, if they exist, over the indicated interval. Where no interval is specified, use the real line.

13. $f(x) = x(3-x)$

13. _____

14. $f(x) = -8x^3 + 9x^2 + 6x - 2; [-1, 1]$

14. _____

15. $f(x) = -x^2 + 4.2x + 6$

15. _____

16. $f(x) = 2x - 3; [-2, 2]$

16. _____

17. $f(x) = 2x - 3$

17. _____

18. $f(x) = 2x^2 - 5x + 3$

18. _____

19. $f(x) = x^2 - \dfrac{250}{x}; (-\infty, 0)$

19. _____

20. Of all numbers whose difference is 18, find the two that have the minimum product.

20. _____

21. Minimize $Q = x^2 + 2y^2$, where $x - y = 4$.

21. _____

22. *Business: Maximum profit.* Find the maximum profit and the number of units that must be produced and sold in order to yield the maximum profit.

$$R(x) = x^2 + 100x + 50$$
$$C(x) = 1.2x^2 + 8x + 60$$

22. _____

23. From a thin piece of cardboard 100 in. by 100 in., square corners are cut out so that the sides can be folded up to make a box. What dimensions will yield a box of maximum volume? What is the maximum volume?

23. _____

24. *Business: Minimizing Inventory Costs.* An appliance store sells 192 washing machines per year. It costs $40 to store one washing machine for one year. To reorder, there is a fixed cost of $15, plus $2 for each washing machine. How many times per year should the store order washing machines and in what lot size, in order to minimize inventory costs?

24. _____

25. For $y = f(x) = x^2 - 6$, $x = -2$, and $\triangle x = 0.1$, find $\triangle y$ and $f'(x) \triangle x$.

25. _____

26. Approximate $\sqrt{50}$ using $\triangle y \approx f'(x) \triangle x$.

26. _____

27. For $y = \sqrt{2x^2 - 1}$:

(a) Find dy.

(b) Find dy when $x = 5$ and $dx = 0.01$.

27. (a) _____

(b) _____

28. Differentiate the following implicitly to find dy/dx. Then find the slope of the curve at the given point.

$$2x^3 - y^3 = -11; \ (2, 3)$$

28. _____

29. A board 26 ft long leans against a vertical wall. If the lower end is being moved away from the wall at a rate of 0.4 ft/sec, how fast is the upper end coming down when the lower end is 10 ft from the wall?

29. _____

30. Find the absolute maximum and minimum values of the function, if they exist, over the indicated interval.

$$f(x) = \frac{-5x^2}{1 + x^3}; \ [0, \infty)$$

30. _____

31. *Business: Minimizing Average Cost.* The total cost of producing x units of a product is given by

$$C(x) = 180x + 180\sqrt{x} + \frac{\sqrt{x^3}}{180}.$$

(a) Find the average cost $A(x)$.

(b) Find the minimum value of $A(x)$.

31. (a) _____

(b) _____

32. Use a grapher to estimate the relative extrema of the function.

$$f(x) = 2x^3 - 15x^2 + 10x + 10\sqrt{x}$$

32. _____

CALCULUS AND ITS APPLICATIONS

Name:

Chapter 3, Form E

Find the relative extrema of the function. List your answers in terms of ordered pairs. Then sketch a graph of the function.

1. $f(x) = x^2 + 7x + 12$

2. $f(x) = x^4 - 6x^2 + 9$

1. _____

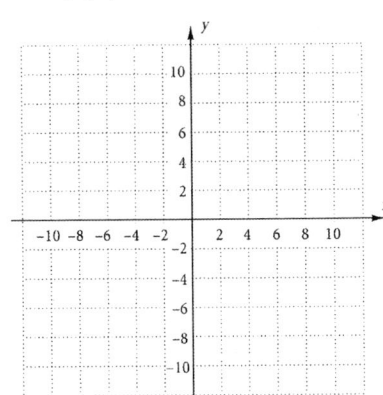
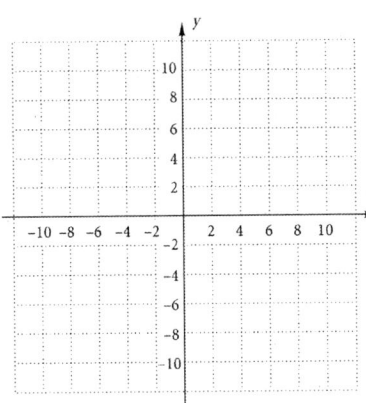

2. _____

3. $f(x) = (x-1)^{2/3} + 2$

4. $f(x) = \dfrac{-6}{x^2 + 2}$

3. _____

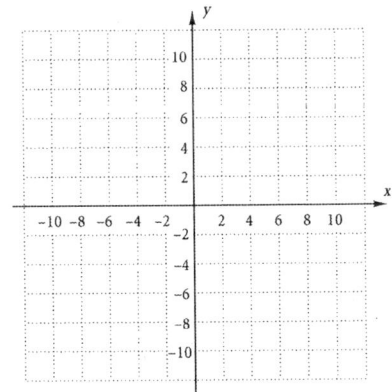
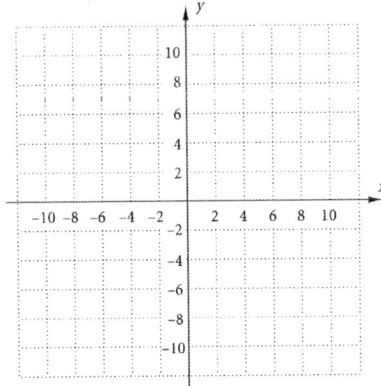

4. _____

5. $f(x) = \dfrac{x^3}{3} + x^2 - 3x + 1$

6. $f(x) = 2 + 3x - x^3$

5. _____

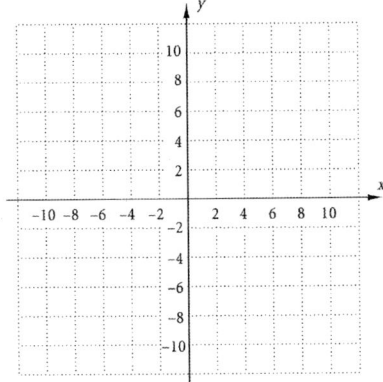
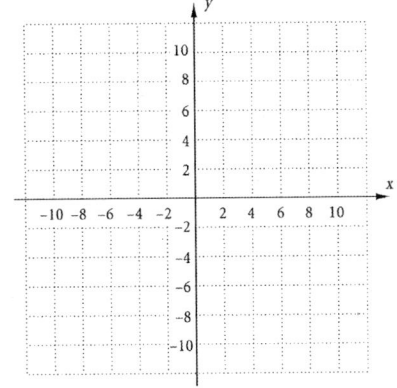

6. _____

77

CALCULUS AND ITS APPLICATIONS   Chapter 3, Form E

Find the relative extrema of the function. List your answers in terms of ordered pairs. Then sketch a graph of the function.

7. $f(x) = (x+3)^3$

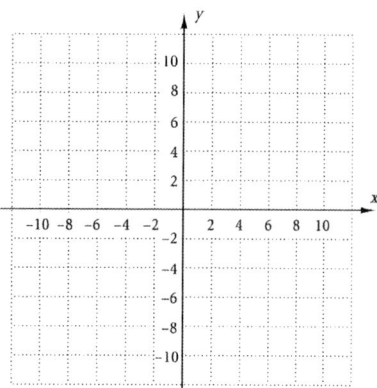

8. $f(x) = x\sqrt{49 - 16x^2}$

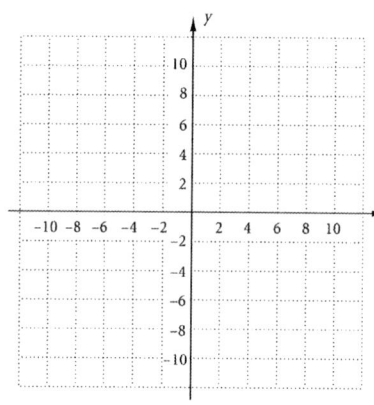

7. _____

8. _____

Sketch a graph of the function.

9. $f(x) = \dfrac{8}{x+5}$

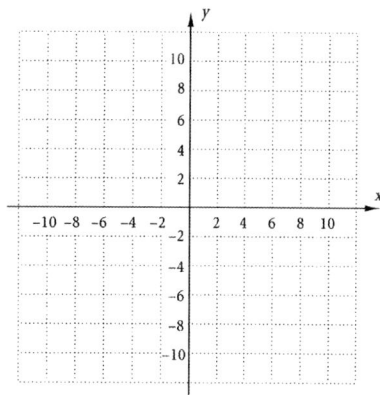

10. $f(x) = \dfrac{-8}{x^2 + 6x + 9}$

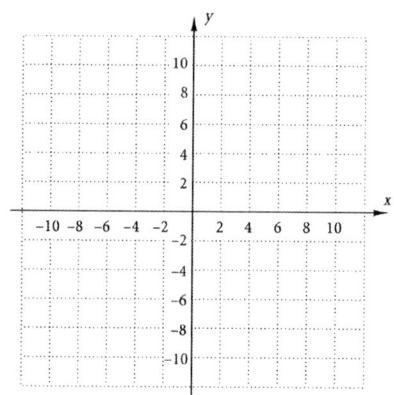

11. $f(x) = \dfrac{x^2 - 16}{x}$

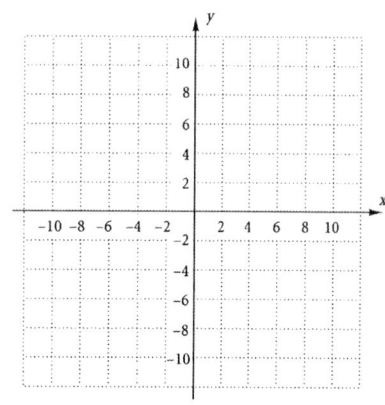

12. $f(x) = \dfrac{x+3}{x-2}$

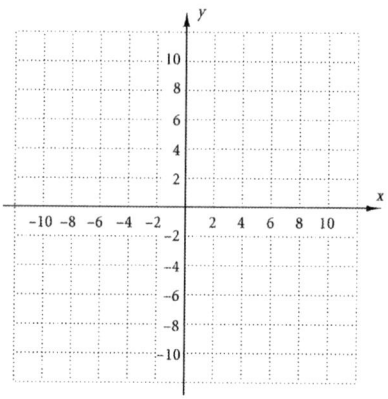

CALCULUS AND ITS APPLICATIONS   Chapter 3, Form E

Find the absolute maximum and minimum values of the function, if they exist, over the indicated interval. Where no interval is specified, use the real line.

13. $f(x) = x(10-x)$

14. $f(x) = \dfrac{x^3}{3} + x^2 - 3x + 1;\ [0,4]$

15. $f(x) = -x^2 + 2.6x - 5$

16. $f(x) = -x + 5;\ [-2,2]$

17. $f(x) = -x + 5$

18. $f(x) = 2x^2 - 3x - 4$

19. $f(x) = x^2 + \dfrac{1024}{x};\ (0, \infty)$

20. Of all numbers whose difference is 8, find the two that have the minimum product.

21. Minimize $Q = 2x^2 + 2y^2$, where $x + y = 10$.

22. *Business: Maximum profit.* Find the maximum profit and the number of units that must be produced and sold in order to yield the maximum profit.
$$R(x) = 0.9x^2 + 100x + 40$$
$$C(x) = x^2 + 10x + 50$$

23. From a thin piece of cardboard 108 in. by 108 in., square corners are cut out so that the sides can be folded up to make a box. What dimensions will yield a box of maximum volume? What is the maximum volume?

24. *Business: Minimizing Inventory Costs.* A furniture store sells 30 divans per year. It costs $20 to store one divan for one year. To reorder, there is a fixed cost of $12, plus $4 for each divan. How many times per year should the store order divans and in what lot size, in order to minimize inventory costs?

25. For $y = f(x) = x^2 - 10$, $x = 5$, and $\triangle x = 0.1$, find $\triangle y$ and $f'(x) \triangle x$.

25. _____

26. Approximate $\sqrt{40}$ using $\triangle y \approx f'(x) \triangle x$.

26. _____

27. For $y = \sqrt{3x^2 + 2}$:

(a) Find dy.

(b) Find dy when $x = 4$ and $dx = 0.01$.

27. (a) _____

(b) _____

28. Differentiate the following implicitly to find dy/dx. Then find the slope of the curve at the given point.

$$3x^3 + y^3 = 3; \ (-2, 3)$$

28. _____

29. A board 17 ft long leans against a vertical wall. If the lower end is being moved away from the wall at a rate of 0.1 ft/sec, how fast is the upper end coming down when the lower end is 8 ft from the wall?

29. _____

30. Find the absolute maximum and minimum values of the function, if they exist, over the indicated interval.

$$f(x) = \frac{4x^2}{1 - x^3}; \ (-\infty, 0]$$

30. _____

31. *Business: Minimizing Average Cost.* The total cost of producing x units of a product is given by

$$C(x) = 40x + 40\sqrt{x} + \frac{\sqrt{x^3}}{40}.$$

(a) Find the average cost $A(x)$.

(b) Find the minimum value of $A(x)$.

31. (a) _____

(b) _____

32. Use a grapher to estimate the relative extrema of the function.

$$f(x) = 3x^3 - 20x^2 + 25x + \sqrt{x}$$

32. _____

CALCULUS AND ITS APPLICATIONS   Name:

Chapter 3, Form F

Find the relative extrema of the function. List your answers in terms of ordered pairs. Then sketch a graph of the function.

1. $f(x) = x^2 + 2x - 8$

2. $f(x) = x^4 - 2x^2 + 3$

1. _____

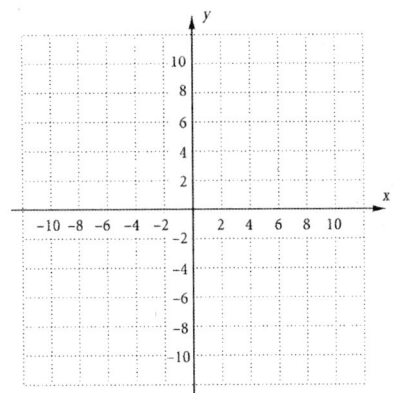 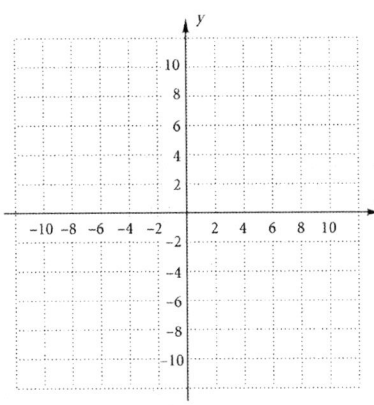

2. _____

3. $f(x) = (x-3)^{2/3} - 1$

4. $f(x) = \dfrac{10}{x^2 + 4}$

3. _____

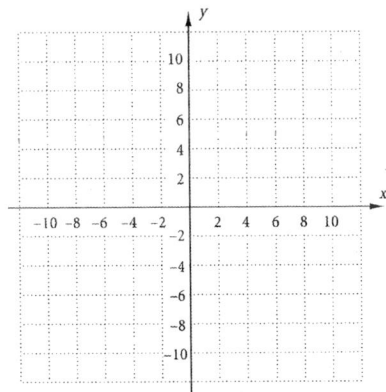 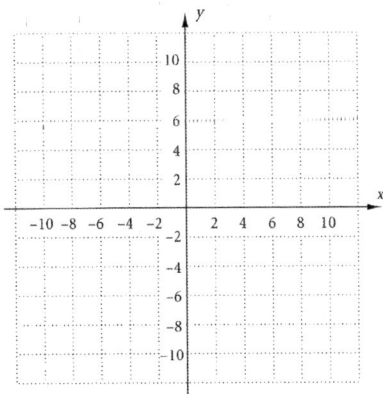

4. _____

5. $f(x) = -8x^3 + 9x^2 + 6x - 5$

6. $f(x) = 5 + 2x - \dfrac{2}{3}x^3$

5. _____

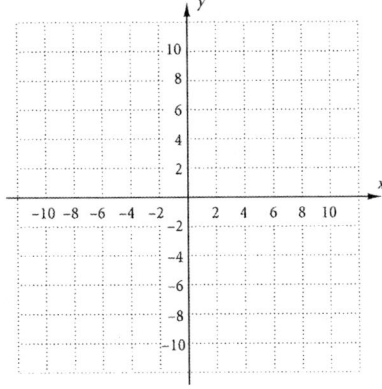 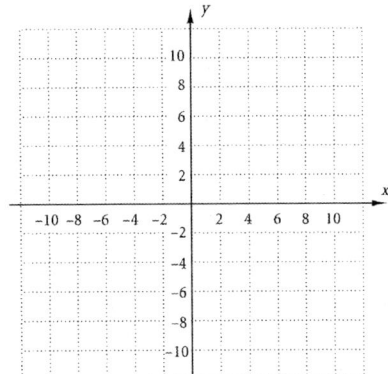

6. _____

CALCULUS AND ITS APPLICATIONS Chapter 3, Form F

Find the relative extrema of the function. List your answers in terms of ordered pairs. Then sketch a graph of the function.

7. $f(x) = (x-2)^3$

8. $f(x) = x\sqrt{25 - 16x^2}$

7. _____

8. _____

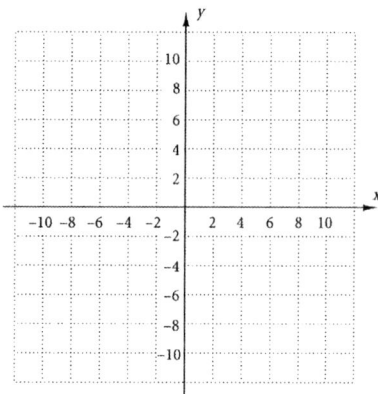
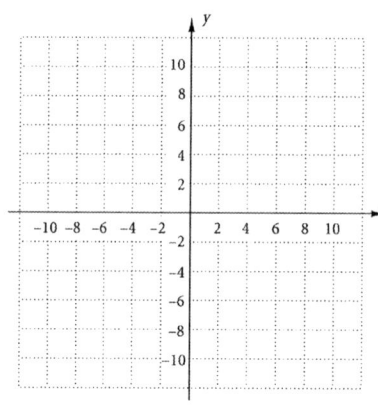

Sketch a graph of the function.

9. $f(x) = \dfrac{3}{x-5}$

10. $f(x) = \dfrac{-6}{x^2 - 25}$

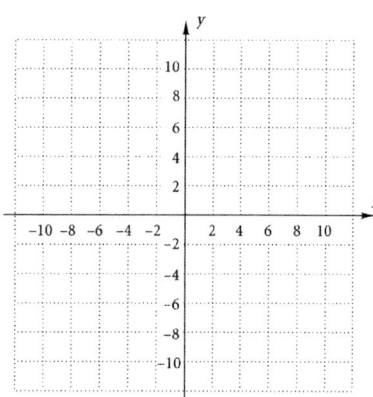
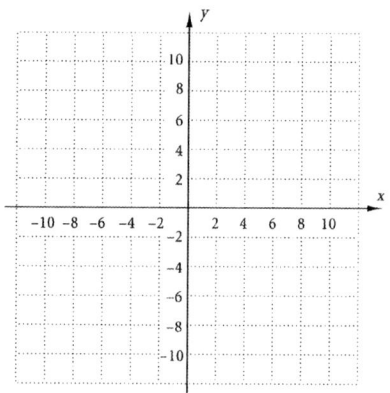

11. $f(x) = \dfrac{x^2 - 16}{x}$

12. $f(x) = \dfrac{x+1}{x-2}$

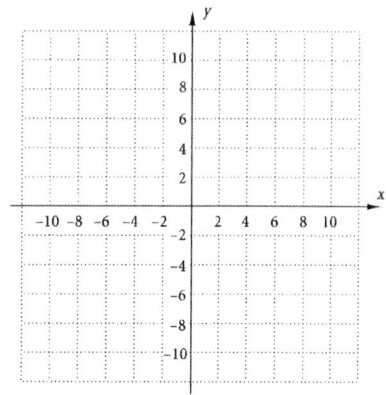
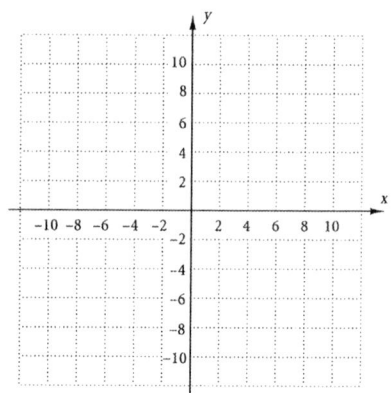

CALCULUS AND ITS APPLICATIONS Chapter 3, Form F

Find the absolute maximum and minimum values of the function, if they exist, over the indicated interval. Where no interval is specified, use the real line.

13. $f(x) = x(2-x)$

13. _____

14. $f(x) = -8x^3 + 9x^2 + 6x - 5; \; [-2, 0]$

14. _____

15. $f(x) = -x^2 + 5.4x + 9$

15. _____

16. $f(x) = -2x + 3; \; [-3, 3]$

16. _____

17. $f(x) = -2x + 3$

17. _____

18. $f(x) = 5x^2 - 4x + 2$

18. _____

19. $f(x) = x^2 + \dfrac{16}{x}; \; (0, \infty)$

19. _____

20. Of all numbers whose difference is 16, find the two that have the minimum product.

20. _____

21. Minimize $Q - x^2 + 3y^2$, where $x - y - 6$.

21. _____

22. *Business: Maximum profit.* Find the maximum profit and the number of units that must be produced and sold in order to yield the maximum profit.

$$R(x) = 0.8x^2 + 90x + 50$$
$$C(x) = 0.9x^2 + 7x + 70$$

22. _____

23. From a thin piece of cardboard 81 in. by 81 in., square corners are cut out so that the sides can be folded up to make a box. What dimensions will yield a box of maximum volume? What is the maximum volume?

23. _____

24. *Business: Minimizing Inventory Costs.* A department store sells 300 laundry baskets per year. It costs $0.54 to store one laundry basket for one year. To reorder, there is a fixed cost of $5, plus $0.05 for each laundry basket. How many times per year should the store order laundry baskets and in what lot size, in order to minimize inventory costs?

24. _____

25. For $y = f(x) = x^2 - 8$, $x = 4$, and $\triangle x = -0.1$, find $\triangle y$ and $f'(x) \triangle x$.

25. _____

26. Approximate $\sqrt{75}$ using $\triangle y \approx f'(x) \triangle x$.

26. _____

27. For $y = \sqrt{x^2 + 7}$:

(a) Find dy.

(b) Find dy when $x = 3$ and $dx = 0.01$.

27. (a) _____

(b) _____

28. Differentiate the following implicitly to find dy/dx. Then find the slope of the curve at the given point.

$$-2x^3 + y^3 = -17;\ (2, -1)$$

28. _____

29. A board 13 ft long leans against a vertical wall. If the lower end is being moved away from the wall at a rate of 0.1 ft/sec, how fast is the upper end coming down when the lower end is 12 ft from the wall?

29. _____

30. Find the absolute maximum and minimum values of the function, if they exist, over the indicated interval.

$$f(x) = \frac{x^2}{4 - x^3};\ (-\infty, 0]$$

30. _____

31. *Business: Minimizing Average Cost.* The total cost of producing x units of a product is given by

$$C(x) = 120x + 120\sqrt{x} + \frac{\sqrt{x^3}}{120}.$$

(a) Find the average cost $A(x)$.

(b) Find the minimum value of $A(x)$.

31. (a) _____

(b) _____

32. Use a grapher to estimate the relative extrema of the function.

$$f(x) = 4x^3 - 22x^2 + 30x + 4\sqrt{x}$$

32. _____

CALCULUS AND ITS APPLICATIONS

Name:

Chapter 4, Form A

Differentiate.

1. $y = 3e^x$

    1. _____

2. $y = \ln x$

    2. _____

3. $f(x) = e^{-x^4}$

    3. _____

4. $f(x) = \ln \dfrac{x}{5}$

    4. _____

5. $f(x) = e^x - 6x^3$

    5. _____

6. $f(x) = 2e^x \ln x$

    6. _____

7. $y = \ln(e^x - x^2)$

    7. _____

8. $y = \dfrac{\ln x}{5e^x}$

    8. _____

Given $\log_b 6 = 1.2925$ and $\log_b 12 = 1.7925$, find each of the following.

9. $\log_b 36$

    9. _____

10. $\log_b 2$

    10. _____

11. $\log_b 72$

    11. _____

12. Find the function that satisfies $dE/dt = kE$. List the answer in terms of E_0.

12. _____

13. The doubling time of a certain bacteria culture is 8 hr. What is the growth rate? Round to the nearest tenth of a percent.

13. _____

14. *Business: Interest Compounded Continuously.* An investment is made at 6.5% per year, compounded continuously. What is the doubling time? Round to the nearest year.

14. _____

15. *Business: Cost of Pizza.* The cost C of a slice of pizza at a certain shop was 5¢ in 1955. In 1998 the cost was $1 per slice. Assuming the exponential-growth model applies:

(a) Find the exponential-growth rate, and write the equation.

15. (a) _____

_____

(b) Find the cost of a slice of this pizza in 2005.

(b) _____

16. *Life Science: Bacterial Population.* After beginning antiseptic treatment, the number of bacteria in a population decreases at the rate of 12% per hour, that is,

$$\frac{dN}{dt} = -0.12N,$$

where N is the number of bacteria and t is the time, in hours.

(a) When $t = 0$, there are 100,000 bacteria present. Find a function that satisfies the equation.

16. (a) _____

(b) How many bacteria will remain after 8 hr?

(b) _____

(c) After how long do half of the original bacteria remain?

(c) _____

CALCULUS AND ITS APPLICATIONS   Chapter 4, Form A

17. *Life Science: Decay Rate.* The decay rate of iodine-131 is 8.6% per day. What is its half-life?

17. _____

18. *Life Science: Half-Life.* The half-life of radium-226 is 1600 years. What is its decay rate? As a percent, round to three decimal places.

18. _____

19. *Business: Effect of Advertising.* A company introduces a new product on a trial run in a city. They advertised the product on television and found the percentage P of people who bought the product after t ads had run satisfied the function

$$P(t) = \frac{100\%}{1 + 20e^{-0.15t}}$$

(a) What percentage buys the product after the ad has been run 1 time, 5 times, 20 times?

(b) Find the rate of change $P'(t)$.

(c) Sketch a graph of the function.

19. (a) _____

(b) _____

(c)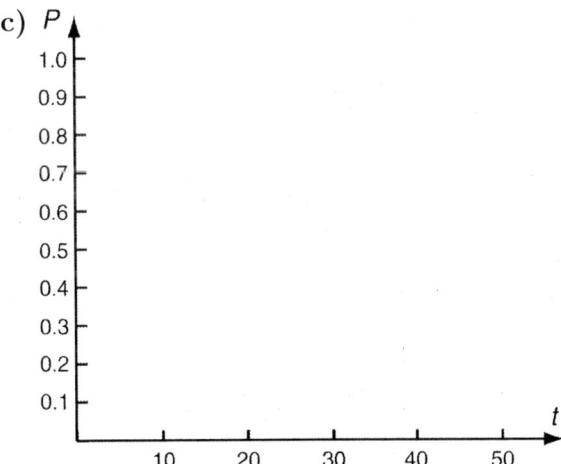

20. *Business: Present Value.* Find the present value of $50,000 due 10 yr later at 4.3%, compounded continuously.

20. _____

Differentiate.

21. $f(x) = 5^x$

21. _____

22. $y = \log_{15} x$

22. _____

23. *Economics: Elasticity of Demand.* Consider the demand function

$$x = D(p) = 280e^{-0.24p}.$$

(a) Find the elasticity.

(b) Find the elasticity at $p = \$5$, stating whether the demand is elastic or inelastic.

(c) At a price of $5, will a small increase in price cause the total revenue to increase or decrease?

(d) Find the value for p for which the total revenue is a maximum.

23. (a) _____

(b) _____

(c) _____

(d) _____

24. Differentiate: $y = x(\ln x)^3 - x \ln x + 3x$.

24. _____

25. Find the maximum and minimum values of $f(x) = x^2 e^{-x}$ over $[0, 10]$.

25. _____

26. Use your grapher to graph: $f(x) = \dfrac{e^{3x} - e^{-x}}{e^{3x} + e^{-x}}$.
Then sketch the graph.

26.

27. Find $\lim\limits_{x \to 0} \dfrac{e^{3x} - e^{-x}}{e^{3x} + e^{-x}}$.

27. _____

CALCULUS AND ITS APPLICATIONS

Name:

Chapter 4, Form B

Differentiate.

1. $y = e^x$

2. $y = 4\ln x$

3. $f(x) = e^{-x^3}$

4. $f(x) = \ln \dfrac{x}{5}$

5. $f(x) = e^x + 3x^4$

6. $f(x) = 10e^x \ln x$

7. $y = \ln(x^2 + e^x)$

8. $y = \dfrac{6\ln x}{e^x}$

Given $\log_b 3 = 0.5283$ and $\log_b 18 = 1.3900$, find each of the following.

9. $\log_b \dfrac{1}{3}$

10. $\log_b 6$

11. $\log_b 54$

12. Find the function that satisfies $dJ/dt = kJ$. List the answer in terms of J_0.

12. _____

13. The doubling time of a certain bacteria culture is 8 hr. What is the growth rate? Round to the nearest tenth of a percent.

13. _____

14. *Business: Interest Compounded Continuously.* An investment is made at 7.625% per year, compounded continuously. What is the doubling time? Round to the nearest year.

14. _____

15. *Business: Cost of Icecream.* The cost C of an icecream bar was \$0.10 in 1957. In 1997 the cost was \$1.50. Assuming the exponential-growth model applies:

(a) Find the exponential-growth rate, and write the equation.

15. (a) _____

_____

(b) Find the cost of an icecream bar in 2010.

(b) _____

16. *Life Science: Bacterial Population.* After beginning antiseptic treatment, the number of bacteria in a population decreases at the rate of 25% per day, that is,

$$\frac{dN}{dt} = -0.25N,$$

where N is the number of bacteria and t is the time, in days.

(a) When $t = 0$, there are 80,000 bacteria present. Find a function that satisfies the equation.

16. (a) _____

(b) How many bacteria will remain after 7 days?

(b) _____

(c) After how long do half of the original bacteria remain?

(c) _____

CALCULUS AND ITS APPLICATIONS Chapter 4, Form B

17. *Life Science: Decay Rate.* The decay rate of krypton-85 is 6.3% per year. What is its half-life?

17. _____

18. *Life Science: Half-Life.* The half-life of phosphorus-32 is 14.3 days. What is its decay rate? As a percent, round to two decimal places.

18. _____

19. *Business: Effect of Advertising.* A company introduces a new product on a trial run in a city. They advertised the product on television and found the percentage P of people who bought the product after t ads had run satisfied the function

$$P(t) = \frac{100\%}{1 + 40e^{-0.32t}}$$

(a) What percentage buys the product after the ad has been run 1 time, 5 times, 20 times?

(b) Find the rate of change $P'(t)$.

(c) Sketch a graph of the function.

19. (a) _____

(b) _____

(c)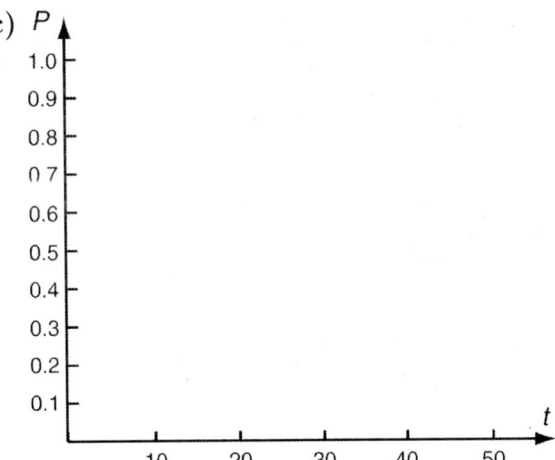

20. *Business: Present Value.* Find the present value of $75,000 due 10 yr later at 4.7%, compounded continuously.

20. _____

Differentiate.

21. $f(x) = 13^x$

21. _____

22. $y = \log_{25} x$

22. _____

23. *Economics: Elasticity of Demand.* Consider the demand function

$$x = D(p) = 200e^{-0.15p}.$$

(a) Find the elasticity.

(b) Find the elasticity at $p = \$6$, stating whether the demand is elastic or inelastic.

(c) At a price of $\$6$, will a small increase in price cause the total revenue to increase or decrease?

(d) Find the value for p for which the total revenue is a maximum.

23. (a) _____

(b) _____

(c) _____

(d) _____

---

24. Differentiate: $y = x(\ln x)^3 - 2x^2 \ln x + x^2$.

24. _____

25. Find the maximum and minimum values of $f(x) = x^5 e^{-x}$ over $[0, 10]$.

25. _____

---

26. Use your grapher to graph: $f(x) = \dfrac{e^{-5x} - e^{5x}}{e^{5x} + e^{-5x}}$.
Then sketch the graph.

26.

27. Find $\lim\limits_{x \to 0} \dfrac{e^{2x} - e^{-2x}}{e^{2x} + e^{-2x}}$.

27. _____

CALCULUS AND ITS APPLICATIONS

Name:

Chapter 4, Form C

Differentiate.

1. $y = e^x$

2. $y = 2\ln x$

3. $f(x) = e^{2x}$

4. $f(x) = \ln \dfrac{4x}{3}$

5. $f(x) = e^x - 4x^2$

6. $f(x) = 5e^x \ln x$

7. $y = \ln(e^x - x)$

8. $y = \dfrac{\ln x}{2e^x}$

Given $\log_b 5 = 0.7740$ and $\log_b 3 = 0.5283$, find each of the following.

9. $\log_b 15$

10. $\log_b 25$

11. $\log_b \dfrac{1}{3}$

12. Find the function that satisfies $dQ/dt = kQ$. List the answer in terms of Q_0.

12. _____

13. The doubling time of a certain bacteria culture is 6 hr. What is the growth rate? Round to the nearest tenth of a percent.

13. _____

14. *Business: Interest Compounded Continuously.* An investment is made at 5.768% per year, compounded continuously. What is the doubling time? Round to the nearest year.

14. _____

15. *Business: Cost of Candy.* The cost C of a certain candy bar in 1974 was 25¢. In 1999 the cost was 65¢. Assuming the exponential-growth model applies:

   (a) Find the exponential-growth rate, and write the equation.

   15. (a) _____

   (b) Find the cost of this candy bar in 2010.

   (b) _____

16. *Life Science: Bacterial Population.* After beginning antiseptic treatment, the number of bacteria in a population decreases at the rate of 10% per hour, that is,

$$\frac{dN}{dt} = -0.1N,$$

where N is the number of bacteria and t is the time, in hours.

   (a) When $t = 0$, there are 80,000 bacteria present. Find a function that satisfies the equation.

   16. (a)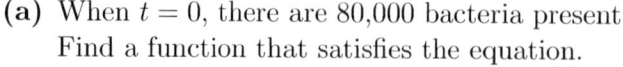

   (b) How many bacteria will remain after 24 hr?

   (b) _____

   (c) After how long do half of the original bacteria remain?

   (c) _____

CALCULUS AND ITS APPLICATIONS Chapter 4, Form C

17. *Life Science: Decay Rate.* The decay rate of phosphorus-32 is 4.85% per day. What is its half-life?

17. _____

18. *Life Science: Half-Life.* The half-life of uranium-238 is 4.5 billion years. What is its decay rate? As a percent, round to nine decimal places.

18. _____

19. *Business: Effect of Advertising.* A company introduces a new product on a trial run in a city. They advertised the product on television and found the percentage P of people who bought the product after t ads had run satisfied the function

$$P(t) = \frac{100\%}{1 + 20e^{-0.25t}}$$

(a) What percentage buys the product after the ad has been run 1 time, 5 times, 20 times?

(b) Find the rate of change $P'(t)$.

(c) Sketch a graph of the function.

19. (a) _____

(b) _____

(c)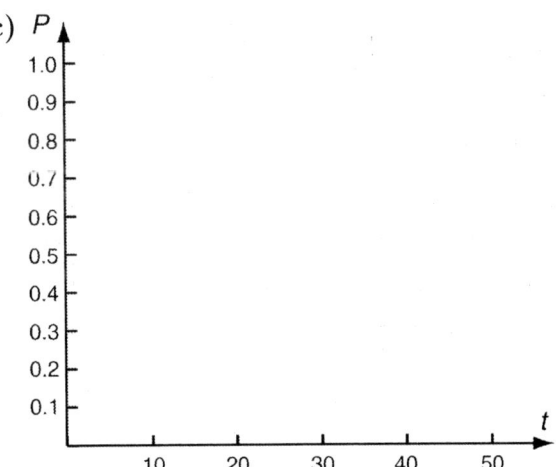

20. *Business: Present Value.* Find the present value of $50,000 due 10 yr later at 6.25%, compounded continuously.

20. _____

Differentiate.

21. $f(x) = 20^x$

21. _____

22. $y = \log_{15} x$

22. _____

23. *Economics: Elasticity of Demand.* Consider the demand function

$$x = D(p) = 500e^{-0.2p}.$$

(a) Find the elasticity.

(b) Find the elasticity at $p = \$4$, stating whether the demand is elastic or inelastic.

(c) At a price of $4, will a small increase in price cause the total revenue to increase or decrease?

(d) Find the value for p for which the total revenue is a maximum.

23. (a) _____

(b) _____

(c) _____

(d) _____

---

24. Differentiate: $y = x\sqrt{\ln x} + 3x \ln x - 3x$.

24. _____

25. Find the maximum and minimum values of $f(x) = x^3 e^{-3x}$ over $[0, 10]$.

25. _____

---

26. Use your grapher to graph: $f(x) = \dfrac{e^{3x} - e^{-3x}}{e^{3x} + e^{-3x}}$. Then sketch the graph.

26.

27. Find $\lim\limits_{x \to 0} \dfrac{e^{3x} - e^{-3x}}{e^{3x} + e^{-3x}}$.

27. _____

CALCULUS AND ITS APPLICATIONS

Name:

Chapter 4, Form D

Differentiate.

1. $y = 5e^x$

1. _____

2. $y = \ln x$

2. _____

3. $f(x) = e^{3x^2}$

3. _____

4. $f(x) = \ln 4x$

4. _____

5. $f(x) = e^x + 4x^5$

5. _____

6. $f(x) = 5e^x \ln x$

6. _____

7. $y = \ln(x^5 - e^x)$

7. _____

8. $y = \dfrac{e^x}{\ln x}$

8. _____

Given $\log_b 6 = 1.2925$ and $\log_b 10 = 1.6610$, find each of the following.

9. $\log_b \dfrac{1}{6}$

9. _____

10. $\log_b 60$

10. _____

11. $\log_b 100$

11. _____

12. Find the function that satisfies $dF/dt = kF$. List the answer in terms of F_0.

12. _____

13. The doubling time of a certain bacteria culture is 12 hr. What is the growth rate? Round to the nearest tenth of a percent.

13. _____

14. *Business: Interest Compounded Continuously.* An investment is made at 8.085% per year, compounded continuously. What is the doubling time? Round to the nearest year.

14. _____

15. *Business: Cost of Bread.* The cost C of a loaf of bread at a particular bakery was 25¢ in 1950. In 2000 the cost was $3.00 per loaf. Assuming the exponential-growth model applies:

(a) Find the exponential-growth rate, and write the equation.

15. (a) _____

(b) Find the cost of a loaf of bread at this bakery in 2010.

(b) _____

16. *Life Science: Bacterial Population.* After beginning antiseptic treatment, the number of bacteria in a population decreases at the rate of 40% per day, that is,

$$\frac{dN}{dt} = -0.40N,$$

where N is the number of bacteria and t is the time, in days.

(a) When $t = 0$, there are 5,000,000 bacteria present. Find a function that satisfies the equation.

16. (a)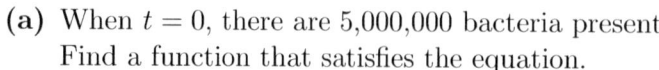

(b) How many bacteria will remain after 3 days?

(b) _____

(c) After how long do half of the original bacteria remain?

(c) _____

CALCULUS AND ITS APPLICATIONS   Chapter 4, Form D

17. *Life Science: Decay Rate.* The decay rate of hydrogen-3 is 5.6% per year. What is its half-life?

17. _____

18. *Life Science: Half-Life.* The half-life of cobalt-60 is 5.3 years. What is its decay rate? As a percent, round to two decimal places.

18. _____

19. *Business: Effect of Advertising.* A company introduces a new product on a trial run in a city. They advertised the product on television and found the percentage P of people who bought the product after t ads had run satisfied the function

$$P(t) = \frac{100\%}{1 + 20e^{-0.15t}}$$

(a) What percentage buys the product after the ad has been run 1 time, 5 times, 20 times?

(b) Find the rate of change $P'(t)$.

(c) Sketch a graph of the function.

19. (a) _____

(b) _____

(c)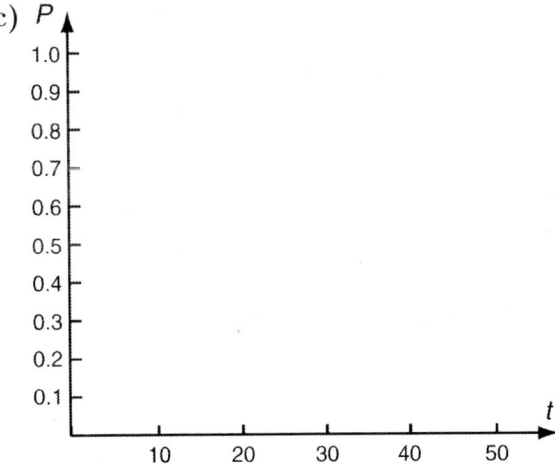

20. *Business: Present Value.* Find the present value of $70,000 due 15 yr later at 8.9%, compounded continuously.

20. _____

Differentiate.

21. $f(x) = 20^x$

21. _____

22. $y = \log_8 x$

22. _____

23. *Economics: Elasticity of Demand.* Consider the demand function

$$x = D(p) = 800e^{-0.02p}.$$

(a) Find the elasticity.

(b) Find the elasticity at $p = \$25$, stating whether the demand is elastic or inelastic.

(c) At a price of $25, will a small increase in price cause the total revenue to increase or decrease?

(d) Find the value for p for which the total revenue is a maximum.

24. Differentiate: $y = x\sqrt{\ln x} - 3x \ln x + 3x$.

25. Find the maximum and minimum values of $f(x) = x^4 e^{-4x}$ over $[0, 5]$.

26. Use your grapher to graph: $f(x) = \dfrac{e^{-3x} - e^{3x}}{e^{-3x} + e^{3x}}$.
Then sketch the graph.

27. Find $\lim\limits_{x \to 0} \dfrac{e^{-3x} - e^{3x}}{e^{-3x} + e^{3x}}$.

CALCULUS AND ITS APPLICATIONS

Chapter 4, Form E

Differentiate.

1. $y = e^x$

2. $y = 9 \ln x$

3. $f(x) = e^{x^5}$

4. $f(x) = \ln \dfrac{x}{2}$

5. $f(x) = -3x^4 + e^x$

6. $f(x) = 5e^x \ln x$

7. $y = \ln(x^2 - e^x)$

8. $y = \dfrac{4 \ln x}{e^x}$

Given $\log_b 6 = 1.2925$ and $\log_b 8 = 1.5$, find each of the following.

9. $\log_b 48$

10. $\log_b 2$

11. $\log_b \dfrac{1}{8}$

CALCULUS AND ITS APPLICATIONS Chapter 4, Form E

12. Find the function that satisfies $dN/dt = kN$. List the answer in terms of N_0.

12. _____

13. The doubling time of a certain bacteria culture is 18 hr. What is the growth rate? Round to the nearest tenth of a percent.

13. _____

14. *Business: Interest Compounded Continuously.* An investment is made at 3.55% per year, compounded continuously. What is the doubling time? Round to the nearest year.

14. _____

15. *Business: Cost of a Movie Ticket.* The cost C of a movie ticket at a certain theatre was $3.75 in 1982. In 1999 the cost was $7.50 per ticket. Assuming the exponential-growth model applies:

(a) Find the exponential-growth rate, and write the equation.

15. (a) _____

(b) Find the cost of a movie ticket at this theatre in 2010.

(b) _____

16. *Life Science: Bacterial Population.* After beginning antiseptic treatment, the number of bacteria in a population decreases at the rate of 20% per hour, that is,

$$\frac{dN}{dt} = -0.20N,$$

where N is the number of bacteria, t is the time, in hours.

(a) When $t = 0$, there are 150,000 bacteria present. Find a function that satisfies the equation.

16. (a) _____

(b) How many bacteria will remain after 6 hr?

(b) _____

(c) After how long do half of the original bacteria remain?

(c) _____

CALCULUS AND ITS APPLICATIONS Chapter 4, Form E

17. *Life Science: Decay Rate.* The decay rate of cobalt-60 is 13.08% per year. What is its half-life?

17. _____

18. *Life Science: Half-Life.* The half-life of radium-226 is 1600 years. What is its decay rate? As a percent, round to three decimal places.

18. _____

19. *Business: Effect of Advertising.* A company introduces a new product on a trial run in a city. They advertised the product on television and found the percentage P of people who bought the product after t ads had run satisfied the function

$$P(t) = \frac{100\%}{1 + 80e^{-0.32t}}$$

(a) What percentage buys the product after the ad has been run 1 time, 5 times, 20 times?

(b) Find the rate of change $P'(t)$.

(c) Sketch a graph of the function.

19. (a) _____

(b) _____

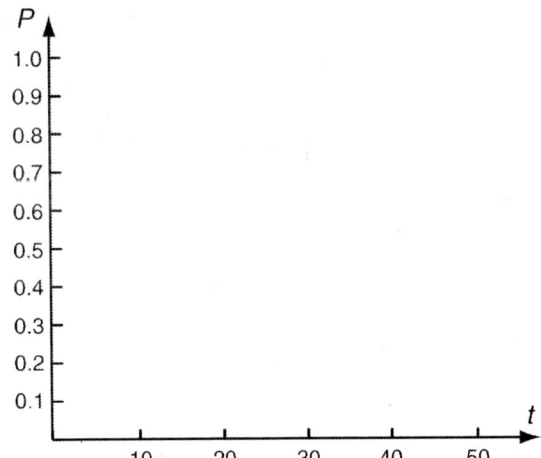

20. *Business: Present Value.* Find the present value of $100,000 due 18 yr later at 5.5%, compounded continuously.

20. _____

Differentiate.

21. $f(x) = 15^x$

21. _____

22. $y = \log_5 x$

22. _____

23. *Economics: Elasticity of Demand.* Consider the demand function

$$x = D(p) = 200e^{-0.5p}.$$

(a) Find the elasticity.

(b) Find the elasticity at $p = \$1.50$, stating whether the demand is elastic or inelastic.

(c) At a price of $1.50, will a small increase in price cause the total revenue to increase or decrease?

(d) Find the value for p for which the total revenue is a maximum.

23. (a) _____

(b) _____

(c) _____

(d) _____

24. Differentiate: $y = 3x(\ln x)^4 - 4x \ln x + 4x$.

24. _____

25. Find the maximum and minimum values of $f(x) = 4x^2 e^{-2x}$ over $[0, 10]$.

25. _____

26. Use your grapher to graph: $f(x) = \dfrac{e^{-5x} - e^{5x}}{e^{-5x} + e^{5x}}$. Then sketch the graph.

26.

27. Find $\lim\limits_{x \to 0} \dfrac{e^{-5x} - e^{5x}}{e^{-5x} + e^{5x}}$.

27. _____

CALCULUS AND ITS APPLICATIONS

Chapter 4, Form F

Differentiate.

1. $y = 2e^x$

2. $y = \ln x$

3. $f(x) = e^{-x^2}$

4. $f(x) = \ln 3x$

5. $f(x) = e^x - 4x^3$

6. $f(x) = 8e^x \ln x$

7. $y = \ln(e^x + 12x)$

8. $y = \dfrac{5e^x}{\ln x}$

Given $\log_b 4 = 0.7124$ and $\log_b 12 = 1.2770$, find each of the following.

9. $\log_b 3$

10. $\log_b \dfrac{1}{12}$

11. $\log_b 8$

105

12. Find the function that satisfies $dS/dt = kS$. List the answer in terms of S_0.

12. _____

13. The doubling time of a certain bacteria culture is 4 hr. What is the growth rate? Round to the nearest tenth of a percent.

13. _____

14. *Business: Interest Compounded Continuously.* An investment is made at 5.676% per year, compounded continuously. What is the doubling time? Round to the nearest year.

14. _____

15. *Business: Cost of Gasoline.* The cost C of a gallon of gasoline at a certain station in 1974 was 49¢. In 2003 the cost was $1.80. Assuming the exponential-growth model applies:

(a) Find the exponential-growth rate, and write the equation.

15. (a) _____

(b) Find the cost of a gallon of gasoline at this station in 2010.

(b) _____

16. *Life Science: Bacterial Population.* After beginning antiseptic treatment, the number of bacteria in a population decreases at the rate of 60% per day, that is,

$$\frac{dN}{dt} = -0.60N,$$

where N is the number of bacteria, t is the time, in days.

(a) When $t = 0$, there are 250,000 bacteria present. Find a function that satisfies the equation.

16. (a) _____

(b) How many bacteria will remain after 5 days?

(b) _____

(c) After how long do half of the original bacteria remain?

(c) _____

CALCULUS AND ITS APPLICATIONS Chapter 4, Form F

17. *Life Science: Decay Rate.* The decay rate of radium-226 is 0.043% per year. What is its half-life?

17. _____

18. *Life Science: Half-Life.* The half-life of iodine-131 is 8.1 days. What is its decay rate? As a percent, round to two decimal places.

18. _____

19. *Business: Effect of Advertising.* A company introduces a new product on a trial run in a city. They advertised the product on television and found the percentage P of people who bought the product after t ads had run satisfied the function

$$P(t) = \frac{100\%}{1 + 50e^{-0.17t}}$$

(a) What percentage buys the product after the ad has been run 1 time, 5 times, 20 times?

(b) Find the rate of change $P'(t)$.

(c) Sketch a graph of the function.

19. (a) _____

(b) _____

(c)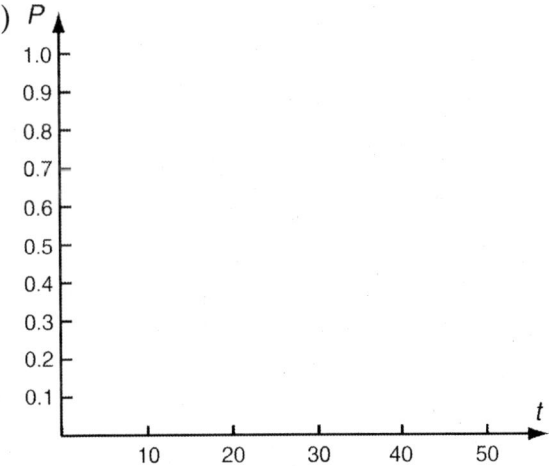

20. *Business: Present Value.* Find the present value of $80,000 due 6 yr later at 6.9%, compounded continuously.

20. _____

Differentiate.

21. $f(x) = 5^x$

21. _____

22. $y = \log_4 x$

22. _____

23. *Economics: Elasticity of Demand.* Consider the demand function

$$x = D(p) = 200e^{-0.05p}.$$

(a) Find the elasticity.

(b) Find the elasticity at $p = \$15$, stating whether the demand is elastic or inelastic.

(c) At a price of $15, will a small increase in price cause the total revenue to increase or decrease?

(d) Find the value for p for which the total revenue is a maximum.

23. (a) _____

(b) _____

(c) _____

(d) _____

24. Differentiate: $y = x^2 (\ln x)^2 - x \ln x + 2x$.

24. _____

25. Find the maximum and minimum values of $f(x) = x^2 e^{-2x}$ over $[0, 5]$.

25. _____

26. Use your grapher to graph: $f(x) = \dfrac{e^{-0.5x} - e^{0.5x}}{e^{-0.5x} + e^{0.5x}}$. Then sketch the graph.

26.

27. Find $\lim\limits_{x \to 0} \dfrac{e^{-0.5x} - e^{0.5x}}{e^{-0.5x} + e^{0.5x}}$.

27. _____

CALCULUS AND ITS APPLICATIONS

Name:

Chapter 5, Form A

Evaluate.

1. $\int dx$

2. $\int 200x^4\, dx$

3. $\int \left(\dfrac{3}{x} + e^x + x^{3/5}\right) dx$

Find the area under the curve over the indicated interval.

4. $y = 3x - x^2$; $[0, 3]$

5. $y = \dfrac{2}{x}$; $[1, 8]$

6. Give two interpretations of the shaded region.

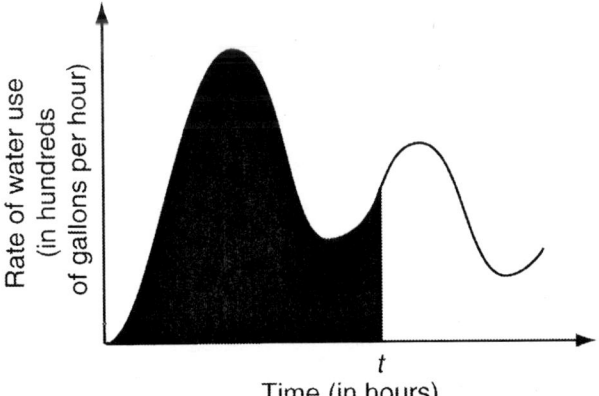

Evaluate.

7. $\displaystyle\int_{-1}^{2} (5x + 4x^3)\, dx$

8. $\displaystyle\int_{0}^{5} e^{-4x}\, dx$

9. $\displaystyle\int_{a}^{2a} \dfrac{dx}{x}$

10. Decide whether $\int_a^b f(x)\,dx$ is positive, negative, or zero.

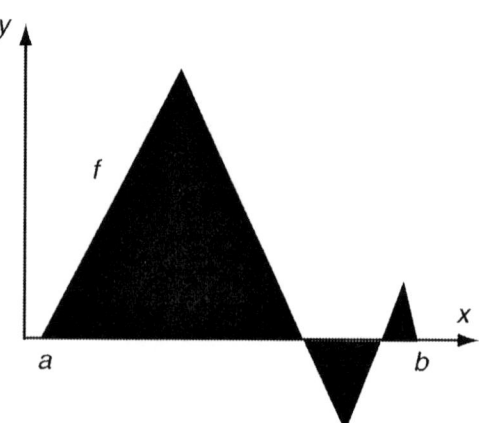

10. _____

Evaluate using substitution. Do not use Table 1.

11. $\displaystyle\int \frac{dx}{x+5}$

11. _____

12. $\displaystyle\int e^{-0.1x}\,dx$

12. _____

13. $\displaystyle\int t\left(t^2 - 5\right)^7 dt$

13. _____

Evaluate using integration by parts. Do not use Table 1.

14. $\displaystyle\int xe^{3x}\,dx$

14. _____

15. $\displaystyle\int x^4 \ln x^5\,dx$

15. _____

Evaluate using Table 1.

16. $\displaystyle\int 3^x\,dx$

16. _____

17. $\displaystyle\int \frac{dx}{(5-x)\,x}$

17. _____

CALCULUS AND ITS APPLICATIONS   Chapter 5, Form A

18. Find the average value of $y = 3t^5 + 2t$ over $[-2, 3]$.

18. _____

19. Find the area of the region bounded by $y = 2x$, $y = x^3$, $x = 0$, and $x = \sqrt{2}$.

19. _____

20. Approximate $\int_1^5 \dfrac{16}{x^2}\, dx$ by computing the area of each rectangle and adding.

20. _____

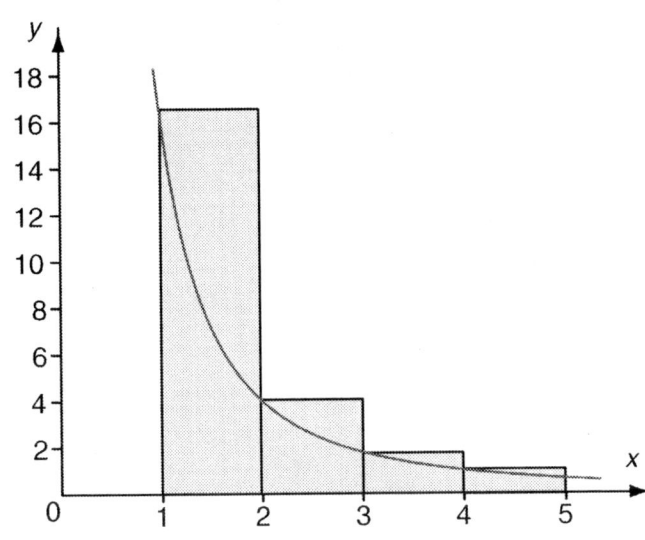

21. *Business: Cost from Marginal Cost.* An appliance company determines that the marginal cost of the xth oven is given by

$$C'(x) = -0.4x + 650,\ C(0) = \$0.$$

Find the total cost of producing 50 ovens.

21. _____

22. *Physical Science: Distance.* A particle is initially at the origin. Its velocity, in meters per second, at any time t, $t \geq 0$, is given by $v(t) = 3t^2 + 2t$. Find the distance that the particle travels in the first 4 seconds (from $t = 0$ to $t = 4$).

22. _____

Integrate using any method.

23. $\displaystyle\int \dfrac{dx}{x(6 - 5x)}$

23. _____

24. $\displaystyle\int 5x^4 e^x\, dx$

24. _____

25. $\displaystyle\int x^3 e^{x^4}\, dx$

25. _____

26. $\displaystyle\int \frac{1}{\sqrt{x}} \ln x\, dx$

26. _____

27. $\displaystyle\int x^3 \sqrt{x^2 + 6}\, dx$

27. _____

28. $\displaystyle\int \frac{dx}{16 - x^2}$

28. _____

29. $\displaystyle\int x^4 e^{-0.5x}\, dx$

29. _____

30. $\displaystyle\int x \ln 12x\, dx$

30. _____

Evaluate using any method.

31. $\displaystyle\int \frac{\left[3 + 2(\ln x)^2 + (\ln x)^5\right]}{x}\, dx$

31. _____

32. $\displaystyle\int \ln\left(\frac{x+9}{x+2}\right) dx$

32. _____

33. Use a grapher to approximate the area between the following curves:

$$y = 4x - x^2,$$
$$y = 3x^3 - x^2 - 8x,$$
$$x = 0,\ x = 2.$$

33. _____

CALCULUS AND ITS APPLICATIONS

Name:

Chapter 5, Form B

Evaluate.

1. $\int dx$

2. $\int 9x^2 \, dx$

3. $\int \left(x^{3/2} + e^x + \frac{4}{x} \right) dx$

Find the area under the curve over the indicated interval.

4. $y = 9x - x^2$; $[0, 9]$

5. $y = \dfrac{3}{x}$; $[1, 3]$

6. Give two interpretations of the shaded area.

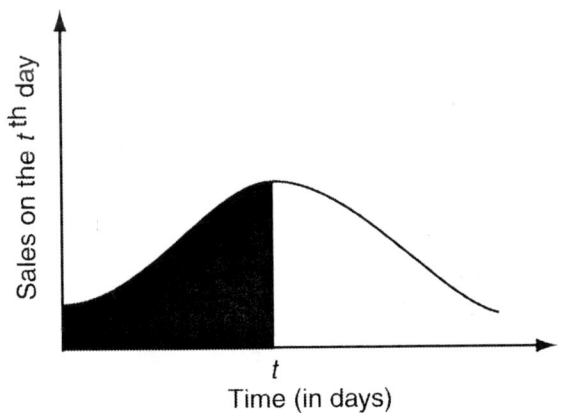

Time (in days)

Evaluate.

7. $\int_{-3}^{2} (4x + 6x^5) \, dx$

8. $\int_{0}^{5} e^{-3x} \, dx$

9. $\int_{0}^{a} \sqrt{x} \, dx$

10. Decide whether $\int_a^b f(x)\,dx$ is positive, negative, or zero.

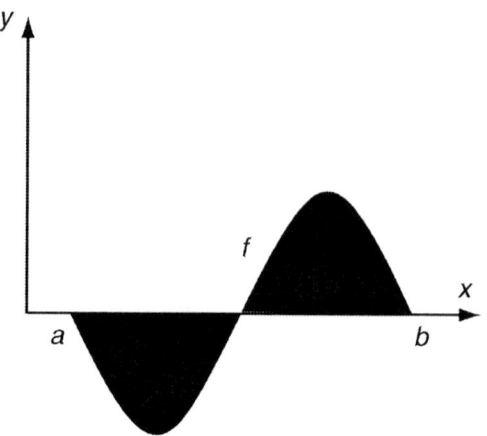

10. _____

Evaluate using substitution. Do not use Table 1.

11. $\int \dfrac{dx}{x-5}$

11. _____

12. $\int e^{-0.8x}\,dx$

12. _____

13. $\int t^4\left(t^5-2\right)^2 dt$

13. _____

Evaluate using integration by parts. Do not use Table 1.

14. $\int xe^{3x}\,dx$

14. _____

15. $\int x^2 \ln x^3\,dx$

15. _____

Evaluate using Table 1.

16. $\int 5^x\,dx$

16. _____

17. $\int \dfrac{dx}{x(3+x)}$

17. _____

CALCULUS AND ITS APPLICATIONS   Chapter 5, Form B

18. Find the average value of $y = 2t - 6t^2$ over $[-5, 4]$.

18. _____

19. Find the area of the region bounded by $y = 4x$, $y = x^3$, $x = 0$, and $x = 2$.

19. _____

20. Approximate $\int_0^3 (9 - x^2)\, dx$ by computing the area of each rectangle and adding.

20. _____

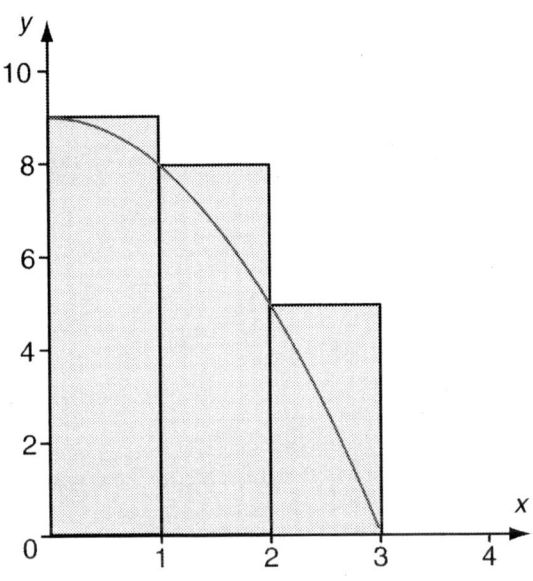

21. *Business: Cost from Marginal Cost.* A shoe company determines that the marginal cost of the xth pair of shoes is given by
$$C'(x) = -0.006x + 40,\ C(0) = \$0.$$
Find the total cost of producing 1000 pairs of shoes.

21. _____

22. *Physical Science: Distance.* A particle is initially at the origin. Its velocity, in meters per second, at any time t, $t \geq 0$, is given by $v(t) = 4t^2 + t$. Find the distance that the particle travels in the first 6 seconds (from $t = 0$ to $t = 6$).

22. _____

Integrate using any method.

23. $\displaystyle\int \frac{dx}{x(5 + 2x)}$

23. _____

24. $\displaystyle\int 3x^3 e^x\, dx$

24. _____

25. $\displaystyle\int x^9 e^{x^{10}}\, dx$

25. _____

26. $\displaystyle\int \sqrt[3]{x}\, \ln x\, dx$

26. _____

27. $\displaystyle\int x^5 \sqrt{x^3 + 2}\, dx$

27. _____

28. $\displaystyle\int \frac{dx}{25 - x^2}$

28. _____

29. $\displaystyle\int x^2 e^{-0.8x}\, dx$

29. _____

30. $\displaystyle\int x \ln 10x\, dx$

30. _____

Evaluate using any method.

31. $\displaystyle\int \frac{\left[3 (\ln x)^2 + 5 \ln x - 2\right]}{x}\, dx$

31. _____

32. $\displaystyle\int \ln\left(\frac{x+4}{x-3}\right) dx$

32. _____

33. Use a grapher to approximate the area between the following curves:

$$y = 4x - x^2,$$
$$y = x^3 + x^2 - 4x,$$
$$x = 0,\ x = 2.$$

33. _____

CALCULUS AND ITS APPLICATIONS

Name:

Chapter 5, Form C

Evaluate.

1. $\int dx$

2. $\int 250x^4\, dx$

3. $\int \left(x^{1/2} + 2e^x + \dfrac{1}{x}\right) dx$

Find the area under the curve over the indicated interval.

4. $y = 4x - x^2;\ [0, 4]$

5. $y = \dfrac{6}{x};\ [1, 4]$

6. Give two interpretations of the shaded area.

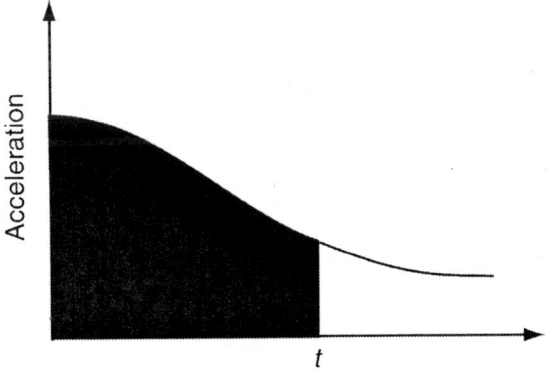

Evaluate.

7. $\displaystyle\int_{-3}^{2} (2x + 4x^3)\, dx$

8. $\displaystyle\int_{0}^{4} e^{-6x}\, dx$

9. $\displaystyle\int_{-a}^{0} \sqrt[3]{x}\, dx$

10. Decide whether $\int_a^b f(x)\,dx$ is positive, negative, or zero.

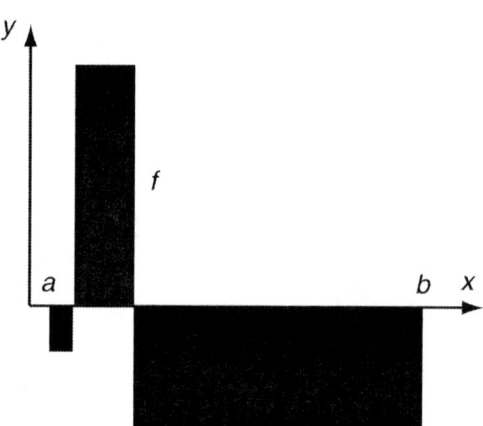

10. _____

Evaluate using substitution. Do not use Table 1.

11. $\int \dfrac{dx}{x+6}$

11. _____

12. $\int e^{-0.2x}\,dx$

12. _____

13. $\int t^5 \left(4 - t^6\right)^2 dt$

13. _____

Evaluate using integration by parts. Do not use Table 1.

14. $\int xe^{2x}\,dx$

14. _____

15. $\int x^8 \ln x^4\,dx$

15. _____

Evaluate using Table 1.

16. $\int 8^x\,dx$

16. _____

17. $\int \dfrac{dx}{\sqrt{x^2 + 25}}\,dx$

17. _____

CALCULUS AND ITS APPLICATIONS Chapter 5, Form C

18. Find the average value of $y = -2t^2 + 3t$ over $[-3, 4]$.

18. _____

19. Find the area of the region bounded by $y = x$, $y = x^5$, $x = 0$, and $x = 1$.

19. _____

20. Approximate $\int_0^4 (x^2 + 2)\, dx$ by computing the area of each rectangle and adding.

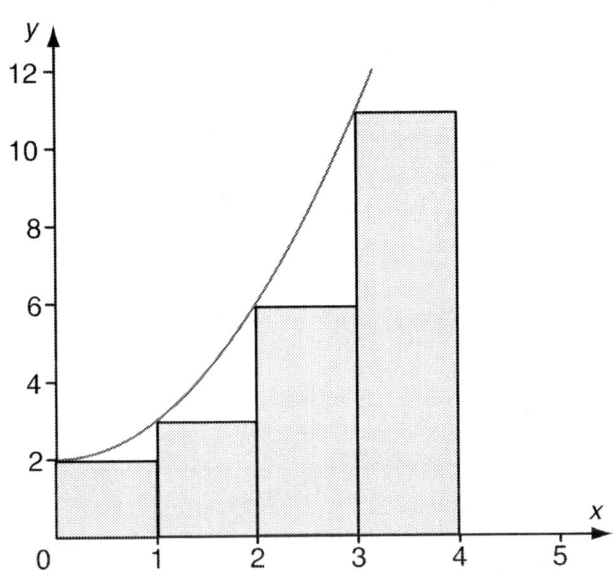

20. _____

21. *Business: Cost from Marginal Cost.* A cookie company determines that the marginal cost of the xth cookie is given by
$$C'(x) = -0.0002x + 1.25, \quad C(0) = \$0.$$
Find the total cost of producing 1000 cookies.

21. _____

22. *Physical Science: Distance.* A particle is initially at the origin. Its velocity, in meters per second, at any time t, $t \geq 0$, is given by $v(t) = t^2 + 2t$. Find the distance that the particle travels in the first 4 seconds (from $t = 0$ to $t = 4$).

22. _____

Integrate using any method.

23. $\int \dfrac{dx}{x(3-x)}$

23. _____

24. $\int 10x^4 e^x\, dx$

24. _____

25. $\displaystyle\int x^6 e^{x^7}\, dx$

25. _____

26. $\displaystyle\int \frac{1}{x} \ln x\, dx$

26. _____

27. $\displaystyle\int x^5 \sqrt{x^3+1}\, dx$

27. _____

28. $\displaystyle\int \frac{dx}{49-x^2}$

28. _____

29. $\displaystyle\int x^4 e^{-1.1x}\, dx$

29. _____

30. $\displaystyle\int x \ln 3x\, dx$

30. _____

Evaluate using any method.

31. $\displaystyle\int \frac{\left[10 + 5(\ln x)^5 - 2(\ln x)^{10}\right]}{x}\, dx$

31. _____

32. $\displaystyle\int \ln\left(\frac{x+8}{x-2}\right) dx$

32. _____

33. Use a grapher to approximate the area between the following curves:

$$y = -x^2 - 2x,$$
$$y = -2x^3 - x^2 + 6x,$$
$$x = 0,\ x = 2.$$

33. _____

CALCULUS AND ITS APPLICATIONS

Name:

Chapter 5, Form D

Evaluate.

1. $\int dx$

1. _____

2. $\int 60x^5 \, dx$

2. _____

3. $\int \left(3e^x + \dfrac{5}{x} + x^{5/6}\right) dx$

3. _____

Find the area under the curve over the indicated interval.

4. $y = 3x - x^2;\ [0, 3]$

4. _____

5. $y = \dfrac{2}{x};\ [1, 4]$

5. _____

6. Give two interpretations of the shaded area.

6. _____

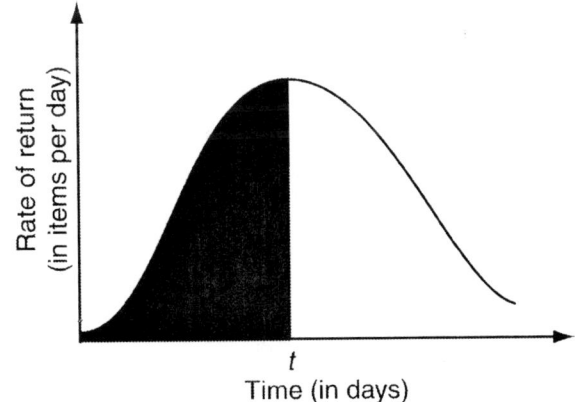

Evaluate.

7. $\int_{-2}^{3} (4x^3 + 10x) \, dx$

7. _____

8. $\int_{1}^{5} e^{-2x} \, dx$

8. _____

9. $\int_{a}^{3a} \dfrac{3}{x} \, dx$

9. _____

10. Decide whether $\int_a^b f(x)\,dx$ is positive, negative, or zero.

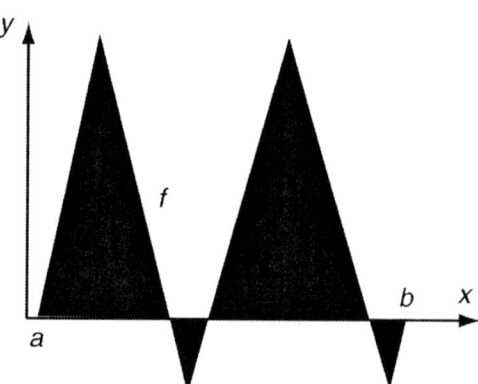

10. _____

Evaluate using substitution. Do not use Table 1.

11. $\int \dfrac{dx}{x+2}$

11. _____

12. $\int e^{-0.25x}\,dx$

12. _____

13. $\int t^3 \sqrt{t^4+3}\,dt$

13. _____

Evaluate using integration by parts. Do not use Table 1.

14. $\int xe^{8x}\,dx$

14. _____

15. $\int x^5 \ln x^6\,dx$

15. _____

Evaluate using Table 1.

16. $\int 11^x\,dx$

16. _____

17. $\int \dfrac{dx}{x^2-9}$

17. _____

CALCULUS AND ITS APPLICATIONS Chapter 5, Form D

18. Find the average value of $y = 5t^4 + 2t$ over $[-3, 4]$.

18. _____

19. Find the area of the region bounded by $y = 9x$, $y = x^3$, $x = -3$, and $x = 0$.

19. _____

20. Approximate $\int_0^4 (16 - x^2)\, dx$ by computing the area of each rectangle and adding.

20. _____

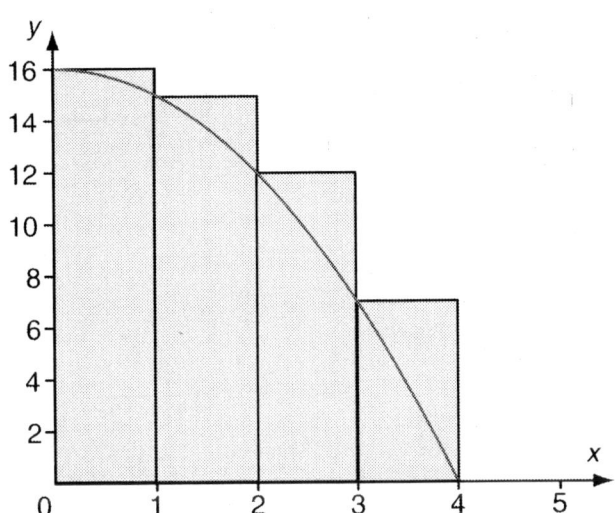

21. *Business: Cost from Marginal Cost.* A clothing company determines that the marginal cost of the xth suit is given by

$$C'(x) = -0.18x + 400,\ C(0) = \$0.$$

Find the total cost of producing 200 suits.

21. _____

22. *Physical Science: Distance.* A particle is initially at the origin. Its velocity, in meters per second, at any time t, $t \geq 0$, is given by $v(t) = 6t^2 + 2t$. Find the distance that the particle travels in the first 4 seconds (from $t = 0$ to $t = 4$).

22. _____

Integrate using any method.

23. $\displaystyle \int \frac{dx}{x(12-x)}$

23. _____

24. $\displaystyle \int x^4 e^x\, dx$

24. _____

25. $\displaystyle\int x^8 e^{x^9}\, dx$

25. _____

26. $\displaystyle\int \frac{1}{x^2} \ln x\, dx$

26. _____

27. $\displaystyle\int x^5 \sqrt{x^3 + 3}\, dx$

27. _____

28. $\displaystyle\int \frac{dx}{81 - x^2}$

28. _____

29. $\displaystyle\int x^2 e^{-0.8x}\, dx$

29. _____

30. $\displaystyle\int x \ln 5x\, dx$

30. _____

Evaluate using any method.

31. $\displaystyle\int \frac{\left[(\ln x)^4 - 3(\ln x)^2 + 4\right]}{x}\, dx$

31. _____

32. $\displaystyle\int \ln(x - 5)(x + 3)\, dx$

32. _____

---

33. Use a grapher to approximate the area between the following curves:
$$y = 4x - 2x^2 + 4,$$
$$y = 4x^3 - 6x^2 + x + 4,$$
$$x = 0,\ x = 1.5.$$

33. _____

CALCULUS AND ITS APPLICATIONS

Name:

Chapter 5, Form E

Evaluate.

1. $\int dx$

2. $\int 180x^8\, dx$

3. $\int \left(\dfrac{3}{x} + e^x + x^{3/4}\right) dx$

Find the area under the curve over the indicated interval.

4. $y = 6x - x^2;\ [0, 6]$

5. $y = \dfrac{5}{x};\ [1, 10]$

6. Give two interpretations of the shaded area.

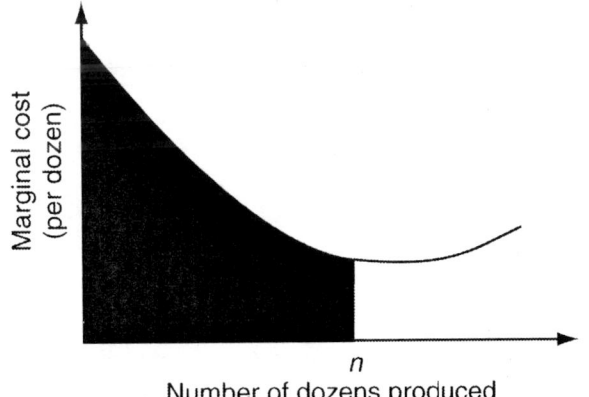

Marginal cost (per dozen)

n
Number of dozens produced

Evaluate.

7. $\displaystyle\int_{-1}^{2} (2x + 12x^5)\, dx$

8. $\displaystyle\int_{0}^{6} e^{-12x}\, dx$

9. $\displaystyle\int_{1}^{a^2} \dfrac{dx}{x^2}$

10. Decide whether $\int_a^b f(x)\,dx$ is positive, negative, or zero.

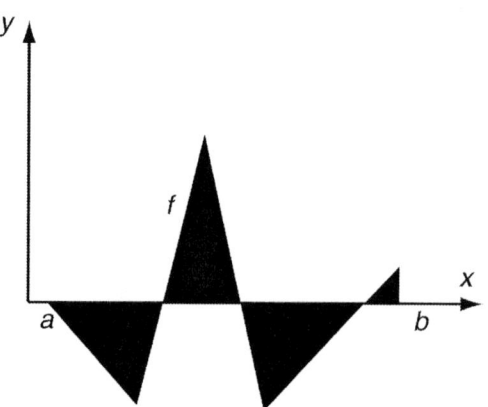

10. _____

Evaluate using substitution. Do not use Table 1.

11. $\displaystyle\int \frac{dx}{x-21}$

11. _____

12. $\displaystyle\int e^{-0.2x}\,dx$

12. _____

13. $\displaystyle\int t^4 \left(t^5+6\right)^3 dt$

13. _____

Evaluate using integration by parts. Do not use Table 1.

14. $\displaystyle\int xe^{3x}\,dx$

14. _____

15. $\displaystyle\int x^8 \ln x^7\,dx$

15. _____

Evaluate using Table 1.

16. $\displaystyle\int 11^x\,dx$

16. _____

17. $\displaystyle\int \frac{x}{5+2x}\,dx$

17. _____

CALCULUS AND ITS APPLICATIONS   Chapter 5, Form E

18. Find the average value of $y = 4t^3 + 6t^2$ over $[-3, -1]$.

18. _____

19. Find the area of the region bounded by $y = x$, $y = x^4$, $x = -1$, and $x = 0$.

19. _____

20. Approximate $\int_1^5 \frac{4}{x}\,dx$ by computing the area of each rectangle and adding.

20. _____

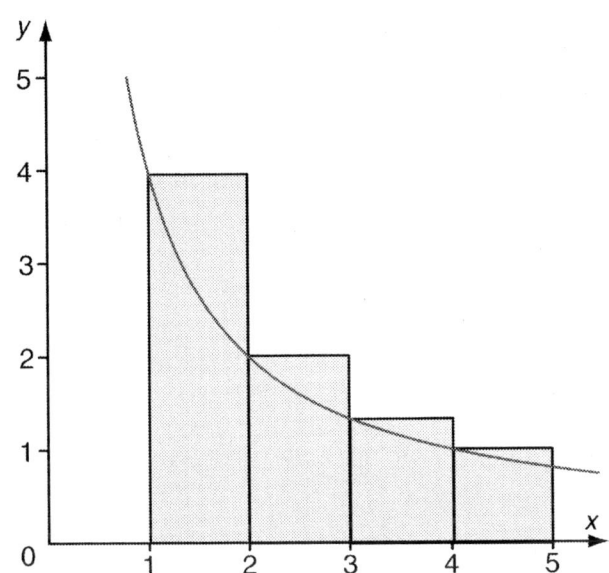

21. *Business: Cost from Marginal Cost.* A tool company determines that the marginal cost of the xth hammer is given by
$$C'(x) = -0.001x + 20, \ C(0) = \$0.$$
Find the total cost of producing 250 hammers.

21. _____

22. *Physical Science: Distance.* A particle is initially at the origin. Its velocity, in meters per second, at any time t, $t \geq 0$, is given by $v(t) = 2t^2 + 6t$. Find the distance that the particle travels in the first 3 seconds (from $t = 0$ to $t = 3$).

22. _____

Integrate using any method.

23. $\int \dfrac{dx}{x(2-x)}$

23. _____

24. $\int x^3 e^x\,dx$

24. _____

25. $\int x^4 e^{x^5}\, dx$ 25. _____

26. $\int \dfrac{1}{\sqrt{x}} \ln x \, dx$ 26. _____

27. $\int x^5 \sqrt{x^3 + 5}\, dx$ 27. _____

28. $\int \dfrac{dx}{16 - x^2}$ 28. _____

29. $\int x^3 e^{0.25x}\, dx$ 29. _____

30. $\int x \ln 25x \, dx$ 30. _____

Evaluate using any method.

31. $\int \dfrac{\left[8(\ln x)^3 - 2(\ln x)^2 - 6\right]}{x}\, dx$ 31. _____

32. $\int \ln \dfrac{x-5}{x+2}\, dx$ 32. _____

33. Use a grapher to approximate the area between the following curves:
$$y = 3x - 2x^2 + 2,$$
$$y = 4x^3 - 2x^2 - x + 2,$$
$$x = 0,\ x = 1.$$
33. _____

CALCULUS AND ITS APPLICATIONS

Name:

Chapter 5, Form F

Evaluate.

1. $\int dx$

1. _____

2. $\int 100x^4 \, dx$

2. _____

3. $\int \left(3e^x + \dfrac{1}{x} + x^{3/2}\right) dx$

3. _____

Find the area under the curve over the indicated interval.

4. $y = 8x - x^2$; $[0, 8]$

4. _____

5. $y = \dfrac{4}{x}$; $[1, 10]$

5. _____

6. Give two interpretations of the shaded area.

6. _____

Evaluate.

7. $\int_{-2}^{3} (6x^2 + 2x) \, dx$

7. _____

8. $\int_0^4 e^{-8x} \, dx$

8. _____

9. $\int_0^a \sqrt{x} \, dx$

9. _____

10. Decide whether $\int_a^b f(x)\,dx$ is positive, negative, or zero.

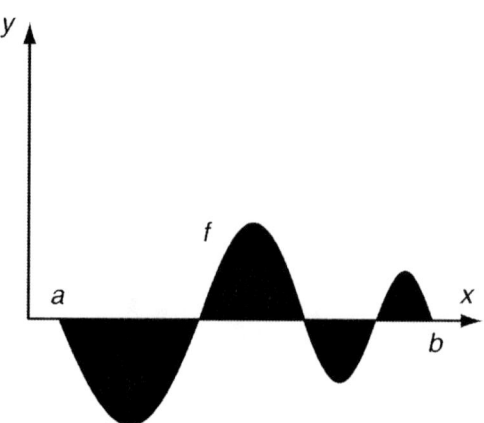

10. _____

Evaluate using substitution. Do not use Table 1.

11. $\int \dfrac{dx}{x-1}$

11. _____

12. $\int e^{-0.4x}\,dx$

12. _____

13. $\int t^4 \sqrt{t^5 + 5}\,dt$

13. _____

Evaluate using integration by parts. Do not use Table 1.

14. $\int x e^{5x}\,dx$

14. _____

15. $\int x^9 \ln x^{10}\,dx$

15. _____

Evaluate using Table 1.

16. $\int 2^x\,dx$

16. _____

17. $\int \dfrac{dx}{x\sqrt{1+x^2}}$

17. _____

CALCULUS AND ITS APPLICATIONS   Chapter 5, Form F

18. Find the average value of $y = -3t^2 + 5t$ over $[-2, 1]$.

18. _____

19. Find the area of the region bounded by $y = x$, $x = 0$, $y = x^5$, and $x = -1$.

19. _____

20. Approximate $\int_0^4 (x^2 + 10)\, dx$ by computing the area of each rectangle and adding.

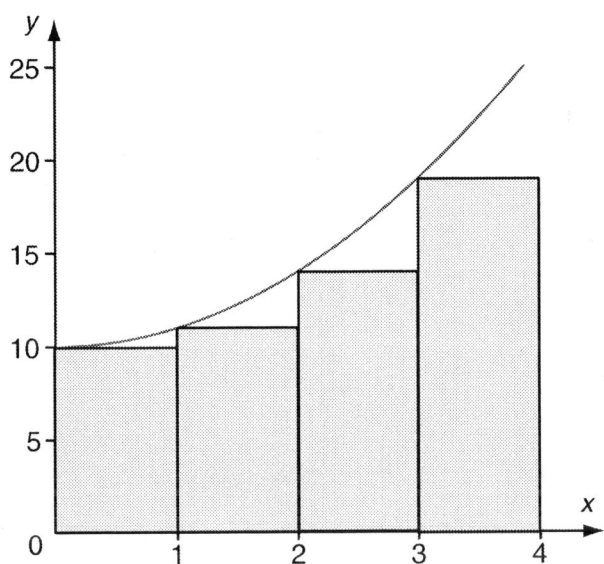

20. _____

21. *Business: Cost from Marginal Cost.* An office equipment company determines that the marginal cost of the xth desk chair is given by
$$C'(x) = -0.12x + 80,\ C(0) = \$0.$$
Find the total cost of producing 200 desk chairs.

21. _____

22. *Physical Science: Distance.* A particle is initially at the origin. Its velocity, in meters per second, at any time t, $t \geq 0$, is given by $v(t) = t^2 + 2t$. Find the distance that the particle travels in the first 3 seconds (from $t = 0$ to $t = 3$).

22. _____

Integrate using any method.

23. $\int \dfrac{dx}{x(4-x)}$

23. _____

24. $\int 3x^5 e^x\, dx$

24. _____

25. $\displaystyle\int x^2 e^{x^3}\,dx$ 25. _____

26. $\displaystyle\int \sqrt[4]{x}\ln x\,dx$ 26. _____

27. $\displaystyle\int x^7\sqrt{x^4+2}\,dx$ 27. _____

28. $\displaystyle\int \frac{dx}{25-x^2}$ 28. _____

29. $\displaystyle\int x^2 e^{-0.5x}\,dx$ 29. _____

30. $\displaystyle\int x\ln 6x\,dx$ 30. _____

Evaluate using any method.

31. $\displaystyle\int \frac{\left[3-2(\ln x)^2+4(\ln x)^3\right]}{x}\,dx$ 31. _____

32. $\displaystyle\int \ln(x-1)(x+6)\,dx$ 32. _____

33. Use a grapher to approximate the area between the following curves:

$$y = 10x + x^2,$$
$$y = x^3 + x^2 + x,$$
$$x = 0,\ x = 3.$$

33. _____

CALCULUS AND ITS APPLICATIONS

Name:

Chapter 6, Form A

Given the demand and supply functions

$$p = D(x) = (x - 8)^2 \text{ and } p = S(x) = x^2 + 5x + 1,$$

in dollars, find each of the following.

1. The equilibrium point

    1. _____

2. The consumer's surplus at the equilibrium point

    2. _____

3. The producer's surplus at the equilibrium point

    3. _____

4. *Business: Amount of Continuous Money Flow.* Find the amount of continuous money flow if $2500 per year is being invested at 5%, compounded continuously for 8 yr.

    4. _____

5. *Business: Continuous Money Flow.* Consider a continuous money flow into an investment at the rate of P_0 dollars per year. What should P_0 be so that the amount of a continuous money flow over 20 yr at 5%, compounded continuously, will be $20,000?

    5. _____

6. *Physical Science: Demand for a Gem.* In 1985 ($t = 0$), the world use of a certain gemstone was 3800 tons, and the demand for newly mined gems of this kind was growing exponentially at the rate of 4% per year. If the demand continues to grow at this rate, how many tons of this gem will the world use from 1985 to 2000?

    6. _____

7. *Physical Science: Depletion of a Gem.* The world reserves of the gemstone in Question 6 were 1,500,000 tons in 1985. Assuming that the growth rate of 4% per year continues and that no new reserves are discovered, when will the world reserves of this gemstone be exhausted?

    7. _____

8. *Business: Present Value.* Following the birth of a child, a parent wants to make an initial investment P_0 that will grow to $25,000 by the child's 18th birthday. Interest is compounded continuously at 6%. What should the initial investment be?

    8. _____

9. **Business: Accumulated Present Value.** Find the accumulated present value of an investment over a 15-yr period if there is a continuous money flow of $1000 per year and the current rate is 4%.

9. _____

10. **Business: Accumulated Present Value.** Find the accumulated present value of an investment for which the continuous money flow in Question 9 is perpetual.

10. _____

Determine whether the improper integral is convergent or divergent, and calculate its value if it is convergent.

11. $\int_{1}^{\infty} \frac{2dx}{x^4}$

11. _____

12. $\int_{0}^{\infty} \frac{5}{3+x} dx$

12. _____

13. Find k such that $f(x) = kx^4$ is a probability density function over the interval $[0, 4]$. Then find the probability density function.

13. _____

14. **Business: Length of Wait.** A grocery store determines that the length of time t, in minutes, that a customer must wait in line is an exponentially distributed random variable with probability density function

$$f(t) = \frac{1}{4}e^{-0.25t}, \ 0 \leq t < \infty.$$

Find the probability that a customer will wait in line no more than 2 min.

14. _____

Given the probability density function $f(x) = \frac{1}{6}x$ over $[2, 4]$, find each of the following.

15. $E(x)$

15. _____

16. $E(x^2)$

16. _____

17. The mean

17. _____

CALCULUS AND ITS APPLICATIONS   Chapter 6, Form A   135

Given the probability density function $f(x) = \frac{1}{6}x$ over $[2, 4]$, find each of the following.

18. The variance   18. _____

19. The standard deviation   19. _____

Let x be a continuous random variable with standard normal density. Using Table 2, find each of the following.

20. $P(0 \leq x \leq 2.1)$ 20. _____

21. $P(-0.56 \leq x \leq -0.2)$ 21. _____

22. $P(-2.2 \leq x \leq 1.83)$ 22. _____

23. The price per pound p of coffee at various stores in a certain city is normally distributed with mean $4.50 and standard deviation $0.30. What is the probability that the price per pound is $5.00 or more?   23. _____

Find the volume generated by revolving about the x-axis the regions bounded by the following graphs.

24. $y = \dfrac{2}{\sqrt[4]{x}}$, $x = 4$, $x = 9$ 24. _____

25. $y = \sqrt{3x - 1}$, $x = 0$, $x = 4$ 25. _____

Solve the differential equation.

26. $\dfrac{dy}{dx} = 6x^2 y$ 26. _____

27. $\dfrac{dy}{dx} = \dfrac{8}{y}$ 27. _____

28. $\dfrac{dy}{dt} = 2y$; $y = 4$ when $t = 0$ 28. _____

29. $y' = 6x^3 - x^3 y$ 29. _____

Solve the differential equation.

30. $\dfrac{dr}{dt} = 5r^{-3}$

31. $y' = 5y + xy$

32. *Economics: Elasticity.* Find the demand function given the elasticity condition
$$E(p) = 3 \text{ for all } p > 0.$$

33. *Business: Stock Growth.* The growth rate of a stock for Hanson Electric is modeled by
$$\dfrac{dV}{dt} = k(L - V),$$
where V is the value of the stock per share, after time t (in months), $L = \$14$, the limiting value of the stock, k is a constant, and $V(0) = \$0$.

(a) Write the solution $V(t)$ in terms of L and k.

(b) If $V(5) = \$2.50$, determine k to the nearest hundredth.

(c) Rewrite $V(t)$ in terms of t and k using the value of k found in part (b).

(d) Use the equation in part (c) to find $V(18)$, the value of the stock after 18 months.

(e) In how many months will the value be $10?

34. The function $f(x) = x^6$ is a probability density function over the interval $[0, b]$. What is b?

35. Determine whether the following improper integral is convergent or divergent, and calculate its value if it is convergent:
$$\int_{1}^{\infty} x^4 e^{-x^5} \, dx.$$

36. Approximate the integral.
$$\int_{-\infty}^{\infty} \dfrac{2}{1 + x^2} \, dx$$

CALCULUS AND ITS APPLICATIONS

Name:

Chapter 6, Form B

Given the demand and supply functions

$$p = D(x) = (x-6)^2 \text{ and } p = S(x) = x^2 + 6x,$$

in dollars, find each of the following.

1. The equilibrium point

    1. _____

2. The consumer's surplus at the equilibrium point

    2. _____

3. The producer's surplus at the equilibrium point

    3. _____

4. *Business: Amount of Continuous Money Flow.* Find the amount of continuous money flow if $5000 per year is being invested at 7%, compounded continuously for 10 yr.

    4. _____

5. *Business: Continuous Money Flow.* Consider a continuous money flow into an investment at the rate of P_0 dollars per year. What should P_0 be so that the amount of a continuous money flow over 20 yr at 7%, compounded continuously, will be $50,000?

    5. _____

6. *Physical Science: Demand for a Gem.* In 1995 ($t=0$), the world use of a certain gemstone was 12,000 tons, and the demand for newly mined gems of this kind was growing exponentially at the rate of 5% per year. If the demand continues to grow at this rate, how many tons of this gem will the world use from 1995 to 2005?

    6. _____

7. *Physical Science: Depletion of a Gem.* The world reserves of the gemstone in Question 6 were 20,000,000 tons in 1995. Assuming that the growth rate of 5% per year continues and that no new reserves are discovered, when will the world reserves of this gemstone be exhausted?

    7. _____

8. *Business: Present Value.* Following the birth of a child, a parent wants to make an initial investment P_0 that will grow to $30,000 by the child's 18th birthday. Interest is compounded continuously at 8%. What should the initial investment be?

    8. _____

9. *Business: Accumulated Present Value.* Find the accumulated present value of an investment over a 20-yr period if there is a continuous money flow of $4500 per year and the current rate is 3%.

9. _____

10. *Business: Accumulated Present Value.* Find the accumulated present value of an investment for which the continuous money flow in Question 9 is perpetual.

10. _____

Determine whether the improper integral is convergent or divergent, and calculate its value if it is convergent.

11. $\displaystyle\int_2^\infty \frac{dx}{x^4}$

11. _____

12. $\displaystyle\int_0^\infty \frac{4}{3+x}\, dx$

12. _____

13. Find k such that $f(x) = kx^5$ is a probability density function over the interval $[0, 3]$. Then find the probability density function.

13. _____

14. *Business: Length of Wait.* A print shop determines that the length of time t, in weeks, that a customer must wait for an order is an exponentially distributed random variable with probability density function

$$f(t) = 0.9e^{-0.9t},\ 0 \le t < \infty.$$

Find the probability that a customer will wait for an order no more than 2 weeks.

14. _____

Given the probability density function $f(x) = \frac{1}{3}x$ over $[1, 2]$, find each of the following.

15. $E(x)$

15. _____

16. $E(x^2)$

16. _____

17. The mean

17. _____

CALCULUS AND ITS APPLICATIONS   Chapter 6, Form B

Given the probability density function $f(x) = \frac{1}{3}x$ over $[1, 2]$, find each of the following.

18. The variance   18. _____

19. The standard deviation   19. _____

Let x be a continuous random variable with standard normal density. Using Table 2, find each of the following.

20. $P(0 \leq x \leq 2.3)$ 20. _____

21. $P(0.75 \leq x \leq 2.15)$ 21. _____

22. $P(-1.4 \leq x \leq 2.02)$ 22. _____

23. The price per pound p of cheddar cheese at various stores in a certain city is normally distributed with mean \$4.70 and standard deviation \$0.25. What is the probability that the price per pound is \$4.00 or less?   23. _____

Find the volume generated by revolving about the x-axis the regions bounded by the following graphs.

24. $y = \frac{1}{2}x^4$, $x = 0$, $x = 2$ 24. _____

25. $y = \sqrt{3+x}$, $x = 2$, $x = 4$ 25. _____

Solve the differential equation.

26. $\dfrac{dy}{dx} = 8x^3 y$ 26. _____

27. $\dfrac{dy}{dx} = \dfrac{4}{y}$ 27. _____

28. $\dfrac{dy}{dt} = 8y$; $y = 5$ when $t = 0$ 28. _____

29. $y' = 4x^4 - x^4 y$ 29. _____

Solve the differential equation.

30. $\dfrac{dr}{dt} = 6r^{-5}$

30. _____

31. $y' = 3y + xy$

31. _____

32. *Economics: Elasticity.* Find the demand function given the elasticity condition

$$E(p) = 10 \text{ for all } p > 0.$$

32. _____

33. *Business: Stock Growth.* The growth rate of a stock for Hanson Electric is modeled by

$$\dfrac{dV}{dt} = k(L - V),$$

where V is the value of the stock per share, after time t (in months), $L = \$40$, the limiting value of the stock, k is a constant, and $V(0) = \$0$.

(a) Write the solution $V(t)$ in terms of L and k.

33. (a) _____

(b) If $V(6) = \$8.75$, determine k to the nearest hundredth.

(b) _____

(c) Rewrite $V(t)$ in terms of t and k using the value of k found in part (b).

(c) _____

(d) Use the equation in part (c) to find $V(18)$, the value of the stock after 18 months.

(d) _____

(e) In how many months will the value be $30?

(e) _____

34. The function $f(x) = x^4$ is a probability density function over the interval $[0, b]$. What is b?

34. _____

35. Determine whether the following improper integral is convergent or divergent, and calculate its value if it is convergent:

$$\int_1^\infty 2x^2 e^{-x^3}\, dx.$$

35. _____

36. Approximate the integral.

$$\int_{-\infty}^\infty \dfrac{5}{1 + 3x^2}\, dx$$

36. _____

CALCULUS AND ITS APPLICATIONS

Name:

Chapter 6, Form C

Given the demand and supply functions

$$p = D(x) = (x - 5)^2 \text{ and } p = S(x) = x^2 + 2x + 1,$$

in dollars, find each of the following.

1. The equilibrium point

2. The consumer's surplus at the equilibrium point

3. The producer's surplus at the equilibrium point

4. *Business: Amount of Continuous Money Flow.* Find the amount of continuous money flow if $2000 per year is being invested at 6%, compounded continuously for 5 yr.

5. *Business: Continuous Money Flow.* Consider a continuous money flow into an investment at the rate of P_0 dollars per year. What should P_0 be so that the amount of a continuous money flow over 15 yr at 6%, compounded continuously, will be $25,000?

6. *Physical Science: Demand for a Gem.* In 1985 ($t = 0$), the world use of a certain gemstone was 10,000 tons, and the demand for newly mined gems of this kind was growing exponentially at the rate of 3% per year. If the demand continues to grow at this rate, how many tons of this gem will the world use from 1985 to 2005?

7. *Physical Science: Depletion of a Gem.* The world reserves of the gemstone in Question 6 were 3,000,000 tons in 1985. Assuming that the growth rate of 3% per year continues and that no new reserves are discovered, when will the world reserves of this gemstone be exhausted?

8. *Business: Present Value.* Following the birth of a child, a parent wants to make an initial investment P_0 that will grow to $40,000 by the child's 18th birthday. Interest is compounded continuously at 8%. What should the initial investment be?

9. **Business: Accumulated Present Value.** Find the accumulated present value of an investment over a 15-yr period if there is a continuous money flow of $2100 per year and the current rate is 6%.

9. _____

10. **Business: Accumulated Present Value.** Find the accumulated present value of an investment for which the continuous money flow in Question 9 is perpetual.

10. _____

Determine whether the improper integral is convergent or divergent, and calculate its value if it is convergent.

11. $\displaystyle\int_4^\infty \frac{5\,dx}{x^5}$

11. _____

12. $\displaystyle\int_0^\infty \frac{3}{x+6}\,dx$

12. _____

13. Find k such that $f(x) = kx^2$ is a probability density function over the interval $[0, 3]$. Then find the probability density function.

13. _____

14. **Business: Length of Wait.** A restaurant determines that the length of time t, in minutes, that a customer must wait for an order is an exponentially distributed random variable with probability density function

$$f(t) = 0.06e^{-0.06t},\ 0 \le t < \infty.$$

Find the probability that a customer will wait for an order no more than 30 min.

14. _____

Given the probability density function $f(x) = \frac{1}{10}x$ over $[4, 6]$, find each of the following.

15. $E(x)$

15. _____

16. $E(x^2)$

16. _____

17. The mean

17. _____

CALCULUS AND ITS APPLICATIONS   Chapter 6, Form C

Given the probability density function $f(x) = \frac{1}{10}x$ over $[4, 6]$, find each of the following.

18. The variance

18. _____

19. The standard deviation

19. _____

Let x be a continuous random variable with standard normal density. Using Table 2, find each of the following.

20. $P(0 \leq x \leq 0.8)$

20. _____

21. $P(-0.9 \leq x \leq -0.15)$

21. _____

22. $P(-2.2 \leq x \leq 1.8)$

22. _____

23. The price per dozen eggs at various stores in a certain city is normally distributed with mean $1.29 and standard deviation $0.10. What is the probability that the price per dozen is $1.50 or more?

23. _____

Find the volume generated by revolving about the x-axis the regions bounded by the following graphs.

24. $y = \dfrac{1}{x},\ x = 1,\ x = 6$

24. _____

25. $y = e^x,\ x = -3,\ x = 5$

25. _____

Solve the differential equation.

26. $\dfrac{dy}{dx} = 6x^2 y$

26. _____

27. $\dfrac{dy}{dx} = \dfrac{3}{y}$

27. _____

28. $\dfrac{dy}{dt} = 8y;\ y = 4$ when $t = 0$

28. _____

29. $y' = 5x^{10} - x^{10} y$

29. _____

Solve the differential equation.

30. $\dfrac{dr}{dt} = 6r^{-3}$

30. _____

31. $y' = 12y - xy$

31. _____

32. *Economics: Elasticity.* Find the demand function given the elasticity condition

$$E(p) = 2 \text{ for all } p > 0.$$

32. _____

33. *Business: Stock Growth.* The growth rate of a stock for Hanson Electric is modeled by

$$\dfrac{dV}{dt} = k(L - V),$$

where V is the value of the stock per share, after time t (in months), $L = \$50$, the limiting value of the stock, k is a constant, and $V(0) = \$0$.

(a) Write the solution $V(t)$ in terms of L and k.

33. (a) _____

(b) If $V(6) = \$24$, determine k to the nearest hundredth.

(b) _____

(c) Rewrite $V(t)$ in terms of t and k using the value of k found in part (b).

(c) _____

(d) Use the equation in part (c) to find $V(18)$, the value of the stock after 18 months.

(d) _____

(e) In how many months will the value be $40?

(e) _____

34. The function $f(x) = x^8$ is a probability density function over the interval $[0, b]$. What is b?

34. _____

35. Determine whether the following improper integral is convergent or divergent, and calculate its value if it is convergent:

$$\int_1^\infty 2x^4 e^{x^5}\, dx.$$

35. _____

36. Approximate the integral.

$$\int_{-\infty}^\infty \dfrac{4}{x^2 + 2}\, dx$$

36. _____

CALCULUS AND ITS APPLICATIONS

Name:

Chapter 6, Form D

Given the demand and supply functions

$$p = D(x) = (x-4)^2 \text{ and } p = S(x) = x^2,$$

in dollars, find each of the following.

1. The equilibrium point

2. The consumer's surplus at the equilibrium point

3. The producer's surplus at the equilibrium point

4. *Business: Amount of Continuous Money Flow.* Find the amount of continuous money flow if $1700 per year is being invested at 5%, compounded continuously for 6 yr.

5. *Business: Continuous Money Flow.* Consider a continuous money flow into an investment at the rate of P_0 dollars per year. What should P_0 be so that the amount of a continuous money flow over 18 yr at 6%, compounded continuously, will be $16,000?

6. *Physical Science: Demand for a Gem.* In 1990 ($t = 0$), the world use of a certain gemstone was 4000 tons, and the demand for newly mined gems of this kind was growing exponentially at the rate of 1.5% per year. If the demand continues to grow at this rate, how many tons of this gem will the world use from 1990 to 2010?

7. *Physical Science: Depletion of a Gem.* The world reserves of the gemstone in Question 6 were 5,000,000 tons in 1990. Assuming that the growth rate of 1.5% per year continues and that no new reserves are discovered, when will the world reserves of this gemstone be exhausted?

8. *Business: Present Value.* Following the birth of a child, a parent wants to make an initial investment P_0 that will grow to $35,000 by the child's 18th birthday. Interest is compounded continuously at 8%. What should the initial investment be?

1. _____

2. _____

3. _____

4. _____

5. _____

6. _____

7. _____

8. _____

9. **Business: Accumulated Present Value.** Find the accumulated present value of an investment over a 20-yr period if there is a continuous money flow of $3000 per year and the current rate is 4%.

9. _____

10. **Business: Accumulated Present Value.** Find the accumulated present value of an investment for which the continuous money flow in Question 9 is perpetual.

10. _____

Determine whether the improper integral is convergent or divergent, and calculate its value if it is convergent.

11. $\displaystyle\int_{10}^{\infty} \frac{3dx}{x^5}$

11. _____

12. $\displaystyle\int_{0}^{\infty} \frac{1}{2x+3}\,dx$

12. _____

13. Find k such that $f(x) = kx^6$ is a probability density function over the interval $[0, 2]$. Then find the probability density function.

13. _____

14. **Business: Length of Wait.** A hardware store determines that the length of time t, in minutes, that a customer must wait in line is an exponentially distributed random variable with probability density function

$$f(t) = 0.2e^{-0.2t},\ 0 \leq t < \infty.$$

Find the probability that a customer will wait in line no more than 4 min.

14. _____

Given the probability density function $f(x) = \frac{1}{8}x$ over $[1, 4]$, find each of the following.

15. $E(x)$

15. _____

16. $E(x^2)$

16. _____

17. The mean

17. _____

CALCULUS AND ITS APPLICATIONS Chapter 6, Form D

Given the probability density function $f(x) = \frac{1}{8}x$ over $[1, 4]$, find each of the following.

18. The variance

18. _____

19. The standard deviation

19. _____

Let x be a continuous random variable with standard normal density. Using Table 2, find each of the following.

20. $P(0 \le x \le 0.84)$

20. _____

21. $P(0.46 \le x \le 1.43)$

21. _____

22. $P(-1.2 \le x \le 2.05)$

22. _____

23. The price per gallon of gas at various stores in a certain city is normally distributed with mean $1.69 and standard deviation $0.05. What is the probability that the price per gallon is $1.65 or less?

23. _____

Find the volume generated by revolving about the x-axis the regions bounded by the following graphs.

24. $y = 3x$, $x = 1$, $x = 6$

24. _____

25. $y = \sqrt{x-3}$, $x = 2$, $x = 8$

25. _____

Solve the differential equation.

26. $\dfrac{dy}{dx} = 12x^5 y$

26. _____

27. $\dfrac{dy}{dx} = \dfrac{5}{y}$

27. _____

28. $\dfrac{dy}{dt} = 6y$; $y = 8$ when $t = 0$

28. _____

29. $y' = 5x^5 - x^5 y$

29. _____

Solve the differential equation.

30. $\dfrac{dr}{dt} = 2r^{-10}$

31. $y' = 10y - xy$

32. *Economics: Elasticity.* Find the demand function given the elasticity condition

$$E(p) = 8 \text{ for all } p > 0.$$

33. *Business: Stock Growth.* The growth rate of a stock for Hanson Electric is modeled by

$$\dfrac{dV}{dt} = k(L - V),$$

where V is the value of the stock per share, after time t (in months), $L = \$25$, the limiting value of the stock, k is a constant, and $V(0) = \$0$.

(a) Write the solution $V(t)$ in terms of L and k.

(b) If $V(6) = \$12.50$, determine k to the nearest hundredth.

(c) Rewrite $V(t)$ in terms of t and k using the value of k found in part (b).

(d) Use the equation in part (c) to find $V(18)$, the value of the stock after 18 months.

(e) In how many months will the value be $10?

34. The function $f(x) = x^5$ is a probability density function over the interval $[0, b]$. What is b?

35. Determine whether the following improper integral is convergent or divergent, and calculate its value if it is convergent:

$$\int_1^\infty 8x^3 e^{-x^4}\, dx.$$

36. Approximate the integral.

$$\int_{-\infty}^\infty \dfrac{5}{2x^2 + 1}\, dx$$

CALCULUS AND ITS APPLICATIONS

Name:

Chapter 6, Form E

Given the demand and supply functions

$$p = D(x) = (x-9)^2 \text{ and } p = S(x) = x^2 + 2x + 1,$$

in dollars, find each of the following.

1. The equilibrium point

   1. _____

2. The consumer's surplus at the equilibrium point

   2. _____

3. The producer's surplus at the equilibrium point

   3. _____

4. *Business: Amount of Continuous Money Flow.* Find the amount of continuous money flow if $2200 per year is being invested at 8%, compounded continuously for 10 yr.

   4. _____

5. *Business: Continuous Money Flow.* Consider a continuous money flow into an investment at the rate of P_0 dollars per year. What should P_0 be so that the amount of a continuous money flow over 18 yr at 8%, compounded continuously, will be $20,000?

   5. _____

6. *Physical Science: Demand for a Gem.* In 1985 ($t = 0$), the world use of a certain gemstone was 1200 tons, and the demand for newly mined gems of this kind was growing exponentially at the rate of 4% per year. If the demand continues to grow at this rate, how many tons of this gem will the world use from 1985 to 2005?

   6. _____

7. *Physical Science: Depletion of a Gem.* The world reserves of the gemstone in Question 6 were 3,000,000 tons in 1985. Assuming that the growth rate of 4% per year continues and that no new reserves are discovered, when will the world reserves of this gemstone be exhausted?

   7. _____

8. *Business: Present Value.* Following the birth of a child, a parent wants to make an initial investment P_0 that will grow to $50,000 by the child's 18th birthday. Interest is compounded continuously at 6%. What should the initial investment be?

   8. _____

9. **Business: Accumulated Present Value.** Find the accumulated present value of an investment over a 15-yr period if there is a continuous money flow of $2100 per year and the current rate is 7%.

9. _____

10. **Business: Accumulated Present Value.** Find the accumulated present value of an investment for which the continuous money flow in Question 9 is perpetual.

10. _____

Determine whether the improper integral is convergent or divergent, and calculate its value if it is convergent.

11. $\int_{1}^{\infty} \dfrac{dx}{x^2}$

11. _____

12. $\int_{0}^{\infty} \dfrac{5}{5+x}\,dx$

12. _____

13. Find k such that $f(x) = kx^4$ is a probability density function over the interval $[0, 2]$. Then find the probability density function.

13. _____

14. **Business: Length of Wait.** A grocery store determines that the length of time t, in minutes, that a customer must wait in line is an exponentially distributed random variable with probability density function

$$f(t) = 0.09e^{-0.09t},\ 0 \leq t < \infty.$$

Find the probability that a customer will wait in line no more than 5 min.

14. _____

Given the probability density function $f(x) = \tfrac{1}{5}x$ over $[1, 3]$, find each of the following.

15. $E(x)$

15. _____

16. $E(x^2)$

16. _____

17. The mean

17. _____

CALCULUS AND ITS APPLICATIONS   Chapter 6, Form E

Given the probability density function $f(x) = \frac{1}{5}x$ over $[1, 3]$, find each of the following.

18. The variance   18. _____

19. The standard deviation   19. _____

Let x be a continuous random variable with standard normal density. Using Table 2, find each of the following.

20. $P(0 \leq x \leq 1.5)$ 20. _____

21. $P(0.42 \leq x \leq 2.86)$ 21. _____

22. $P(-1.4 \leq x \leq 2.15)$ 22. _____

23. The price per pound of brie at various stores in a certain city is normally distributed with mean $9.00 and standard deviation $0.75. What is the probability that the price per pound is $10.00 or more?   23. _____

Find the volume generated by revolving about the x-axis the regions bounded by the following graphs.

24. $y = 5x$, $x = 1$, $x = 2$ 24. _____

25. $y = \sqrt{3 + x}$, $x = 4$, $x = 5$ 25. _____

Solve the differential equation.

26. $\dfrac{dy}{dx} = 5x^4 y$ 26. _____

27. $\dfrac{dy}{dx} = \dfrac{6}{y}$ 27. _____

28. $\dfrac{dy}{dt} = 4y$; $y = 6$ when $t = 0$ 28. _____

29. $y' = 2x^3 - x^3 y$ 29. _____

Solve the differential equation.

30. $\dfrac{dr}{dt} = -2r^{-4}$

30. _____

31. $y' = 2y + xy$

31. _____

32. *Economics: Elasticity.* Find the demand function given the elasticity condition

$$E(p) = 6 \text{ for all } p > 0.$$

32. _____

33. *Business: Stock Growth.* The growth rate of a stock for Hanson Electric is modeled by

$$\dfrac{dV}{dt} = k(L - V),$$

where V is the value of the stock per share, after time t (in months), $L = \$8$, the limiting value of the stock, k is a constant, and $V(0) = \$0$.

(a) Write the solution $V(t)$ in terms of L and k.

33. (a) _____

(b) If $V(6) = \$3.00$, determine k to the nearest hundredth.

(b) _____

(c) Rewrite $V(t)$ in terms of t and k using the value of k found in part (b).

(c) _____

(d) Use the equation in part (c) to find $V(18)$, the value of the stock after 18 months.

(d) _____

(e) In how many months will the value be $7?

(e) _____

34. The function $f(x) = x^7$ is a probability density function over the interval $[0, b]$. What is b?

34. _____

35. Determine whether the following improper integral is convergent or divergent, and calculate its value if it is convergent:

$$\int_{-\infty}^{0} 3x^4 e^{x^5}\, dx.$$

35. _____

36. Approximate the integral.

$$\int_{-\infty}^{\infty} \dfrac{12}{1 + x^2}\, dx$$

36. _____

CALCULUS AND ITS APPLICATIONS

Name:

Chapter 6, Form F

Given the demand and supply functions

$$p = D(x) = (x - 6)^2 \text{ and } p = S(x) = x^2 + 12,$$

in dollars, find each of the following.

1. The equilibrium point

   1. _____

2. The consumer's surplus at the equilibrium point

   2. _____

3. The producer's surplus at the equilibrium point

   3. _____

4. *Business: Amount of Continuous Money Flow.* Find the amount of continuous money flow if $3200 per year is being invested at 6%, compounded continuously for 12 yr.

   4. _____

5. *Business: Continuous Money Flow.* Consider a continuous money flow into an investment at the rate of P_0 dollars per year. What should P_0 be so that the amount of a continuous money flow over 20 yr at 6%, compounded continuously, will be $40,000?

   5. _____

6. *Physical Science: Demand for a Gem.* In 1985 ($t = 0$), the world use of a certain gemstone was 1900 tons, and the demand for newly mined gems of this kind was growing exponentially at the rate of 3% per year. If the demand continues to grow at this rate, how many tons of this gem will the world use from 1985 to 2000?

   6. _____

7. *Physical Science: Depletion of a Gem.* The world reserves of the gemstone in Question 6 were 2,000,000 tons in 1985. Assuming that the growth rate of 3% per year continues and that no new reserves are discovered, when will the world reserves of this gemstone be exhausted?

   7. _____

8. *Business: Present Value.* Following the birth of a child, a parent wants to make an initial investment P_0 that will grow to $60,000 by the child's 18th birthday. Interest is compounded continuously at 5%. What should the initial investment be?

   8. _____

9. **Business: Accumulated Present Value.** Find the accumulated present value of an investment over a 20-yr period if there is a continuous money flow of $2500 per year and the current rate is 3%.

9. _____

10. **Business: Accumulated Present Value.** Find the accumulated present value of an investment for which the continuous money flow in Question 9 is perpetual.

10. _____

Determine whether the improper integral is convergent or divergent, and calculate its value if it is convergent.

11. $\displaystyle\int_{2}^{\infty} \frac{3dx}{x^2}$

11. _____

12. $\displaystyle\int_{0}^{\infty} \frac{6}{2x+3} dx$

12. _____

13. Find k such that $f(x) = kx^3$ is a probability density function over the interval $[1, 4]$. Then find the probability density function.

13. _____

14. **Business: Length of Wait.** A sub shop determines that the length of time t, in minutes, that a customer must wait in line is an exponentially distributed random variable with probability density function

$$f(t) = 0.1e^{-0.1t}, \ 0 \leq t < \infty.$$

Find the probability that a customer will wait in line no more than 10 min.

14. _____

Given the probability density function $f(x) = \frac{1}{14}x$ over $[6, 8]$, find each of the following.

15. $E(x)$

15. _____

16. $E(x^2)$

16. _____

17. The mean

17. _____

CALCULUS AND ITS APPLICATIONS   Chapter 6, Form F

Given the probability density function $f(x) = \frac{1}{14}x$ over $[6, 8]$, find each of the following.

18. The variance

18. _____

19. The standard deviation

19. _____

Let x be a continuous random variable with standard normal density. Using Table 2, find each of the following.

20. $P(0 \leq x \leq 1.8)$

20. _____

21. $P(-1.2 \leq x \leq -0.35)$

21. _____

22. $P(-1.1 \leq x \leq 0.78)$

22. _____

23. The price per dozen doughnuts at various stores in a certain city is normally distributed with mean $5.00 and standard deviation $0.50. What is the probability that the price per dozen is $3.80 or more?

23. _____

Find the volume generated by revolving about the x-axis the regions bounded by the following graphs.

24. $y = \dfrac{5}{x}$, $x = 1$, $x = 9$

24. _____

25. $y = \sqrt{x + 5}$, $x = 1$, $x = 8$

25. _____

Solve the differential equation.

26. $\dfrac{dy}{dx} = 12x^5 y$

26. _____

27. $\dfrac{dy}{dx} = \dfrac{3}{y}$

27. _____

28. $\dfrac{dy}{dt} = 8y$; $y = 4$ when $t = 0$

28. _____

29. $y' = 10x^4 - x^4 y$

29. _____

Solve the differential equation.

30. $\dfrac{dr}{dt} = -9r^{-3}$

31. $y' = 2y - xy$

32. *Economics: Elasticity.* Find the demand function given the elasticity condition
$$E(p) = 8 \text{ for all } p > 0.$$

33. *Business: Stock Growth.* The growth rate of a stock for Hanson Electric is modeled by
$$\dfrac{dV}{dt} = k(L - V),$$
where V is the value of the stock per share, after time t (in months), $L = \$10$, the limiting value of the stock, k is a constant, and $V(0) = \$0$.

  (a) Write the solution $V(t)$ in terms of L and k.

  (b) If $V(12) = \$7.75$, determine k to the nearest hundredth.

  (c) Rewrite $V(t)$ in terms of t and k using the value of k found in part (b).

  (d) Use the equation in part (c) to find $V(18)$, the value of the stock after 18 months.

  (e) In how many months will the value be $\$5$?

34. The function $f(x) = 2x$ is a probability density function over the interval $[0, b]$. What is b?

35. Determine whether the following improper integral is convergent or divergent, and calculate its value if it is convergent:
$$\int_{1}^{\infty} \dfrac{1}{2} x^4 e^{-x^5}\, dx.$$

36. Approximate the integral.
$$\int_{-\infty}^{\infty} \dfrac{7}{x^2 + 3}\, dx$$

CALCULUS AND ITS APPLICATIONS

Name:

Chapter 7, Form A

Given $f(x, y) = 2y + 4x^5 y - e^x$, find each of the following.

1. $f(2, -1)$

2. $\dfrac{\partial f}{\partial x}$

3. $\dfrac{\partial f}{\partial y}$

4. $\dfrac{\partial^2 f}{\partial x^2}$

5. $\dfrac{\partial^2 f}{\partial x \, \partial y}$

6. $\dfrac{\partial^2 f}{\partial y \, \partial x}$

7. $\dfrac{\partial^2 f}{\partial y^2}$

1. _____
2. _____
3. _____
4. _____
5. _____
6. _____
7. _____

Find the relative maximum and minimum values.

8. $f(x, y) = x^2 + xy + y^3 + 2x$

9. $f(x, y) = 3x^2 + 7y^2 - 12xy$

8. _____

9. _____

10. *Business: Predicting Total Revenue.* Consider the data in the following table regarding the total revenue of a company during the first three years of operation.

Year, x	Revenue, y (in millions)
1	$8
2	12
3	15

(a) Find the regression line $y = mx + b$.

(b) Use the regression line to predict sales in the fourth year.

10. (a) _____

(b) _____

157

11. Find the maximum value of
$$f(x, y) = 4xy - 3x^2 + 5y^2$$
subject to the constraint $x + 2y = 60$.

11. _____

12. Evaluate
$$\int_0^1 \int_2^x (x^3 + 3y)\, dy\, dx.$$

12. _____

13. *Business: Maximizing Production.* An automotive company has the following Cobb-Douglas production function for a certain product,
$$p(x, y) = 40x^{3/5}y^{2/5},$$
where x is labor, measured in dollars, and y is capital, measured in dollars. Suppose that a company can make a total investment in labor and capital of \$500,000. How should it allocate the investment between labor and capital in order to maximize production?

13. _____

14. Find f_x and f_t:
$$f(x, t) = \frac{x^2 + 5t}{x^3 - 5t}.$$

14. _____

15. Use a 3D grapher to graph
$$f(x, y) = x + \frac{1}{3}y^2 - \frac{1}{2}x^3.$$

Sketch the graph.

15.

CALCULUS AND ITS APPLICATIONS

Name:

Chapter 7, Form B

Given $f(x, y) = e^x + 4x^2y + 2y$, find each of the following.

1. $f(1, -2)$

2. $\dfrac{\partial f}{\partial x}$

3. $\dfrac{\partial f}{\partial y}$

4. $\dfrac{\partial^2 f}{\partial x^2}$

5. $\dfrac{\partial^2 f}{\partial x \partial y}$

6. $\dfrac{\partial^2 f}{\partial y \partial x}$

7. $\dfrac{\partial^2 f}{\partial y^2}$

1. _____

2. _____

3. _____

4. _____

5. _____

6. _____

7. _____

Find the relative maximum and minimum values.

8. $f(x, y) = x^2 + 2y^2 - 4xy + y$

9. $f(x, y) = 3x^2 + y^2$

8. _____

9. _____

10. *Business: Predicting Total Revenue.* Consider the data in the following table regarding the total revenue of a company during the first three years of operation.

Year, x	Revenue, y (in millions)
1	$12
2	13
3	16

(a) Find the regression line $y = mx + b$.

(b) Use the regression line to predict sales in the fourth year.

10. (a) _____

(b) _____

159

11. Find the maximum value of

$$f(x, y) = 2xy - x^2 - 2y^2$$

subject to the constraint $x - 4y = 11$.

11. _____

12. Evaluate

$$\int_0^2 \int_1^x (x^4 + 4y)\, dy\, dx.$$

12. _____

13. *Business: Maximizing Production.* A clothing company has the following Cobb-Douglas production function for a certain product,

$$p(x, y) = 80x^{5/8} y^{3/8},$$

where x is labor, measured in dollars, and y is capital, measured in dollars. Suppose that a company can make a total investment in labor and capital of \$250,000. How should it allocate the investment between labor and capital in order to maximize production?

13. _____

14. Find f_x and f_t:

$$f(x, t) = \frac{x^4 - 5t}{x^3 + 5t}.$$

14. _____

15. Use a 3D grapher to graph

$$f(x, y) = 2x + \frac{1}{4}y^2 - \frac{1}{5}x^2.$$

Sketch the graph.

15.

CALCULUS AND ITS APPLICATIONS

Name:

Chapter 7, Form C

Given $f(x, y) = \ln x + 4x^2 y + y$, find each of the following.

1. $f(2, 3)$

2. $\dfrac{\partial f}{\partial x}$

3. $\dfrac{\partial f}{\partial y}$

4. $\dfrac{\partial^2 f}{\partial x^2}$

5. $\dfrac{\partial^2 f}{\partial x \, \partial y}$

6. $\dfrac{\partial^2 f}{\partial y \, \partial x}$

7. $\dfrac{\partial^2 f}{\partial y^2}$

1. _____

2. _____

3. _____

4. _____

5. _____

6. _____

7. _____

Find the relative maximum and minimum values.

8. $f(x, y) = y^3 - 6xy + x^2 - 48y$

9. $f(x, y) = 2x^2 + 3y^2$

8. _____

9. _____

10. *Business: Predicting Total Revenue.* Consider the data in the following table regarding the total revenue of a company during the first three years of operation.

Year, x	Revenue, y (in millions)
1	$ 5
2	10
3	12

(a) Find the regression line $y = mx + b$.

(b) Use the regression line to predict sales in the fourth year.

10. (a) _____

(b) _____

161

11. Find the maximum value of

$$f(x,y) = 4xy - 2x^2 - 5y^2$$

subject to the constraint $x + 2y = 4$.

11. _____

12. Evaluate

$$\int_0^2 \int_1^x (3x^2 - y)\, dy\, dx.$$

12. _____

13. *Business: Maximizing Production.* A software company has the following Cobb-Douglas production function for a certain product,

$$p(x,y) = 30x^{4/5}y^{1/5},$$

where x is labor, measured in dollars, and y is capital, measured in dollars. Suppose that a company can make a total investment in labor and capital of $150,000. How should it allocate the investment between labor and capital in order to maximize production?

13. _____

14. Find f_x and f_t:

$$f(x,t) = \frac{t - 3x^3}{2t + 3x^3}.$$

14. _____

15. Use a 3D grapher to graph

$$f(x,y) = 2x^2 - 3y^2 + x.$$

Sketch the graph.

15.

CALCULUS AND ITS APPLICATIONS

Name:

Chapter 7, Form D

Given $f(x, y) = 3x^2y + 2e^x - y$, find each of the following.

1. $f(2, -1)$

2. $\dfrac{\partial f}{\partial x}$

3. $\dfrac{\partial f}{\partial y}$

4. $\dfrac{\partial^2 f}{\partial x^2}$

5. $\dfrac{\partial^2 f}{\partial x\, \partial y}$

6. $\dfrac{\partial^2 f}{\partial y\, \partial x}$

7. $\dfrac{\partial^2 f}{\partial y^2}$

1. _____
2. _____
3. _____
4. _____
5. _____
6. _____
7. _____

Find the relative maximum and minimum values.

8. $f(x, y) = 3x^2 + y^2 + x - xy$

9. $f(x, y) = 8y^2 - 2x^2$

8. _____

9. _____

10. *Business: Predicting Total Revenue.* Consider the data in the following table regarding the total revenue of a company during the first three years of operation.

Year, x	Revenue, y (in millions)
1	$ 5
2	9
3	11

(a) Find the regression line $y = mx + b$.

(b) Use the regression line to predict sales in the fourth year.

10. (a) _____

(b) _____

163

11. Find the maximum value of

$$f(x,y) = 2xy - x^2 - 2y^2$$

subject to the constraint $2x + y = 13$.

11. _____

12. Evaluate

$$\int_0^3 \int_1^x (x^3 - 4y) \, dy \, dx.$$

12. _____

13. *Business: Maximizing Production.* An entertainment company has the following Cobb-Douglas production function for a certain product,

$$p(x, y) = 40x^{7/10} y^{3/10},$$

where x is labor, measured in dollars, and y is capital, measured in dollars. Suppose that a company can make a total investment in labor and capital of $300,000. How should it allocate the investment between labor and capital in order to maximize production?

13. _____

14. Find f_x and f_t:

$$f(x, t) = \frac{4x^2 - t}{4x^3 + t}.$$

14. _____

15. Use a 3D grapher to graph

$$f(x, y) = \frac{1}{2}x^3 - 8x^2 + x.$$

Sketch the graph.

15.

CALCULUS AND ITS APPLICATIONS

Name:

Chapter 7, Form E

Given $f(x, y) = 2e^x + 4x^2y - y$, find each of the following.

1. $f(1, -2)$

2. $\dfrac{\partial f}{\partial x}$

3. $\dfrac{\partial f}{\partial y}$

4. $\dfrac{\partial^2 f}{\partial x^2}$

5. $\dfrac{\partial^2 f}{\partial x \, \partial y}$

6. $\dfrac{\partial^2 f}{\partial y \, \partial x}$

7. $\dfrac{\partial^2 f}{\partial y^2}$

1. _____
2. _____
3. _____
4. _____
5. _____
6. _____
7. _____

Find the relative maximum and minimum values.

8. $f(x, y) = 2x^2 + 4xy - y^3 + x$

9. $f(x, y) = 2x^2 - 5y^2$

8. _____

9. _____

10. *Business: Predicting Total Revenue.* Consider the data in the following table regarding the total revenue of a company during the first three years of operation.

Year, x	Revenue, y (in millions)
1	$1
2	2
3	5

(a) Find the regression line $y = mx + b$.

(b) Use the regression line to predict sales in the fourth year.

10. (a) _____

(b) _____

165

11. Find the minimum value of

$$f(x, y) = 2xy - 4x^2 + 6y^2$$

subject to the constraint $2x - y = 24$.

11. _____

12. Evaluate

$$\int_0^2 \int_1^x (4x - 3y^2)\, dy\, dx.$$

12. _____

13. *Business: Maximizing Production.* A sales company has the following Cobb-Douglas production function for a certain product,

$$p(x, y) = 50x^{3/4} y^{1/4},$$

where x is labor, measured in dollars, and y is capital, measured in dollars. Suppose that a company can make a total investment in labor and capital of $200{,}000. How should it allocate the investment between labor and capital in order to maximize production?

13. _____

14. Find f_x and f_t:

$$f(x, t) = \frac{2t + 3x^2}{2t - 3x^3}.$$

14. _____

15. Use a 3D grapher to graph

$$f(x, y) = 5x - \frac{1}{5}y^2 - \frac{1}{2}x^2.$$

Sketch the graph.

15.

CALCULUS AND ITS APPLICATIONS

Name:

Chapter 7, Form F

Given $f(x, y) = 2y + 6x^2 y - \dfrac{1}{x}$, find each of the following.

1. $f(1, -2)$

2. $\dfrac{\partial f}{\partial x}$

3. $\dfrac{\partial f}{\partial y}$

4. $\dfrac{\partial^2 f}{\partial x^2}$

5. $\dfrac{\partial^2 f}{\partial x \, \partial y}$

6. $\dfrac{\partial^2 f}{\partial y \, \partial x}$

7. $\dfrac{\partial^2 f}{\partial y^2}$

1. _____
2. _____
3. _____
4. _____
5. _____
6. _____
7. _____

Find the relative maximum and minimum values.

8. $f(x, y) = 2y^2 - 3x^2$

9. $f(x, y) = x^3 + 2xy + y^2 + y$

8. _____

9. _____

10. *Business: Predicting Total Revenue.* Consider the data in the following table regarding the total revenue of a company during the first three years of operation.

Year, x	Revenue, y (in millions)
1	$ 8
2	10
3	11

(a) Find the regression line $y = mx + b$.

(b) Use the regression line to predict sales in the fourth year.

10. (a) _____

(b) _____

11. Find the maximum value of

$$f(x,y) = 2xy - 2x^2 + y^2$$

subject to the constraint $x + 2y = 11$.

11. _____

12. Evaluate

$$\int_0^2 \int_1^x (y^3 - 2x)\, dy\, dx.$$

12. _____

13. *Business: Maximizing Production.* A scientific testing company has the following Cobb-Douglas production function for a certain product,

$$p(x,y) = 50x^{1/3}y^{2/3},$$

where x is labor, measured in dollars, and y is capital, measured in dollars. Suppose that a company can make a total investment in labor and capital of \$4,200,000. How should it allocate the investment between labor and capital in order to maximize production?

13. _____

14. Find f_x and f_t:

$$f(x,t) = \frac{x^3 - t}{x^4 + t}.$$

14. _____

15. Use a 3D grapher to graph

$$f(x,y) = y - \frac{2}{3}x^2 - \frac{3}{4}y^2.$$

Sketch the graph.

15.

CALCULUS AND ITS APPLICATIONS

Name:

Final Exam, Form A

1. Write an equation of the line with slope $-\frac{3}{4}$ and containing the point $(5, -2)$.

    1. _____

2. For $f(x) = 5x^2 - 2$, find the simplified difference quotient.

    2. _____

3. (a) Graph:
$$f(x) = \begin{cases} x^2 + 1, & \text{for } x \neq 0, \\ 0, & \text{for } x = 0. \end{cases}$$

    3. (a)

    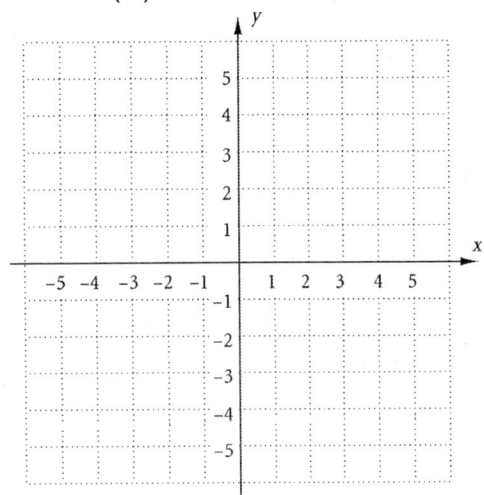

   (b) Find $\lim_{x \to 0} f(x)$.

   (b) _____

   (c) Find $f(0)$.

   (c) _____

   (d) Is f continuous at 0?

   (d) _____

Find the limit, if it exists.

4. $\lim_{x \to -2} (2x^5 + x^3 + 4)$

    4. _____

5. $\lim_{x \to 0} \frac{5}{x}$

    5. _____

6. $\lim_{x \to 3} \frac{x - 3}{x^2 + x - 12}$

    6. _____

7. $\lim_{x \to \infty} \frac{1}{x^2}$

    7. _____

Differentiate.

8. $y = x^3 - 4x + 5$

8. _____

9. $y = x^{5/6}$

9. _____

10. $f(x) = x^{-6}$

10. _____

11. $f(x) = (x+2)(x-3)^4$

11. _____

12. $f(x) = \dfrac{3x+4}{x^2-2}$

12. _____

13. $y = \ln(x^3 + x + 1)$

13. _____

14. $y = 5e^x$

14. _____

15. $y = e^{x^2 + 2x}$

15. _____

16. $f(x) = 4(2x+1)^5$

16. _____

17. For $y = 10x^3 + 4x$, find $\dfrac{d^2y}{dx^2}$.

17. _____

18. Differentiate implicitly to find $\dfrac{dy}{dx}$ if $x^4 - 2xy^3 = 3$.

18. _____

19. Find an equation of the tangent line to the graph of $y = 2e^x + x^3 - 4$ at the point $(0, -2)$.

19. _____

CALCULUS AND ITS APPLICATIONS   Final Exam, Form A

Find the relative extrema of the function. List your answer in terms of an ordered pair. Then sketch a graph of the function.

20. $f(x) = x^3 + 6x^2 + 9x + 1$

20.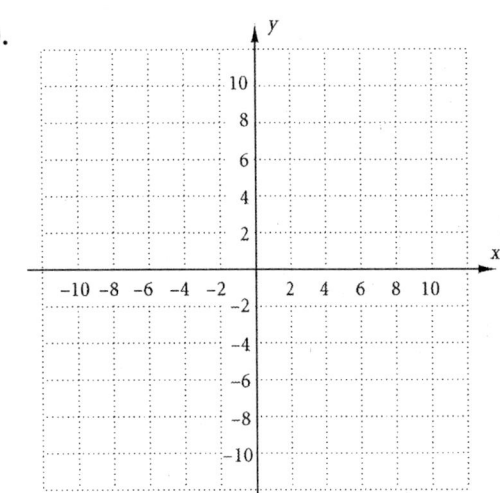

21. $f(x) = 4 - 5x^2 + x^4$

21.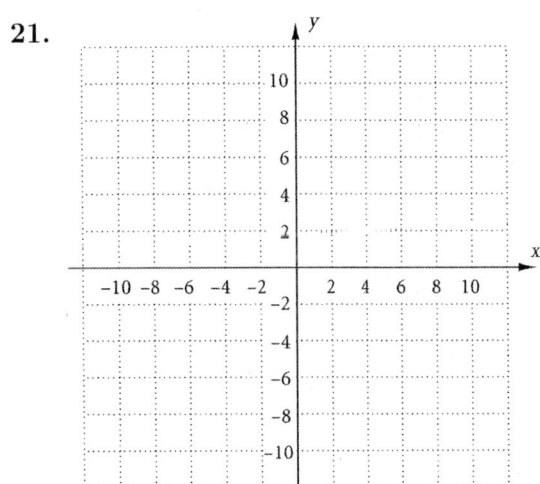

22. $f(x) = \dfrac{-3x}{x^2 + 2}$

22.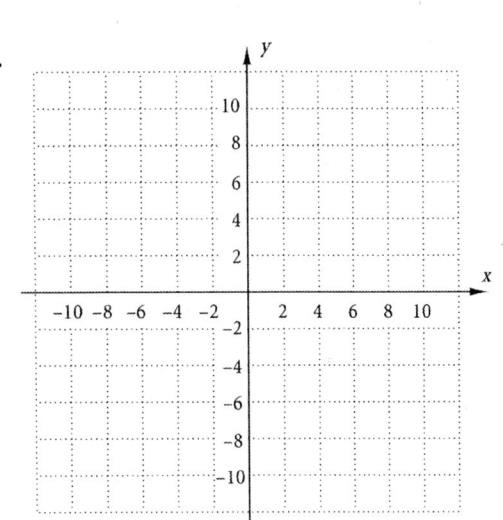

23. $f(x) = \dfrac{4}{x^2 - 4}$ **23.**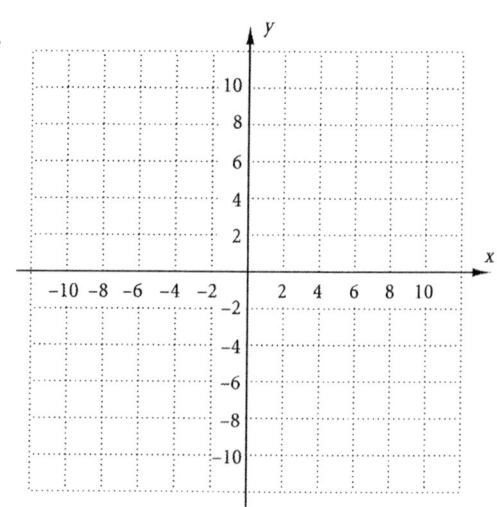

Find the absolute maximum and minimum values, if they exist, over the indicated interval. If no interval is indicated, use the real line.

24. $f(x) = 4x^2 - 3x + 2$ **24.** _____

25. $f(x) = -2x + 7$ **25.** _____

26. $f(x) = \dfrac{1}{3}x^3 + x^2 - 8x + 2;\ [-1, 3]$ **26.** _____

27. *Business: Maximizing Profit.* For a certain product, the total-revenue and total-cost functions are given by

$$R(x) = 30x - 0.3x^2 \text{ and}$$
$$C(x) = 0.2x^2 + 5x.$$

Find the number of units that must be produced and sold in order to maximize profit.   **27.** _____

28. *Business: Minimizing Inventory Costs.* A shoe store sells 800 pairs of sneakers per year. It costs $2 to store one pair for one year. To order sneakers there is a fixed cost of $8 plus $2 for each pair. How many times per year should the store reorder sneakers, and in what lot size, in order to minimize inventory costs?   **28.** _____

CALCULUS AND ITS APPLICATIONS   Final Exam, Form A

29. For $f(x) = 3x^2 - 2$, $x = 5$, and $\Delta x = 0.01$, find Δy and $f'(x)\Delta x$.

29. _____

30. *Social Science: Population Growth.* A town is experiencing a growth pattern of 3% per year in its population P; that is

$$\frac{dP}{dt} = 0.03P,$$

where P is the population of the town and t is the time, in years, from 2000.

(a) Given that the town's population in 2000 was 14,000, find the solution of the equation assuming $P_0 = 14{,}000$ and $k = 0.03$.

30. (a) _____

(b) What will the population be in 2010?

(b) _____

(c) What is the doubling time of the population?

(c) _____

31. *Economics: Elasticity of Demand.* Consider the demand function

$$x = D(p) = 120 - 5p.$$

(a) Find the elasticity.

31. (a) _____

(b) Find the elasticity at $p = \$10$, stating whether the demand is elastic or inelastic.

(b) _____

(c) At a price of $10, will a small increase in price cause the total revenue to increase or decrease?

(c) _____

(d) Find the value of p for which the total revenue is a maximum.

(d) _____

Evaluate.

32. $\int 2x^7\, dx$

32. _____

33. $\int_0^4 (3e^x + x)\, dx$

33. _____

34. $\displaystyle\int \frac{1}{\sqrt{x^2-25}}\, dx$ (Use Table 1.)

34. _____

35. $\displaystyle\int x^2 e^{x^3+1}\, dx$ (Use substitution. Do not use Table 1.)

35. _____

36. $\displaystyle\int (x-2)\ln x\, dx$

36. _____

37. $\displaystyle\int \frac{18}{x}\, dx$

37. _____

38. $\displaystyle\int_1^8 \frac{2}{\sqrt[3]{x}}\, dx$

38. _____

39. Find the area under the graph of $y = x^3 + 2x$ over the interval $[1, 3]$.

39. _____

40. *Business: Present Value.* Find the present value of $140,000 due in 25 yr at 7.8% compounded continuously.

40. _____

41. *Business: Accumulated Present Value.* Find the accumulated present value of an investment over a 30-yr period in which there is a continuous money flow of $4000 and the current interest rate is 6%, compounded continuously.

41. _____

Determine whether the improper integral is convergent or divergent, and calculate its value if it is convergent.

42. $\displaystyle\int_1^\infty x^{-2}\, dx$

42. _____

43. $\displaystyle\int_0^\infty \frac{1}{100} e^x\, dx$

43. _____

CALCULUS AND ITS APPLICATIONS   Final Exam, Form A

44. Given the probability density function
$$f(x) = \frac{x^2}{9} \text{ over } [0, 3]$$
find each of the following.

(a) The mean

(b) The standard deviation

44. (a) _____

(b) _____

45. Let x be a continuous random variable that is normally distributed with mean $\mu = 5$ and standard deviation $\sigma = 0.8$. Using Table 2, find $P(3 \leq x \leq 5.2)$.

45. _____

46. *Economics: Supply and Demand.* Given the demand and supply functions, in dollars,
$$p = D(x) = (x - 10)^2 \text{ and } p = S(x) = x^2 + 5x + 25,$$
find the equilibrium point and the producer's surplus at the equilibrium point.

46. _____

47. Find the volume of the solid of revolution generated by rotating the region under the graph of
$$e^{x/2} \text{ from } x = 0 \text{ to } x - 10$$
about the x-axis.

47. _____

Solve the differential equation.

48. $\dfrac{dy}{dx} = 2x^3 y$

48. _____

49. $y' = \dfrac{8}{y}$

49. _____

50. *Total Profit.* A firm's marginal profit P as a function of its total cost C is given by
$$\frac{dP}{dC} = \frac{-100}{(C+4)^{3/2}}.$$

(a) Find the profit function $P(C)$ if $P = \$5$ when $C = \$60$.

(b) At what value will the firm break even ($P = 0$)?

50. (a) _____

(b) _____

Given $f(x, y) = 6y - 2x^3y + 4e^x$, find each of the following.

51. f_y

51. _____

52. f_{yx}

52. _____

53. Find the relative maximum and minimum values of $f(x, y) = 2x^2 + 5y^2 + 6$.

53. _____

54. Maximize $f(x, y) = 40x + 10y - x^2 + 6$ subject to the constraint $x + 3y = 45$.

54. _____

55. Evaluate:
$$\int_{-1}^{1} \int_{0}^{1} (e^y + 2) \, dx \, dy.$$

55. _____

56. Integrate: $\displaystyle\int \frac{e^x}{1 + e^x} \, dx.$

56. _____

57. Find $\displaystyle\lim_{x \to 3} \frac{x^3 - 27}{x - 3}.$

57. _____

58. Use your grapher to graph $f(x) = 3x^3 - 2x^4$. Then sketch the graph.

58.

CALCULUS AND ITS APPLICATIONS

Name:

Final Exam, Form B

1. Write an equation of the line with slope $-\frac{2}{3}$ and containing the point $(3, -1)$.

   1. _____

2. For $f(x) = 2x^2 + 4$, find the simplified difference quotient.

   2. _____

3. (a) Graph:
   $$f(x) = \begin{cases} x^2 - 5, & \text{for } x \neq 2, \\ 0, & \text{for } x = 2. \end{cases}$$

   3. (a)

   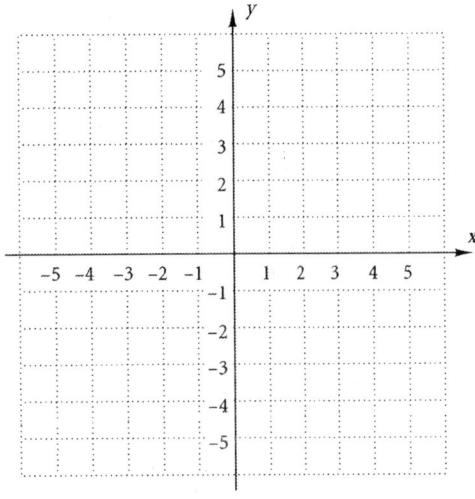

   (b) Find $\lim_{x \to 2} f(x)$.

   (b) _____

   (c) Find $f(2)$.

   (c) _____

   (d) Is f continuous at 2?

   (d) _____

Find the limit, if it exists.

4. $\lim_{x \to -2} (5x^3 - 3x + 4)$

   4. _____

5. $\lim_{x \to -3} \dfrac{8}{x + 3}$

   5. _____

6. $\lim_{x \to -1} \dfrac{x + 1}{x^2 - 4x - 5}$

   6. _____

7. $\lim_{x \to \infty} \dfrac{1}{x^5}$

   7. _____

Differentiate.

8. $y = x^3 + 2x - 10$

    8. _____

9. $y = x^{4/3}$

    9. _____

10. $f(x) = x^{-16}$

    10. _____

11. $f(x) = (x+5)(x-2)^4$

    11. _____

12. $f(x) = \dfrac{8 - 5x}{x^3 + 3}$

    12. _____

13. $y = \ln(x^2 - 6x + 4)$

    13. _____

14. $y = 5e^x$

    14. _____

15. $y = e^{4x^2 - 3x}$

    15. _____

16. $f(x) = 4(5x - 2)^8$

    16. _____

17. For $y = 8x^3 - 2x^5$, find $\dfrac{d^2y}{dx^2}$.

    17. _____

18. Differentiate implicitly to find $\dfrac{dy}{dx}$ if $x^{10} - 5x^2y^2 = 4$.

    18. _____

19. Find an equation of the tangent line to the graph of $y = 3x^2 - 2x + 4e^{-x}$ at the point $(0, 4)$.

    19. _____

CALCULUS AND ITS APPLICATIONS Final Exam, Form B

Find the relative extrema of the function. List your answer in terms of an ordered pair. Then sketch a graph of the function.

20. $f(x) = 2x^3 + 12x^2 + 18x + 5$

20.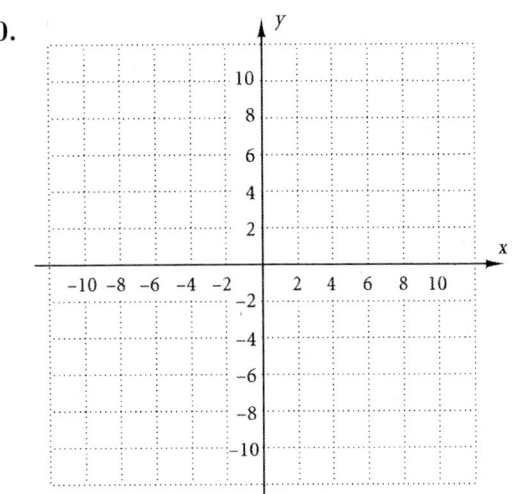

21. $f(x) = x^4 - 6$

21.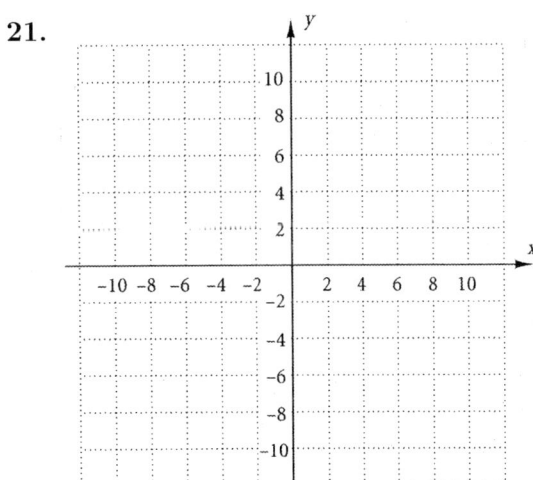

22. $f(x) = \dfrac{-6x}{x^2 + 4}$

22.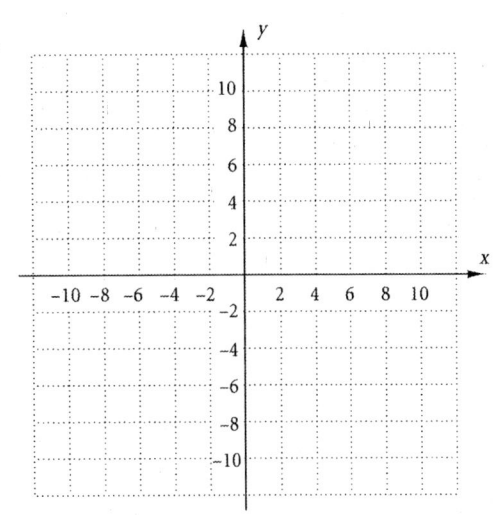

23. $f(x) = \dfrac{-2}{1+x^2}$

23.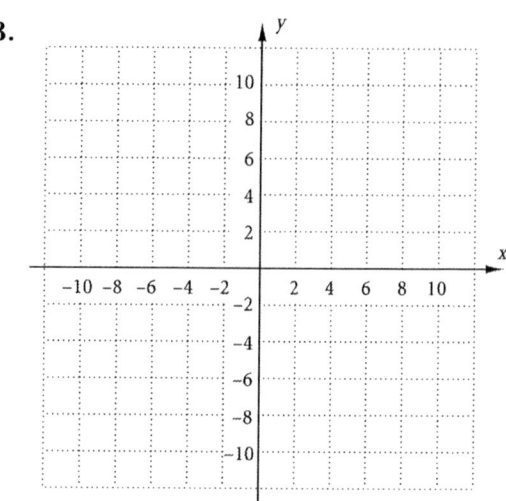

Find the absolute maximum and minimum values, if they exist, over the indicated interval. If no interval is indicated, use the real line.

24. $f(x) = 2x^2 + 3x - 7$

24. _____

25. $f(x) = -5x + 1$

25. _____

26. $f(x) = \dfrac{1}{3}x^3 + x^2 - 8x + 4$; $[0, 4]$

26. _____

27. *Business: Maximizing Profit.* For a certain product, the total-revenue and total-cost functions are given by

$$R(x) = 60x - 0.8x^2 \text{ and}$$
$$C(x) = 1.2x^2 + 200.$$

Find the number of units that must be produced and sold in order to maximize profit.

27. _____

28. *Business: Minimizing Inventory Costs.* A shoe store sells 400 pairs of pumps per year. It costs $2 to store one pair for one year. To order pumps there is a fixed cost of $4 plus $1.25 for each pair. How many times per year should the store reorder pumps, and in what lot size, in order to minimize inventory costs?

28. _____

CALCULUS AND ITS APPLICATIONS   Final Exam, Form B

29. For $f(x) = 3x^2 + 4$, $x = -1$, and $\Delta x = 0.01$, find Δy and $f'(x)\Delta x$.

29. _____

30. *Social Science: Population Growth.* A town is experiencing a growth pattern of 6% per year in its population P; that is

$$\frac{dP}{dt} = 0.06P,$$

where P is the population of the town and t is the time, in years, from 1990.

(a) Given that the town's population in 1990 was 8500, find the solution of the equation assuming $P_0 = 8500$ and $k = 0.06$.

30. (a) _____

(b) What will the population be in 2005?

(b) _____

(c) What is the doubling time of the population?

(c) _____

31. *Economics: Elasticity of Demand.* Consider the demand function

$$x = D(p) = 100 - 8p.$$

(a) Find the elasticity.

31. (a) _____

(b) Find the elasticity at $p = \$8$, stating whether the demand is elastic or inelastic.

(b) _____

(c) At a price of $8, will a small increase in price cause the total revenue to increase or decrease?

(c) _____

(d) Find the value of p for which the total revenue is a maximum.

(d) _____

Evaluate.

32. $\int 4x^5 \, dx$

32. _____

33. $\int_0^2 (4e^x + 3x) \, dx$

33. _____

34. $\displaystyle\int \frac{1}{x(3x+1)}\, dx$ (Use Table 1.)   34. _____

35. $\displaystyle\int 12xe^{x^2+4}\, dx$ (Use substitution. Do not use Table 1.)   35. _____

36. $\displaystyle\int (x+6)\ln x\, dx$ 36. _____

37. $\displaystyle\int \frac{50}{x}\, dx$ 37. _____

38. $\displaystyle\int_1^4 \frac{x\sqrt{x}}{2}\, dx$ 38. _____

39. Find the area under the graph of $y = -x^4 + 8x$ over the interval $[0, 2]$.   39. _____

40. *Business: Present Value.* Find the present value of $100,000 due in 20 yr at 5.6% compounded continuously.   40. _____

41. *Business: Accumulated Present Value.* Find the accumulated present value of an investment over a 20-yr period in which there is a continuous money flow of $2600 and the current interest rate is 7%, compounded continuously.   41. _____

Determine whether the improper integral is convergent or divergent, and calculate its value if it is convergent.

42. $\displaystyle\int_1^\infty x^{-4}\, dx$ 42. _____

43. $\displaystyle\int_1^\infty \frac{1}{3x}\, dx$ 43. _____

CALCULUS AND ITS APPLICATIONS   Final Exam, Form B

44. Given the probability density function
$$f(x) = \frac{3x^2}{125} \text{ over } [0, 5]$$
find each of the following.

(a) The mean

(b) The standard deviation

44. (a) _____

(b) _____

45. Let x be a continuous random variable that is normally distributed with mean $\mu = 10$ and standard deviation $\sigma = 0.5$. Using Table 2, find $P(9.15 \leq x \leq 10)$.

45. _____

46. *Economics: Supply and Demand.* Given the demand and supply functions, in dollars,
$$p = D(x) = (x - 20)^2 \text{ and } p = S(x) = x^2 + 5x + 40,$$
find the equilibrium point and the producer's surplus at the equilibrium point.

46. _____

47. Find the volume of the solid of revolution generated by rotating the region under the graph of
$$y = e^{2x} \text{ from } x - 0 \text{ to } x = 5$$
about the x-axis.

47. _____

Solve the differential equation.

48. $\dfrac{dy}{dx} = 6x^3 y$

48. _____

49. $y' = \dfrac{-10}{y}$

49. _____

50. *Total Profit.* A firm's marginal profit P as a function of its total cost C is given by
$$\frac{dP}{dC} = \frac{-250}{(C+5)^{3/2}}.$$

(a) Find the profit function $P(C)$ if $P = \$8$ when $C = \$76$.

(b) At what value will the firm break even $(P = 0)$?

50. (a) _____

(b) _____

Given $f(x, y) = 6x^2 + xy^5 - 7e^y$, find each of the following.

51. f_y

51. _____

52. f_{yx}

52. _____

53. Find the relative maximum and minimum values of $f(x, y) = 3x^2 - 5y^2 + 4$.

53. _____

54. Maximize $f(x, y) = 12x + 6y - x^2 + y^2 + 2$ subject to the constraint $x - 4y = -12$.

54. _____

55. Evaluate:
$$\int_{-1}^{0} \int_{0}^{2} (e^x - 3) \, dy \, dx.$$

55. _____

56. Find f_x: $f(x, t) = \dfrac{4x^3 - t}{xt + 2}$.

56. _____

57. Integrate: $\displaystyle\int \dfrac{e^x}{e^x + 5} \, dx$.

57. _____

58. Use your grapher to graph $f(x) = 8x^3 - 3x^5$. Then sketch the graph.

58.

CALCULUS AND ITS APPLICATIONS

Name:

Final Exam, Form C

1. Write an equation of the line with slope -3 and containing the point $\left(\frac{2}{3}, -1\right)$.

   1. _____

2. For $f(x) = 4x^2 - x$, find the simplified difference quotient.

   2. _____

3. (a) Graph:
$$f(x) = \begin{cases} -x^2 + 2, & \text{for } x \neq -1, \\ -1, & \text{for } x = -1. \end{cases}$$

   3. (a)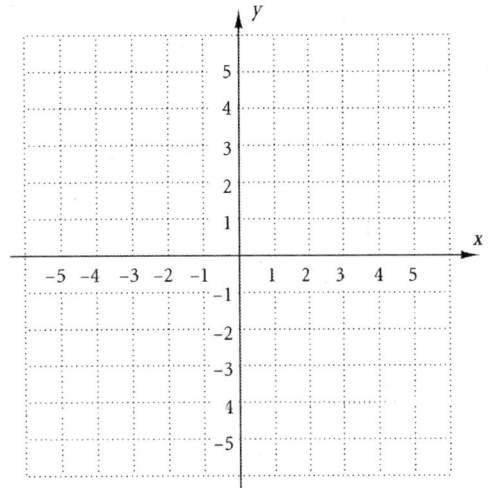

   (b) Find $\lim_{x \to -1} f(x)$.

   (b) _____

   (c) Find $f(-1)$.

   (c) _____

   (d) Is f continuous at -1?

   (d) _____

Find the limit, if it exists.

4. $\lim_{x \to 2} (-x^3 + 3x - 5)$

   4. _____

5. $\lim_{x \to 0} \dfrac{6}{x}$

   5. _____

6. $\lim_{x \to -2} \dfrac{x+2}{x^2 - 4}$

   6. _____

7. $\lim_{x \to \infty} \dfrac{1}{\sqrt{x}}$

   7. _____

Differentiate.

8. $y = x^4 - 3x^2 + 5$

8. _____

9. $y = x^{3/4}$

9. _____

10. $f(x) = x^{-8}$

10. _____

11. $f(x) = (x+4)(x-2)^5$

11. _____

12. $f(x) = \dfrac{3x-4}{x^2+5}$

12. _____

13. $y = \ln(x^2 - 3x + 1)$

13. _____

14. $y = 10e^x$

14. _____

15. $y = e^{x^2 - 4x}$

15. _____

16. $f(x) = -2(3x-8)^6$

16. _____

17. For $y = 2x^6 - 4x^3$, find $\dfrac{d^2y}{dx^2}$.

17. _____

18. Differentiate implicitly to find $\dfrac{dy}{dx}$ if $2xy^4 - 3 = 4x^2$.

18. _____

19. Find an equation of the tangent line to the graph of $y = x^2 - 5 + 3e^x$ at the point $(0, -2)$.

19. _____

CALCULUS AND ITS APPLICATIONS Final Exam, Form C

Find the relative extrema of the function. List your answer in terms of an ordered pair. Then sketch a graph of the function.

20. $f(x) = x^3 + 6x^2 + 9x + 2$

20.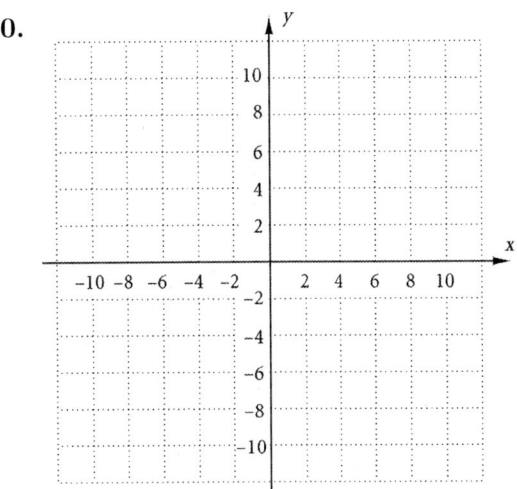

21. $f(x) = -2x^2 - 8 + x^4$

21.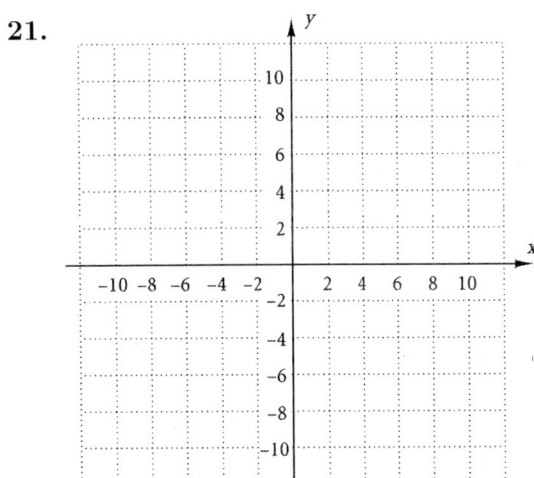

22. $f(x) = \dfrac{12x}{x^2 + 4}$

22.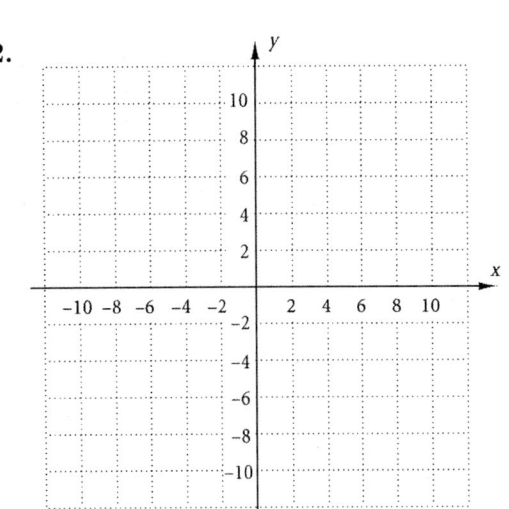

23. $f(x) = \dfrac{12}{x^2 - 4}$ 23.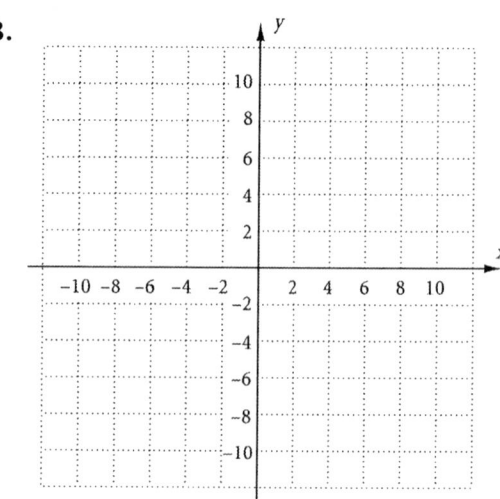

Find the absolute maximum and minimum values, if they exist, over the indicated interval. If no interval is indicated, use the real line.

24. $f(x) = -x^2 + 2x + 1$ 24. _____

25. $f(x) = 3 - 7x$ 25. _____

26. $f(x) = \dfrac{2}{3}x^3 - x^2 - 4x + 6;\ [-3, 3]$ 26. _____

27. *Business: Maximizing Profit.* For a certain product, the total-revenue and total-cost functions are given by
$$R(x) = 40x - 0.5x^2 \text{ and}$$
$$C(x) = 1.5x^2 + 20x.$$
Find the number of units that must be produced and sold in order to maximize profit.   27. _____

28. *Business: Minimizing Inventory Costs.* A shoe store sells 300 pairs of hikers per year. It costs $1 to store one pair for one year. To order hikers there is a fixed cost of $6 plus $0.50 for each pair. How many times per year should the store reorder hikers, and in what lot size, in order to minimize inventory costs?   28. _____

CALCULUS AND ITS APPLICATIONS Final Exam, Form C

29. For $f(x) = 4x^2 + 3$, $x = 3$, and $\Delta x = 0.01$, find Δy and $f'(x)\Delta x$.

29. _____

30. *Social Science: Population Growth.* A town is experiencing a growth pattern of 8% per year in its population P; that is

$$\frac{dP}{dt} = 0.08P,$$

where P is the population of the town and t is the time, in years, from 1995.

(a) Given that the town's population in 1995 was 400, find the solution of the equation assuming $P_0 = 400$ and $k = 0.08$.

30. (a) _____

(b) What will the population be in 2005?

(b) _____

(c) What is the doubling time of the population?

(c) _____

31. *Economics: Elasticity of Demand.* Consider the demand function

$$x = D(p) = 480 - 20p.$$

(a) Find the elasticity.

31. (a) _____

(b) Find the elasticity at $p = \$18$, stating whether the demand is elastic or inelastic.

(b) _____

(c) At a price of $18, will a small increase in price cause the total revenue to increase or decrease?

(c) _____

(d) Find the value of p for which the total revenue is a maximum.

(d) _____

Evaluate.

32. $\int 6x^4 \, dx$

32. _____

33. $\int_0^2 (3x + 4e^x) \, dx$

33. _____

34. $\int \dfrac{x}{(2x+3)^2}\, dx$ (Use Table 1.)

34. _____

35. $\int 2x^3 e^{x^4+2}\, dx$ (Use substitution. Do not use Table 1.)

35. _____

36. $\int (x+6) \ln x\, dx$

36. _____

37. $\int \dfrac{-2}{x}\, dx$

37. _____

38. $\int_0^9 4\sqrt{x}\, dx$

38. _____

39. Find the area under the graph of $y = x^3 - 4x$ over the interval $[-2, 0]$.

39. _____

40. *Business: Present Value.* Find the present value of \$50,000 due in 10 yr at 4.3% compounded continuously.

40. _____

41. *Business: Accumulated Present Value.* Find the accumulated present value of an investment over a 10-yr period in which there is a continuous money flow of \$2500 and the current interest rate is 6%, compounded continuously.

41. _____

Determine whether the improper integral is convergent or divergent, and calculate its value if it is convergent.

42. $\int_0^\infty \dfrac{1}{5x}\, dx$

42. _____

43. $\int_1^\infty x^{-4}\, dx$

43. _____

CALCULUS AND ITS APPLICATIONS   Final Exam, Form C

44. Given the probability density function
$$f(x) = \frac{x^3}{4} \text{ over } [0, 2]$$
find each of the following.

(a) The mean

(b) The standard deviation

44. (a) _____

(b) _____

45. Let x be a continuous random variable that is normally distributed with mean $\mu = 12$ and standard deviation $\sigma = 1.5$. Using Table 2, find $P(9 \leq x \leq 12.6)$.

45. _____

46. *Economics: Supply and Demand.* Given the demand and supply functions, in dollars,
$$p = D(x) = (x - 10)^2 \text{ and } p = S(x) = x^2 + 10x + 10,$$
find the equilibrium point and the producer's surplus at the equilibrium point.

46. _____

47. Find the volume of the solid of revolution generated by rotating the region under the graph of
$$y = e^{x/10} \text{ from } x = 2 \text{ to } x = 5$$
about the x-axis.

47. _____

Solve the differential equation.

48. $\dfrac{dy}{dx} = 4x^5 y$

48. _____

49. $y' = \dfrac{3}{2y}$

49. _____

50. *Total Profit.* A firm's marginal profit P as a function of its total cost C is given by
$$\frac{dP}{dC} = \frac{-200}{(C+8)^{3/2}}.$$

(a) Find the profit function $P(C)$ if $P = \$4$ when $C = \$392$.

(b) At what value will the firm break even $(P = 0)$?

50. (a) _____

(b) _____

CALCULUS AND ITS APPLICATIONS   Final Exam, Form C

Given $f(x, y) = 2y^2 + xy^3 + 2e^x$, find each of the following.

51. f_y

51. _____

52. f_{yx}

52. _____

53. Find the relative maximum and minimum values of $f(x, y) = 5x^2 + 6y^2 + 2$.

53. _____

54. Minimize $f(x, y) = 2x - 3y + x^2 - y^2 + 6$ subject to the constraint $x + 2y = 11$.

54. _____

55. Evaluate:

$$\int_2^4 \int_0^1 (2e^x + 1)\, dy\, dx.$$

55. _____

56. Differentiate: $f(x) = x^2 \ln(xe^x)$.

56. _____

57. Integrate: $\displaystyle\int \frac{2e^x}{1 + e^x}\, dx.$

57. _____

58. Use your grapher to approximate the integral:

$$\int_{-\infty}^{\infty} \frac{3}{x^2 + 2}\, dx.$$

58. _____

CALCULUS AND ITS APPLICATIONS

Name:

Final Exam, Form D

1. Write an equation of the line with slope -3 and containing the point $\left(3, \frac{1}{2}\right)$.

   1. _____

2. For $f(x) = 3x^2 - x$, find the simplified difference quotient.

   2. _____

3. (a) Graph:
$$f(x) = \begin{cases} 4 - x^2, & \text{for } x \neq 1, \\ 1, & \text{for } x = 1. \end{cases}$$

   3. (a)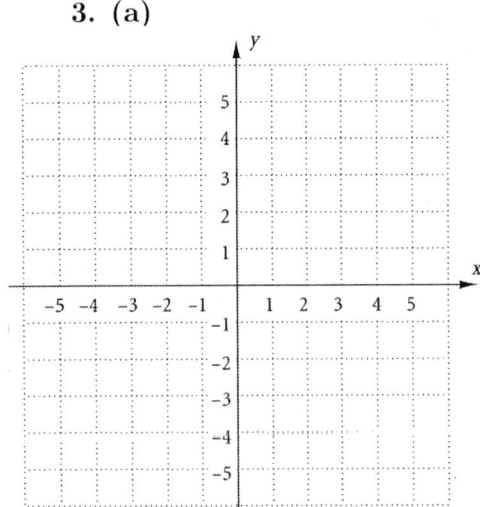

   (b) Find $\lim\limits_{x \to 1} f(x)$.   (b) _____

   (c) Find $f(1)$.   (c) _____

   (d) Is f continuous at 1?   (d) _____

Find the limit, if it exists.

4. $\lim\limits_{x \to -1} (2x^2 + x + 1)$ 4. _____

5. $\lim\limits_{x \to 2} \dfrac{6x}{x - 2}$ 5. _____

6. $\lim\limits_{x \to -3} \dfrac{x + 3}{x^2 - x - 12}$ 6. _____

7. $\lim\limits_{x \to \infty} -\dfrac{5}{x^3}$ 7. _____

Differentiate.

8. $y = x^8 - 3x^2 + 6$

9. $y = x^{3/4}$

10. $f(x) = x^{-1}$

11. $f(x) = (x+3)(x-4)^5$

12. $f(x) = \dfrac{5x-2}{3-x^2}$

13. $y = \ln(x^2 - 7x + 1)$

14. $y = 8e^x$

15. $y = e^{x^2 - x - 3}$

16. $f(x) = -4(6x+5)^3$

17. For $y = 5x^4 + 16x$, find $\dfrac{d^2y}{dx^2}$.

18. Differentiate implicitly to find $\dfrac{dy}{dx}$ if $\sqrt{x} + 8 = 2xy^2$.

19. Find an equation of the tangent line to the graph of $y = \ln x + x^2 + 4$ at the point $(1, 5)$.

CALCULUS AND ITS APPLICATIONS   Final Exam, Form D

Find the relative extrema of the function. List your answer in terms of an ordered pair. Then sketch a graph of the function.

20. $f(x) = x^3 - 6x^2 + 9x - 8$

20.

21. $f(x) = 2x^2 + 5 - x^4$

21.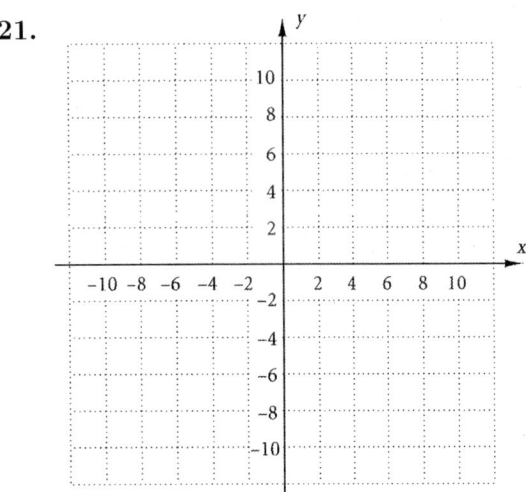

22. $f(x) = \dfrac{-12x}{x^2 + 9}$

22.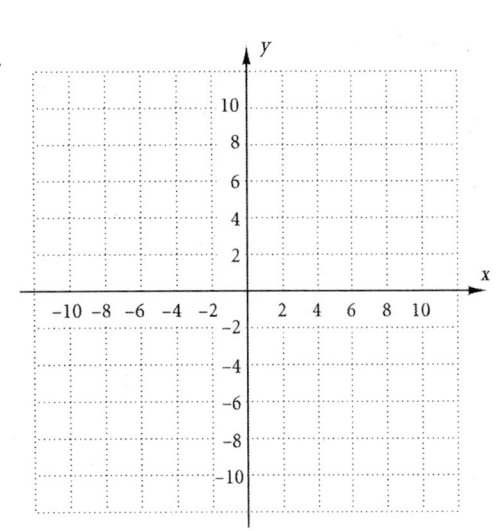

23. $f(x) = \dfrac{-10}{x^2 - 4}$

23.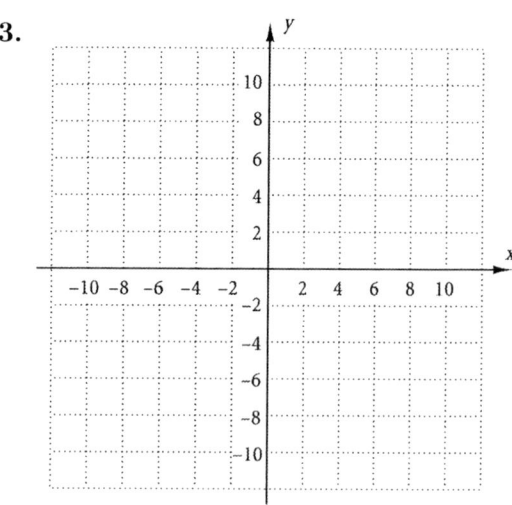

Find the absolute maximum and minimum values, if they exist, over the indicated interval. If no interval is indicated, use the real line.

24. $f(x) = -4x^2 + 5x - 3$

24. _____

25. $f(x) = x + 4$

25. _____

26. $f(x) = \dfrac{2}{3}x^3 - 4x^2 - 10x + 7;\ [1, 10]$

26. _____

27. *Business: Maximizing Profit.* For a certain product, the total-revenue and total-cost functions are given by

$$R(x) = 100x - 0.85x^2 \text{ and}$$
$$C(x) = 0.05x^2 + 19x.$$

Find the number of units that must be produced and sold in order to maximize profit.

27. _____

28. *Business: Minimizing Inventory Costs.* A department store sells 1600 packages of white socks per year. It costs $0.10 to store one package for one year. To order socks there is a fixed cost of $5 plus $0.05 for each package. How many times per year should the store reorder white socks, and in what lot size, in order to minimize inventory costs?

28. _____

CALCULUS AND ITS APPLICATIONS   Final Exam, Form D

29. For $f(x) = 3x^2 + 4$, $x = 5$, and $\Delta x = 0.01$, find Δy and $f'(x)\Delta x$.

29. _____

30. *Biological Science: Population Growth.* A bacteria culture is experiencing a growth pattern of 15% per hour in its population P; that is

$$\frac{dP}{dt} = 0.15P,$$

where P is the population of the culture and t is the time, in hours, from noon.

(a) Given that the culture's population at noon was 8000, find the solution of the equation assuming $P_0 = 8000$ and $k = 0.15$.

30. (a) _____

(b) What will the population be at 5 P.M.?

(b) _____

(c) What is the doubling time of the population?

(c) _____

31. *Economics: Elasticity of Demand.* Consider the demand function

$$x = D(p) = 400 - 10p.$$

(a) Find the elasticity.

31. (a) _____

(b) Find the elasticity at $p = \$25$, stating whether the demand is elastic or inelastic.

(b) _____

(c) At a price of $25, will a small increase in price cause the total revenue to increase or decrease?

(c) _____

(d) Find the value of p for which the total revenue is a maximum.

(d) _____

Evaluate.

32. $\int 4x^3 \, dx$

32. _____

33. $\int_1^2 (4e^x + 6x) \, dx$

33. _____

34. $\int \dfrac{1}{\sqrt{x^2+9}}\, dx$ (Use Table 1.)

34. _____

35. $\int x^3 e^{x^4+6}\, dx$ (Use substitution. Do not use Table 1.)

35. _____

36. $\int (x+2)\ln x\, dx$

36. _____

37. $\int \dfrac{-8\, dx}{x}$

37. _____

38. $\int_0^8 4\sqrt[3]{x}\, dx$

38. _____

39. Find the area under the graph of $y = x^3 + 2x$ over the interval $[0,2]$.

39. _____

40. *Business: Present Value.* Find the present value of \$100,000 due in 20 yr at 8.7% compounded continuously.

40. _____

41. *Business: Accumulated Present Value.* Find the accumulated present value of an investment over a 10-yr period in which there is a continuous money flow of \$4500 and the current interest rate is 3%, compounded continuously.

41. _____

Determine whether the improper integral is convergent or divergent, and calculate its value if it is convergent.

42. $\int_1^\infty \dfrac{1}{10x}\, dx$

42. _____

43. $\int_1^\infty x^{-8}\, dx$

43. _____

CALCULUS AND ITS APPLICATIONS   Final Exam, Form D

44. Given the probability density function

$$f(x) = \frac{4x^3}{625} \text{ over } [0, 5]$$

find each of the following.

(a) The mean

(b) The standard deviation

44. (a) _____

(b) _____

45. Let x be a continuous random variable that is normally distributed with mean $\mu = 3$ and standard deviation $\sigma = 0.25$. Using Table 2, find $P(3.1 \leq x \leq 3.5)$.

45. _____

46. *Economics: Supply and Demand.* Given the demand and supply functions, in dollars,

$$p = D(x) = (x - 10)^2 \text{ and } p = S(x) = x^2 + 4x + 28,$$

find the equilibrium point and the producer's surplus at the equilibrium point.

46. _____

47. Find the volume of the solid of revolution generated by rotating the region under the graph of

$$y = e^{x/4} \text{ from } x = 0 \text{ to } x = 8$$

about the x-axis.

47. _____

Solve the differential equation.

48. $\dfrac{dy}{dx} = -x^3 y$

48. _____

49. $y' = \dfrac{5}{y}$

49. _____

50. *Total Profit.* A firm's marginal profit P as a function of its total cost C is given by

$$\frac{dP}{dC} = \frac{-100}{(C+3)^{3/2}}.$$

(a) Find the profit function $P(C)$ if $P = \$2$ when $C = \$22$.

(b) At what value will the firm break even $(P = 0)$?

50. (a) _____

(b) _____

Given $f(x, y) = 2xy + y^2 + 3e^x$, find each of the following.

51. f_y

51. _____

52. f_{yx}

52. _____

53. Find the relative maximum and minimum values of $f(x, y) = -2x^2 - 5y^2 + 6$.

53. _____

54. Maximize $f(x, y) = 4x + 2y - x^2 - 3y^2 + 2$ subject to the constraint $x + 2y = 5$.

54. _____

55. Evaluate:
$$\int_{-1}^{1} \int_{0}^{x} (2y^2 + e^y) \, dy \, dx.$$

55. _____

56. Differentiate: $f(x) = e^x \ln(xe^x)$.

56. _____

57. Integrate: $\displaystyle\int \frac{5e^x}{5 + e^x} \, dx.$

57. _____

58. Use your grapher to approximate the integral
$$\int_{-\infty}^{\infty} \frac{4}{2 + x^2} \, dx.$$

58. _____

CALCULUS AND ITS APPLICATIONS

Name:

Final Exam, Form E

1. Write an equation of the line with slope $-\frac{1}{2}$ and containing the point $(3, -4)$.

   1. _____

2. For $f(x) = 3x^2 + 4$, find the simplified difference quotient.

   2. _____

3. (a) Graph:
$$f(x) = \begin{cases} x^2 - 4, & \text{for } x \neq 2, \\ 4, & \text{for } x = 2. \end{cases}$$

   3. (a)

   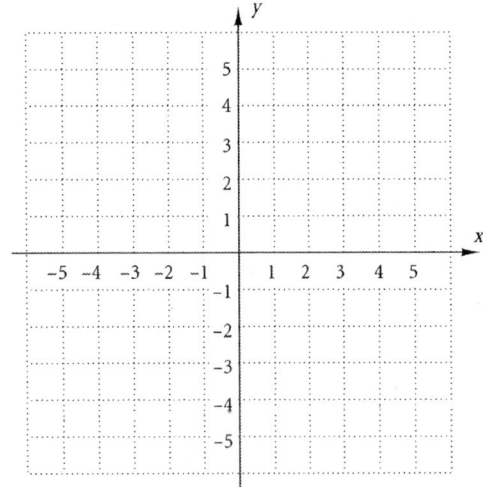

   (b) Find $\lim_{x \to 2} f(x)$.

   (b) _____

   (c) Find $f(2)$.

   (c) _____

   (d) Is f continuous at 2?

   (d) _____

Find the limit, if it exists.

4. $\lim_{x \to -1} 6x^4 - 3x^2 + 8$

   4. _____

5. $\lim_{x \to 0} \dfrac{10}{x}$

   5. _____

6. $\lim_{x \to -3} \dfrac{x-3}{x^2 + x - 12}$

   6. _____

7. $\lim_{x \to \infty} \dfrac{4}{x^4}$

   7. _____

Differentiate.

8. $y = 3x^3 + x + 18$

9. $y = x^{3/5}$

10. $f(x) = x^{-3}$

11. $f(x) = (x-2)(x+3)^4$

12. $f(x) = \dfrac{3x-1}{x^2+6}$

13. $y = \ln(x^4 - 2x)$

14. $y = 5e^x$

15. $y = e^{x^4 - 1}$

16. $f(x) = 4(2x+3)^3$

17. For $y = 13x^3 + 8x$, find $\dfrac{d^2y}{dx^2}$.

18. Differentiate implicitly to find $\dfrac{dy}{dx}$ if $x^2 + 4x = xy$.

19. Find an equation of the tangent line to the graph of $y = 4\ln x + x^2 - 2$ at the point $(1, -1)$.

CALCULUS AND ITS APPLICATIONS   Final Exam, Form E

Find the relative extrema of the function. List your answer in terms of an ordered pair. Then sketch a graph of the function.

20. $f(x) = -x^3 - 6x^2 - 9x + 5$

20.

21. $f(x) = 8 - 8x^2 + x^4$

21.

22. $f(x) = \dfrac{6x}{3 + x^2}$

22.

23. $f(x) = \dfrac{12}{x^2 - 6}$

23.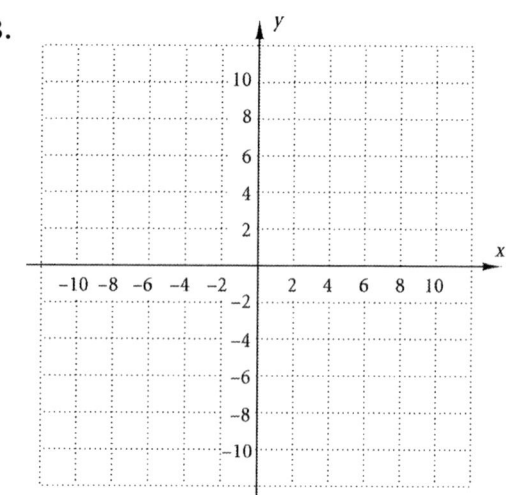

Find the absolute maximum and minimum values, if they exist, over the indicated interval. If no interval is indicated, use the real line.

24. $f(x) = 2x^2 + 8x - 10$

24. _____

25. $f(x) = 5 - x$

25. _____

26. $f(x) = \dfrac{1}{3}x^3 - 2x^2 + 3x + 2;\ [-2, 2]$

26. _____

27. *Business: Maximizing Profit.* For a certain product, the total-revenue and total-cost functions are given by

$$R(x) = 54x - 0.2x^2 \text{ and}$$
$$C(x) = 0.1x^2 + 3x.$$

Find the number of units that must be produced and sold in order to maximize profit.

27. _____

28. *Business: Minimizing Inventory Costs.* A department store sells 240 toaster ovens per year. It costs $4.80 to store one toaster oven for one year. To order toaster ovens there is a fixed cost of $9 plus $2 for each oven. How many times per year should the store reorder toaster ovens, and in what lot size, in order to minimize inventory costs?

28.

CALCULUS AND ITS APPLICATIONS   Final Exam, Form E

29. For $f(x) = 8x^2 - 5$, $x = 3$, and $\Delta x = 0.01$, find Δy and $f'(x)\Delta x$.

29. _____

30. *Biological Science: Population Growth.* A bacterial culture is experiencing a growth pattern of 10% per hour in its population P; that is

$$\frac{dP}{dt} = 0.1P,$$

where P is the population of the culture and t is the time, in hours, from noon.

(a) Given that the culture's population at noon was 10,000, find the solution of the equation assuming $P_0 = 10{,}000$ and $k = 0.1$.

30. (a) _____

(b) What will the population be at 3 P.M.?

(b) _____

(c) What is the doubling time of the population?

(c) _____

31. *Economics: Elasticity of Demand.* Consider the demand function

$$x = D(p) = 200 - 12p.$$

(a) Find the elasticity.

31. (a) _____

(b) Find the elasticity at $p = \$6$, stating whether the demand is elastic or inelastic.

(b) _____

(c) At a price of $6, will a small increase in price cause the total revenue to increase or decrease?

(c) _____

(d) Find the value of p for which the total revenue is a maximum.

(d) _____

Evaluate.

32. $\displaystyle\int 3x^4\, dx$

32. _____

33. $\displaystyle\int_1^2 (8e^x + 3x^2)\, dx$

33. _____

34. $\int 12^x \, dx$ (Use Table 1.)

34. _____

35. $\int x^3 e^{x^4+3} \, dx$ (Use substitution. Do not use Table 1.)

35. _____

36. $\int (x-8) \ln x \, dx$

36. _____

37. $\int \frac{2}{x} \, dx$

37. _____

38. $\int_0^{16} \frac{2}{\sqrt{x}} \, dx$

38. _____

39. Find the area under the graph of $y = 5x - x^3$ over the interval $[0, 2]$.

39. _____

40. *Business: Present Value.* Find the present value of $200,000 due in 30 yr at 6.9% compounded continuously.

40. _____

41. *Business: Accumulated Present Value.* Find the accumulated present value of an investment over a 10-yr period in which there is a continuous money flow of $2500 and the current interest rate is 7%, compounded continuously.

41. _____

Determine whether the improper integral is convergent or divergent, and calculate its value if it is convergent.

42. $\int_1^\infty x^{-19} \, dx$

42. _____

43. $\int_3^\infty \frac{1}{3} e^x \, dx$

43. _____

CALCULUS AND ITS APPLICATIONS Final Exam, Form E

44. Given the probability density function

$$f(x) = \frac{5x^4}{32} \text{ over } [0, 2]$$

find each of the following.

(a) The mean

(b) The standard deviation

45. Let x be a continuous random variable that is normally distributed with mean $\mu = 20$ and standard deviation $\sigma = 2.5$. Using Table 2, find $P(18 \leq x \leq 21)$.

46. *Economics: Supply and Demand.* Given the demand and supply functions, in dollars,

$$p = D(x) = (x-9)^2 \text{ and } p = S(x) = x^2 + 2x + 1,$$

find the equilibrium point and the producer's surplus at the equilibrium point.

47. Find the volume of the solid of revolution generated by rotating the region under the graph of

$$y = e^{-x} \text{ from } x = 1 \text{ to } x = 3$$

about the x-axis.

Solve the differential equation.

48. $\dfrac{dy}{dx} = -8x^7 y$

49. $y' = \dfrac{4}{y}$

50. *Total Profit.* A firm's marginal profit P as a function of its total cost C is given by

$$\frac{dP}{dC} = \frac{-300}{(C+5)^{3/2}}.$$

(a) Find the profit function $P(C)$ if $P = \$14$ when $C = \$251$.

(b) At what value will the firm break even $(P = 0)$?

Given $f(x, y) = \dfrac{y}{x} + 2x - 6y^2$, find each of the following.

51. f_y

51. _____

52. f_{yx}

52. _____

53. Find the relative maximum and minimum values of $f(x, y) = 2x^2 - y^2 + 6$.

53. _____

54. Maximize $f(x, y) = 4x + 2y - x^2 + y^2 + 3$ subject to the constraint $x + 3y = 5$.

54. _____

55. Evaluate:
$$\int_{-1}^{1}\int_{0}^{1} (3e^y + x)\, dy\, dx.$$

55. _____

56. Find f_t: $f(x, t) = \dfrac{2t - 3x^2}{x + 2t}$.

56. _____

57. Find $\displaystyle\lim_{x \to 3} \dfrac{x^4 - 10x^2 + 9}{x - 3}$.

57. _____

58. Use a grapher to approximate the area between the following curves:
$$y = x^3 - 3x^2 - 4x,$$
$$y = x^2 - 4x,$$
$$x = 0,\ x = 4.$$

58. _____

CALCULUS AND ITS APPLICATIONS

Name: _____

Final Exam, Form F

1. Write an equation of the line with slope -4 and containing the point $(-3, 5)$.

   1. _____

2. For $f(x) = 2x^2 + 3x$, find the simplified difference quotient.

   2. _____

3. (a) Graph:
   $$f(x) = \begin{cases} -x^2 + 2, & \text{for } x \neq 1, \\ 0, & \text{for } x = 1. \end{cases}$$

   3. (a)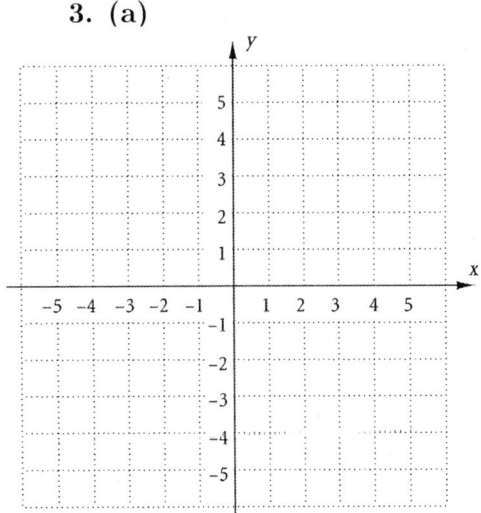

   (b) Find $\lim\limits_{x \to 1} f(x)$.

   (b) _____

   (c) Find $f(1)$.

   (c) _____

   (d) Is f continuous at 1?

   (d) _____

Find the limit, if it exists.

4. $\lim\limits_{x \to -2} (4x^3 - x^2 + x)$

   4. _____

5. $\lim\limits_{x \to -3} \dfrac{12}{x + 3}$

   5. _____

6. $\lim\limits_{x \to -5} \dfrac{x + 5}{x^2 + 3x - 10}$

   6. _____

7. $\lim\limits_{x \to \infty} \dfrac{8}{x}$

   7. _____

Differentiate.

8. $y = 5x^4 - 2x^3 + 6$

9. $y = x^{4/3}$

10. $f(x) = x^{-6}$

11. $f(x) = (x-5)(x+1)^4$

12. $f(x) = \dfrac{3x+2}{x^2-8}$

13. $y = \ln(x^3 - 5x)$

14. $y = -4e^x$

15. $y = e^{x^5 - x}$

16. $f(x) = 8(5x-2)^5$

17. For $y = 4x^4 + 3x^2$, find $\dfrac{d^2y}{dx^2}$.

18. Differentiate implicitly to find $\dfrac{dy}{dx}$ if $4 - \sqrt{x} = 2xy$.

19. Find an equation of the tangent line to the graph of $y = x^4 - \ln x + 1$ at the point $(1, 2)$.

CALCULUS AND ITS APPLICATIONS   Final Exam, Form F

Find the relative extrema of the function. List your answer in terms of an ordered pair. Then sketch a graph of the function.

20. $f(x) = -2x^3 + 12x^2 - 18x + 6$

20.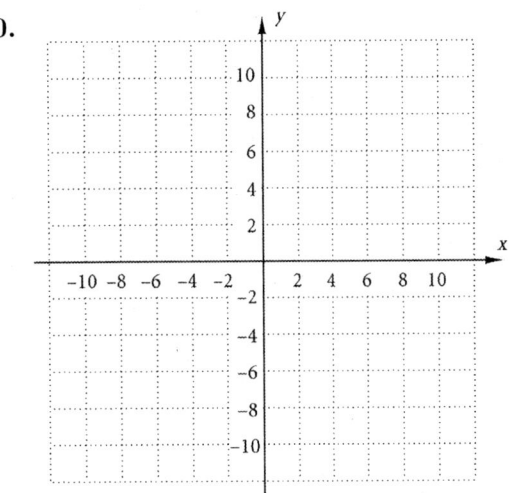

21. $f(x) = 8x^2 - x^4 - 8$

21.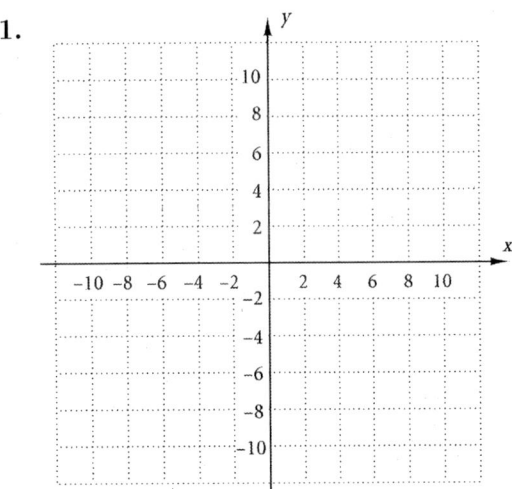

22. $f(x) = \dfrac{20x}{5 + x^2}$

22.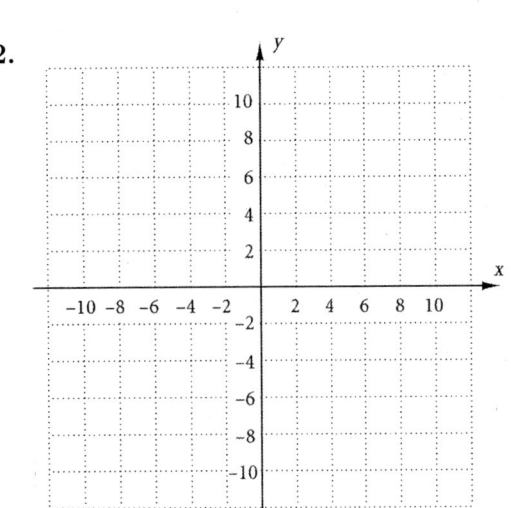

23. $f(x) = \dfrac{-8}{x^2 - 4}$

23.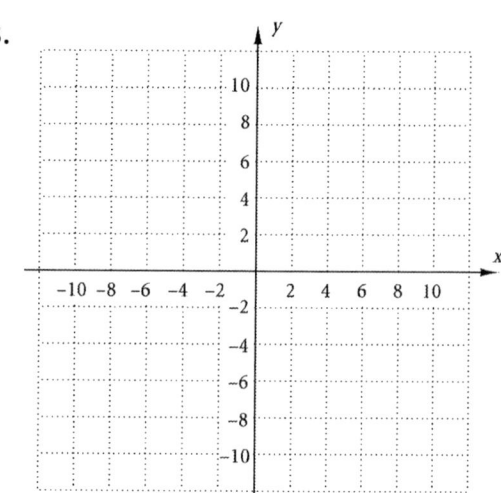

Find the absolute maximum and minimum values, if they exist, over the indicated interval. If no interval is indicated, use the real line.

24. $f(x) = 6x^2 + 3x - 2$

24. _____

25. $f(x) = 2x + 8$

25. _____

26. $f(x) = \dfrac{2}{3}x^3 - x^2 - 4x + 5;\ [-3, 3]$

26. _____

27. *Business: Maximizing Profit.* For a certain product, the total-revenue and total-cost functions are given by

$$R(x) = 60x - 0.3x^2 \text{ and}$$
$$C(x) = 0.2x^2 + 40x.$$

Find the number of units that must be produced and sold in order to maximize profit.

27. _____

28. *Business: Minimizing Inventory Costs.* A department store sells 600 CD players per year. It costs $3 to store one CD player for one year. To order CD players there is a fixed cost of $9 plus $2.50 for each CD player. How many times per year should the store reorder CD players, and in what lot size, in order to minimize inventory costs?

28. _____

CALCULUS AND ITS APPLICATIONS Final Exam, Form F

29. For $f(x) = 6x^2 + 2$, $x = 2$, and $\Delta x = 0.01$, find Δy and $f'(x)\Delta x$.

29. _____

30. *Biological Science: Population Growth.* A bacterial culture is experiencing a growth pattern of 12% per hour in its population P; that is

$$\frac{dP}{dt} = 0.12P,$$

where P is the population of the culture and t is the time, in hours, from noon.

(a) Given that the culture's population at noon was 40,000, find the solution of the equation assuming $P_0 = 40,000$ and $k = 0.12$.

30. (a) _____

(b) What will the population be at 6 P.M.?

(b) _____

(c) What is the doubling time of the population?

(c) _____

31. *Economics: Elasticity of Demand.* Consider the demand function

$$x = D(p) = 200 - 16p.$$

(a) Find the elasticity.

31. (a) _____

(b) Find the elasticity at $p = \$7$, stating whether the demand is elastic or inelastic.

(b) _____

(c) At a price of $7, will a small increase in price cause the total revenue to increase or decrease?

(c) _____

(d) Find the value of p for which the total revenue is a maximum.

(d) _____

Evaluate.

32. $\int 8x^4 \, dx$

32. _____

33. $\int_0^2 (2e^x + 2) \, dx$

33. _____

34. $\displaystyle\int \frac{1}{x(3x+4)}\,dx$ (Use Table 1.) 34. _____

35. $\displaystyle\int 15x^4 e^{x^5+2}\,dx$ (Use substitution. Do not use Table 1.) 35. _____

36. $\displaystyle\int (x-6)\ln x\,dx$ 36. _____

37. $\displaystyle\int \frac{10}{x}\,dx$ 37. _____

38. $\displaystyle\int_4^9 2\sqrt{x}\,dx$ 38. _____

39. Find the area under the graph of $y = 5x - x^2$ over the interval $[2,3]$. 39. _____

40. *Business: Present Value.* Find the present value of $200,000 due in 20 yr at 8.8% compounded continuously. 40. _____

41. *Business: Accumulated Present Value.* Find the accumulated present value of an investment over a 15-yr period in which there is a continuous money flow of $2500 and the current interest rate is 7.6%, compounded continuously. 41. _____

Determine whether the improper integral is convergent or divergent, and calculate its value if it is convergent.

42. $\displaystyle\int_0^\infty \frac{1}{\sqrt{x}}\,dx$ 42. _____

43. $\displaystyle\int_1^\infty x^{-3}\,dx$ 43. _____

CALCULUS AND ITS APPLICATIONS   Final Exam, Form F

44. Given the probability density function
$$f(x) = \frac{x^3}{4} \text{ over } [0, 2]$$
find each of the following.

(a) The mean

(b) The standard deviation

44. (a) _____

(b) _____

45. Let x be a continuous random variable that is normally distributed with mean $\mu = 5$ and standard deviation $\sigma = 1.6$. Using Table 2, find $P(4.6 \leq x \leq 6.2)$.

45. _____

46. *Economics: Supply and Demand.* Given the demand and supply functions, in dollars,
$$p = D(x) = (x - 50)^2 \text{ and } p = S(x) = x^2 + 20x + 700,$$
find the equilibrium point and the producer's surplus at the equilibrium point.

46. _____

47. Find the volume of the solid of revolution generated by rotating the region under the graph of
$$y = e^{3x} \text{ from } x = 0 \text{ to } x = 6$$
about the x-axis.

47. _____

Solve the differential equation.

48. $\dfrac{dy}{dx} = 3x^4 y$

48. _____

49. $y' = \dfrac{110}{y}$

49. _____

50. *Total Profit.* A firm's marginal profit P as a function of its total cost C is given by
$$\frac{dP}{dC} = \frac{-200}{(C+8)^{3/2}}.$$

(a) Find the profit function $P(C)$ if $P = \$10$ when $C = \$316$.

(b) At what value will the firm break even $(P = 0)$?

50. (a) _____

(b) _____

Given $f(x, y) = \dfrac{3x}{y} + e^x + y^2$, find each of the following.

51. f_y

51. _____

52. f_{yx}

52. _____

53. Find the relative maximum and minimum values of $f(x, y) = -x^2 - 2y^2 + 8$.

53. _____

54. Maximize $f(x, y) = x + 2y - 2x^2 + 5y^2 + 1$ subject to the constraint $x + 2y = 3$.

54. _____

55. Evaluate:
$$\int_{-1}^{0} \int_{1}^{2} \frac{1}{x} + y^2 \, dx \, dy.$$

55. _____

56. Find $\displaystyle\lim_{x \to -3} \dfrac{x^4 - 5x^2 - 36}{x + 3}$.

56. _____

57. Differentiate: $\dfrac{\ln(e^x x)}{e^x}$.

57. _____

58. Use a grapher to approximate the area between the following curves:

$$y = x^3 - 3x^2 - 5x,$$
$$y = x - 2x^2,$$
$$x = 0, \, x = 3.$$

58. _____

ANSWERS TO CHAPTER TEST FORMS

CHAPTER 1, FORM A

1. (a) 125;
   (b) 250 mg/dl

2. $1260

3. (a) -12;
   (b) $x^3 + 3x^2h + 3xh^2 + h^3 - 4$

4. $m = -\dfrac{1}{2}$; y-intercept: $(0, -5)$

5. $y = \dfrac{2}{3}x - 8$

6. $\dfrac{2}{3}$

7. $-\$200/\text{yr}$

8. 3 mph

9. $I = \dfrac{1}{3}V$

10. (a) $C(x) = 0.95x + 15{,}000$;
    (b) $R(x) = 3.50x$;
    (c) $P(x) = 2.55x - 15{,}000$;
    (d) 5883

11. $(1, 9)$

12. No

13. Yes

14. (a) 2.2;
    (b) $\{x \mid -4 \leq x \leq 3\}$;
    (c) 2;
    (d) $\{y \mid -5 \leq y \leq 4\}$

15.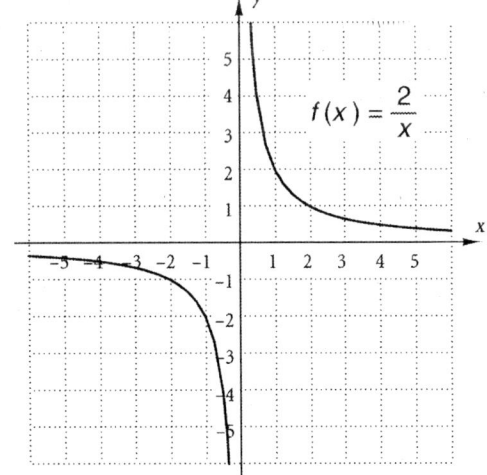

16. $x^{-4/5}$

17. $\dfrac{1}{\sqrt[3]{y^2}}$

217

CHAPTER 1, FORM A (continued)

18.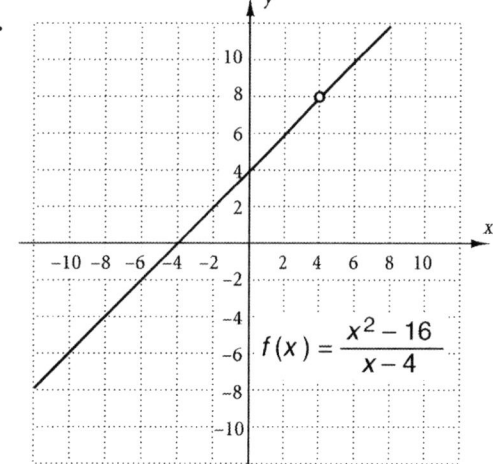

19. $(-\infty, -2) \cup (-2, 3) \cup (3, \infty)$

20. $[3, \infty)$

21. $(a, b]$

22.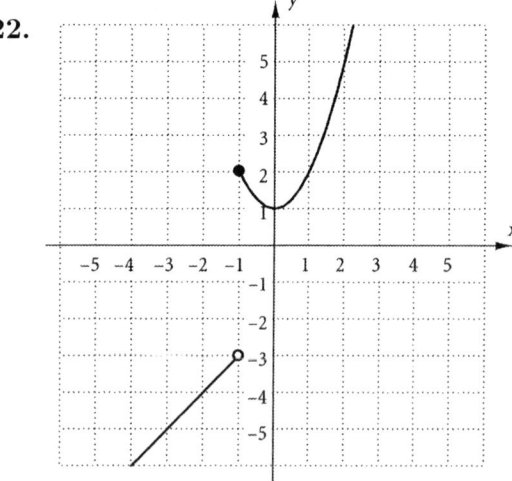

23. (a) $g = 0.25x - 486.75$;
    (b) 10.25 gal

24. (a)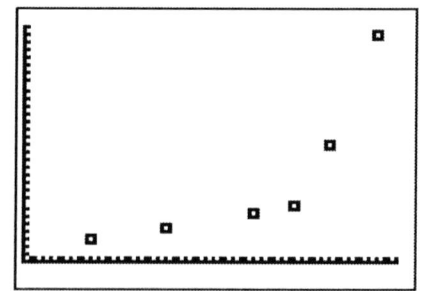
    (b) Yes;
    (c) $f = 0.0402v^2 - 1.8733v + 17.7127$;
    (d) 1.4 newtons

25. (a) 880;
    (b) $5.31

26. $f(x) = x^3 - 4x^2 + 2x - 3$
    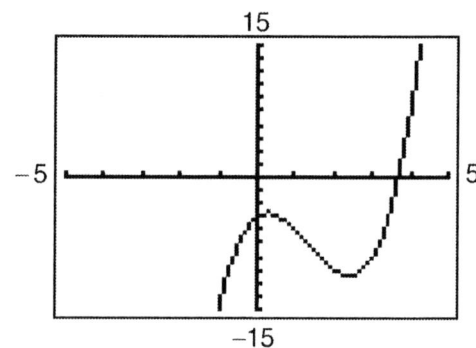

27. $-6, 6$

28. (a) $g = \dfrac{8}{25}x - \dfrac{9376}{15}$, or $g = 0.32x - 625.0\overline{6}$
    (b) $11.09\overline{3}$ gal

ANSWERS TO CHAPTER TEST FORMS

CHAPTER 1, FORM B

1. (a) Approximately 310 mg;
   (b) 0.5 hr and 4.4 hr

2. $800

3. (a) -12;
   (b) $2x^3 + 6x^2a + 6xa^2 + 2a^3 + 4$

4. $m = -4$; y-intercept: $(0, 5)$

5. $y = \dfrac{2}{5}x - 4$

6. $\dfrac{7}{4}$

7. $\dfrac{1}{4}$ bag/student

8. $-6\dfrac{2}{3}$ m/s

9. $N = 240P$

10. (a) $C(x) = 0.76x + 8000$;
    (b) $R(x) = 1.29x$;
    (c) $P(x) = 0.53x - 8000$;
    (d) 15,095

11. Approximately $(1.71, 3.29)$

12. No

13. Yes

14. (a) -4;   (b) $(-\infty, \infty)$;
    (c) $-2, 3$;   (d) $[-6.25, \infty)$

15.

16. $y^{-2/3}$

17. $\dfrac{1}{\sqrt[4]{p^3}}$

CHAPTER 1, FORM B (continued)

18.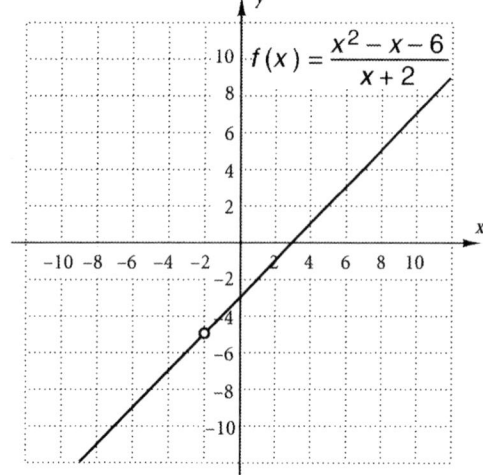

19. $(-\infty, -3) \cup (-3, 4) \cup (4, \infty)$

20. $\left(-\dfrac{3}{2}, \infty\right)$

21. $(a, d]$

22.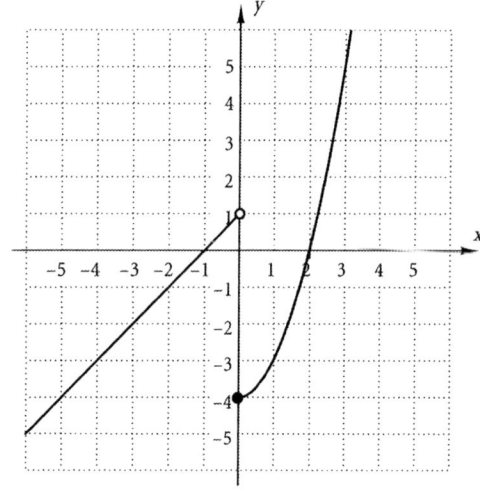

23. (a) $p = -55x + 111{,}200$;
    (b) $1640

24. (a)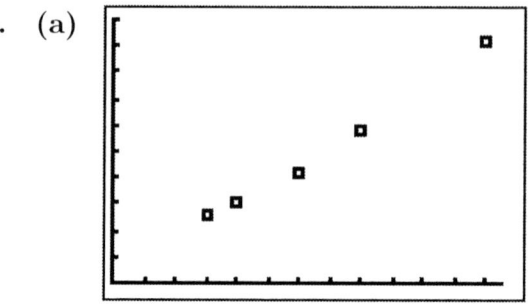
    (b) Yes;
    (c) $p = 0.0125d^2 + 0.375d + 2.3$;
    (d) $7.30

25. (a) 771; (b) $7.94

26. $f(x) = 5x^3 - 6x^2 + x - 2$

    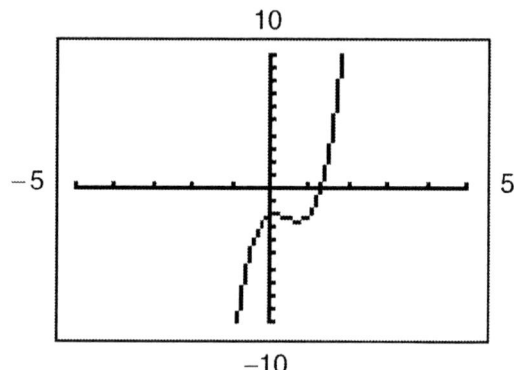

27. -11.18, 11.18

28. (a) $p = -89.\overline{6}x + 180{,}247.\overline{3}$; (b) $1631.33

ANSWERS TO CHAPTER TEST FORMS

CHAPTER 1, FORM C

1. (a) 6 (b) 10.5 min

2. $1352

3. (a) 142;
   (b) $3x^2 + 6x - 2$

4. $m = -\dfrac{2}{3}$; y-intercept: $(0, 8)$

5. $y = -\dfrac{5}{8}x + \dfrac{5}{2}$

6. $\dfrac{15}{7}$

7. Approximately -1600 ft/min

8. $\dfrac{1}{4}$ km/min

9. $E = 6M$

10. (a) $C(x) = 12x + 24{,}000$;
    (b) $R(x) = 18x$;
    (c) $P(x) = 6x - 24{,}000$;
    (d) 4000

11. $\left(\dfrac{25}{11}, \dfrac{900}{121}\right)$, or approximately $(2.27, 7.44)$

12. No

13. Yes

14. (a) -2; (b) $\{x \mid -5 \leq x < 4\}$;
    (c) $\{x \mid 1 \leq x < 4\}$; (d) $\{y \mid -5 \leq y \leq 1\}$

15. [Graph of $f(x) = \dfrac{6}{x-3}$]

16. $5x^{-1/3}$

17. $\dfrac{1}{\sqrt[7]{a^6}}$

CHAPTER 1, FORM C (continued)

18.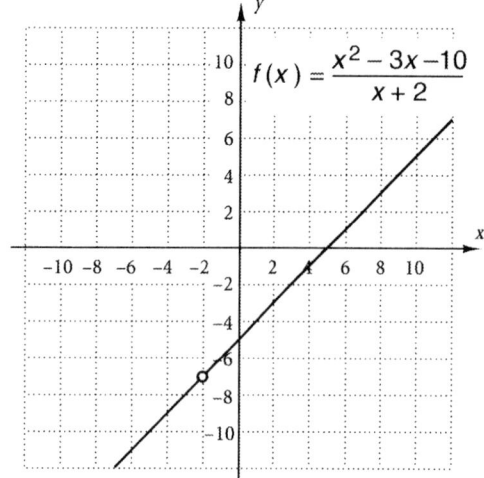

19. $(-\infty, -2) \cup (-2, 6) \cup (6, \infty)$

20. $\left(-\infty, \dfrac{4}{3}\right]$

21. (m, n)

22.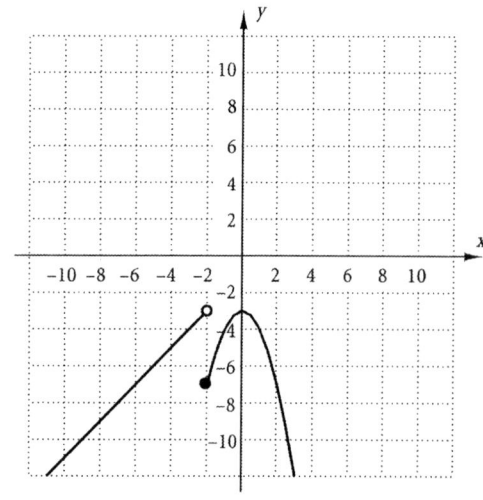

23. (a) $R = 6x - 11{,}879$;
    (b) $133 thousand, or $133,000

24. (a)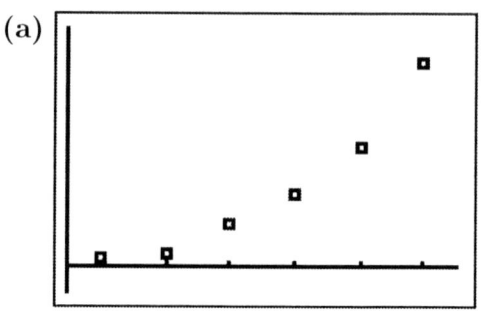
    (b) Yes;
    (c) $w = 6x^2 - 23{,}943x + 23{,}886{,}140$;
    (d) 575 million

25. (a) 376;   (b) $6.30

26. $f(x) = 2x^3 - 5x^2 + x - 4$

    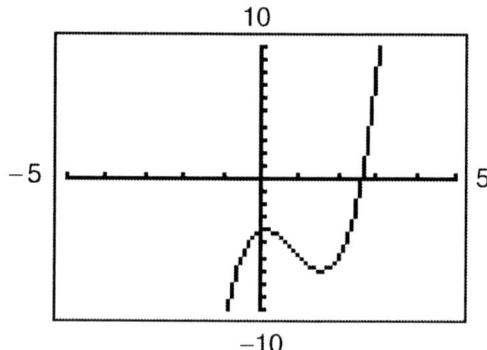

27. $-8.06,\ 8.06$

28. (a) $R = 2.905x - 5704.267$
    (b) $111.5 thousand, or $111,500

ANSWERS TO CHAPTER TEST FORMS

CHAPTER 1, FORM D

1. (a) 48 ft; (b) 0.5 sec, 3.5 sec

2. $750

3. (a) 5; (b) $2a^2 - 12a + 21$

4. $m = 5$; y-intercept: $(0, -4)$

5. $y = \dfrac{5}{8}x - \dfrac{23}{4}$

6. $-\dfrac{3}{2}$

7. $\dfrac{1}{2}$ in./mo

8. $-1666\dfrac{2}{3}$ ft/min

9. $A = 1.07\overline{3}G$

10. (a) $C(x) = 35x + 135{,}000$;
    (b) $R(x) = 70x$;
    (c) $P(x) = 35x - 135{,}000$;
    (d) 3858

11. $(1.92, 9.47)$

12. Yes

13. No

14. (a) -2; (b) $\{x \mid -4 \leq x \leq 2\}$;
    (c) -2; (d) $\{y \mid -3 \leq y \leq 3\}$

15.

16. $5n^{-1/3}$

17. $\dfrac{1}{\sqrt[7]{y^4}}$

CHAPTER 1, FORM D (continued)

18.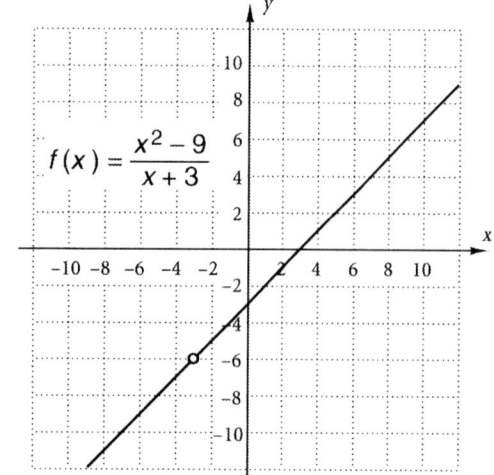

19. $(-\infty, -8) \cup (-8, 4) \cup (4, \infty)$

20. $(-\infty, 3)$

21. $[a, c]$

22.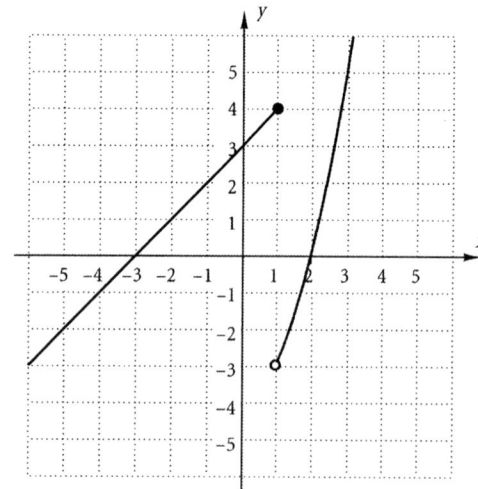

23. (a) $I = 2000x$; (b) $32,000

24. (a)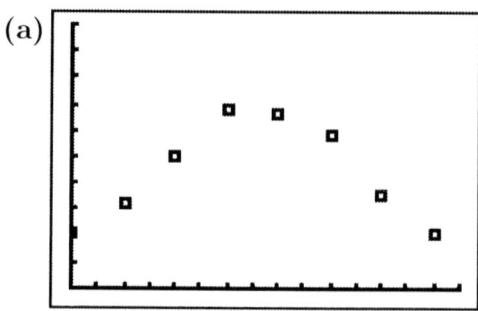

(b) Yes;

(c) $f(x) = -\dfrac{13}{1200}x^2 + \dfrac{17}{120}x + \dfrac{21}{10}$, or $f(x) = -0.0011x^2 + 0.1417x + 2.1$;

(d) 2.2 million

25. (a) 338; (b) $5.31

26. $f(x) = x^3 - 2x^2 - 2x + 3$

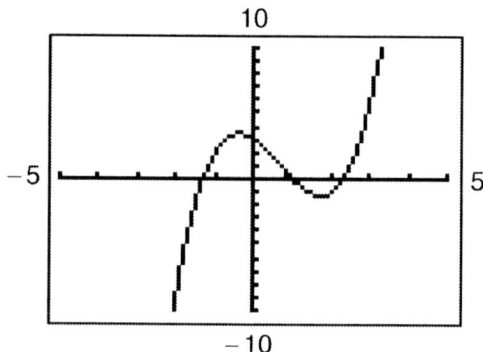

27. $-7.94, 7.94$

28. (a) $f = -0.0009x^2 + 0.1331x + 1.6$;
(b) 3.1 million

CHAPTER 1, FORM E

1. (a) 15 parts per million;
   (b) 4.5 hr, 35.5 hr

2. $1050

3. (a) 60;
   (b) $4x^2 - 15x + 14$

4. $m = -2$; y-intercept: $(0, -6)$

5. $y = -\dfrac{1}{4}x - \dfrac{15}{2}$

6. -3

7. $2500/yr

8. -25 mi/hr

9. $A = 0.042F$

10. (a) $C(x) = 10x + 10{,}000$;
    (b) $R(x) = 30x$;
    (c) $P(x) = 20x - 10{,}000$;
    (d) 500

11. $\left(\dfrac{16}{9}, \dfrac{400}{81}\right)$, or approximately $(1.78, 4.94)$

12. Yes

13. No

14. (a) 2;   (b) $\{x \mid -3 \le x \le 4\}$;
    (c) $-\dfrac{1}{2}$, 3;   (d) $\{y \mid -4.3 \le y \le 4.5\}$

15. [graph of $t(x) = \dfrac{1}{x-4}$]

16. $2p^{-2/3}$

17. $\dfrac{1}{\sqrt{y}}$

CHAPTER 1, FORM E (continued)

18.

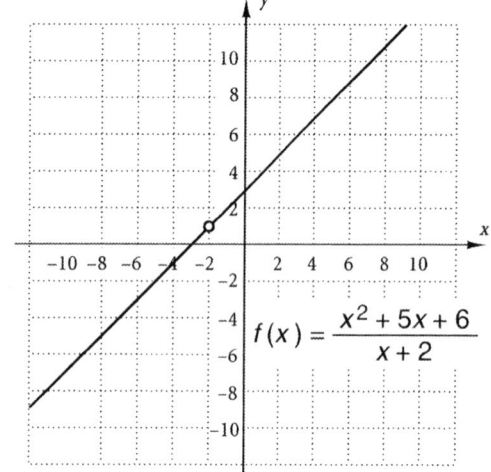

$f(x) = \dfrac{x^2 + 5x + 6}{x + 2}$

19. $(-\infty, -2) \cup (-2, 5) \cup (5, \infty)$

20. $\left(-\infty, \dfrac{3}{2}\right]$

21. $(p, q]$

22.

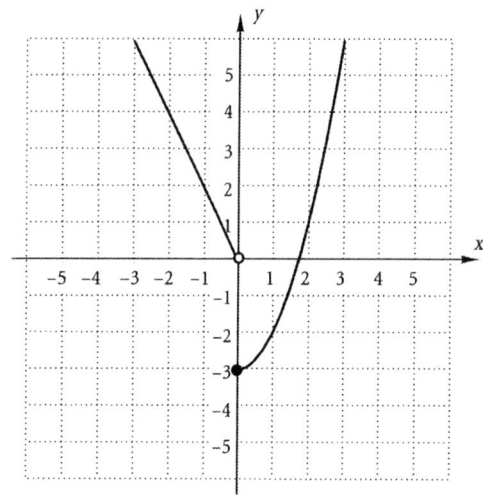

23. (a) $g = 0.05t + 80$; (b) 92

24. (a)

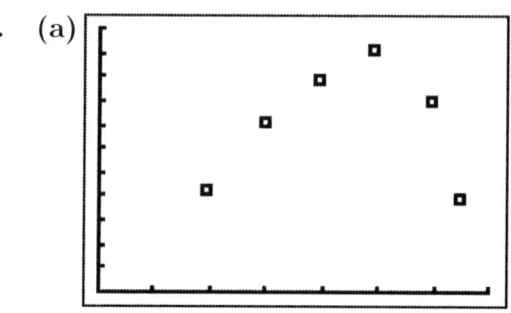

(b) Yes;

(c) $i = -51.6196a^2 + 4414.9491a - 49{,}431.0837$;

(d) $44,712

25. (a) 1125; (b) $5.50

26. $f(x) = 2x^3 - 3x^2 + x + 2$

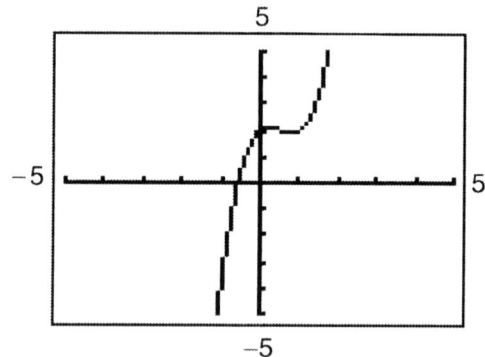

27. $-3.16, 3.16$

28. (a) $g(t) = 0.0630662021t + 77.1184669$

(b) 79

CHAPTER 1, FORM F

1. (a) 3150 parts per billion;
   (b) 1991

2. $1750

3. (a) 52;
   (b) $3x^2 + 6xa + 3a^2 - x - a$

4. $m = -2$; y-intercept: $(0, 3)$

5. $y = -\dfrac{2}{3}x + \dfrac{1}{3}$

6. $\dfrac{12}{7}$

7. $-\$700/\text{yr}$

8. 3 million/yr

9. $d = \dfrac{28}{5}m$

10. (a) $C(x) = 6x + 5760$;
    (b) $R(x) = 10.80x$;
    (c) $P(x) = 4.8x - 5760$;
    (d) 1200

11. $(1.41, 4.24)$

12. Yes   13. No

14. (a) 4;
    (b) $\{-3, -2, -1, 0, 1, 2, 3, 4\}$;
    (c) $-2, -1, 0$;
    (d) $\{-2, 0, 4, 5\}$

15.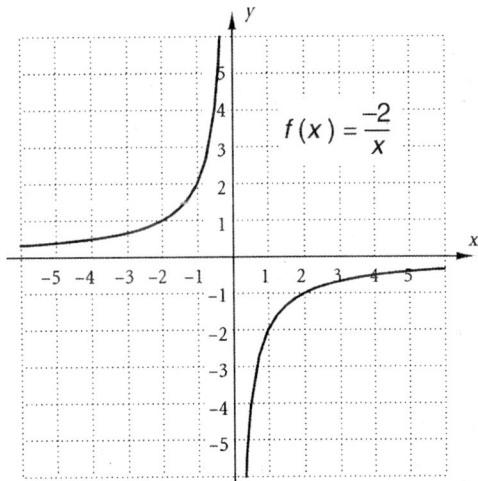

16. $5x^{-1/2}$

17. $\dfrac{1}{\sqrt[4]{m^5}}$, or $\dfrac{1}{m\sqrt[4]{m}}$

CHAPTER 1, FORM F (continued)

18.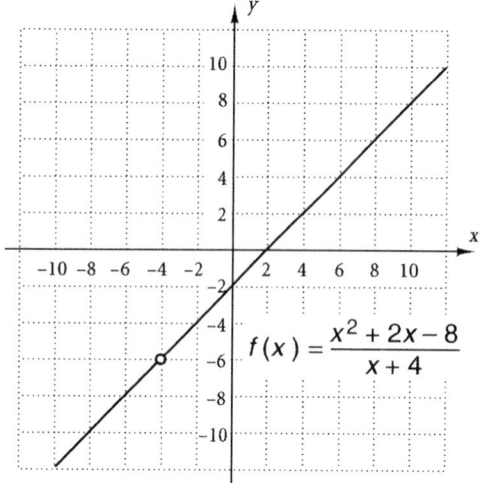

19. $(-\infty, -1) \cup (-1, 2) \cup (2, \infty)$

20. $(-\infty, 5)$

21. $[a, \infty)$

22.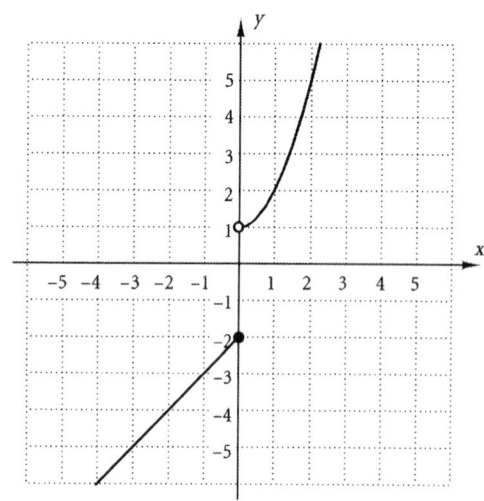

23. (a) $c = \dfrac{10}{3}a + \dfrac{196}{3}$; (b) $192

24. (a)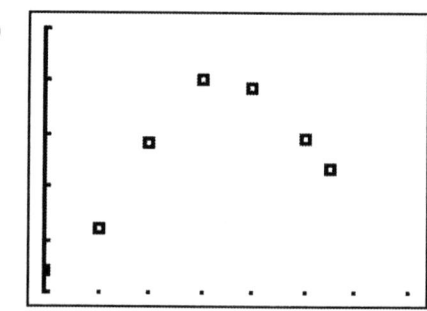

(b) Yes;

(c) $C(x) = -\dfrac{295}{54}x^2 + \dfrac{4165}{54}x + \dfrac{11{,}545}{27}$, or
$C(x) = -5.4630x^2 + 77.1296x + 427.5926$;

(d) 355

25. (a) 344; (b) $5.85

26. $f(x) = -x^3 + 6x - 3$

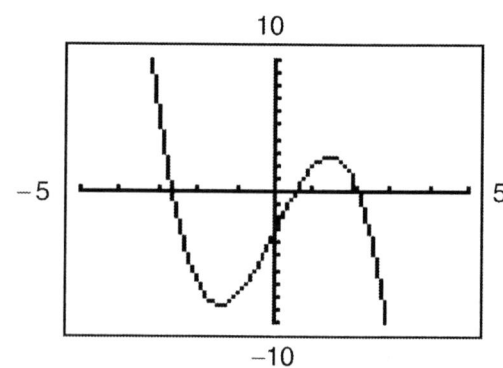

27. $-8.6, 8.6$

28. (a) $C(x) = -3.8207x^2 + 53.1596x + 499.7568$;

(b) 438

ANSWERS TO CHAPTER TEST FORMS

CHAPTER 2, FORM A

1. Does not exist
2. 0
3. Does not exist
4. -3
5. 1
6. 6
7. No
8. Yes
9. 0
10. 0
11. Yes
12. Does not exist
13. 1
14. No
15. 18
16. $-\dfrac{1}{4}$
17. Does not exist
18. $4(2x+h)$
19. $y = -x + 6$
20. $(0,0)$, $(2,-4)$
21. $40x^{39}$
22. $\dfrac{3}{2\sqrt{x}}$
23. $\dfrac{24}{x^4}$
24. $\dfrac{4}{5}x^{-1/5}$, or $\dfrac{4}{5\sqrt[5]{x}}$
25. $-10x + 0.4$
26. $3x^3 - 10x + 4$
27. $\dfrac{-4x - 4}{x^3}$
28. $(x+5)^4 (4-x)(10-7x)$
29. $-20x\left(3x^2 - 2\right)\left(3x^4 - 4x^2 + 6\right)^{-6}$
30. $2x\sqrt{x^3+1} + \dfrac{3x^4}{2\sqrt{x^3+1}}$, or $\dfrac{7x^4 + 4x}{2\sqrt{x^3+1}}$
31. $480x^3 - 54$
32. (a) $\dfrac{dN}{dn} = n - \dfrac{1}{2}$;
    (b) 105;   (c) 14.5 games/team
33. $x^3 - 3x^6$; $x^3 - 9x^4 + 27x^5 - 27x^6$
34. $-\dfrac{8}{3}\left(\dfrac{5-4x}{5+4x}\right)^{1/3}(5+12x)$
35. 12
36. $(0.8610, 15.2837)$, $(3.3413, -15.6609)$

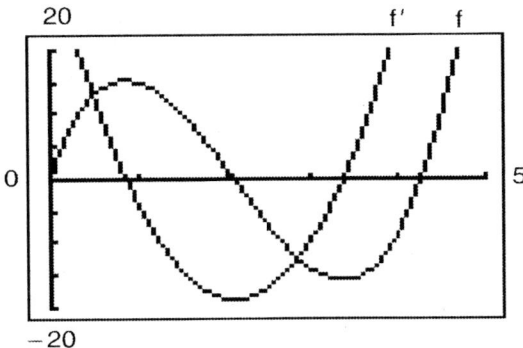

37. -0.5

CHAPTER 2, FORM B

1. 2
2. Does not exist
3. Does not exist
4. 1
5. 2
6. 5
7. No
8. Yes
9. Does not exist
10. 2
11. No
12. 1
13. 1
14. Yes
15. 66
16. $\dfrac{1}{14}$
17. Does not exist
18. $3(2x+h)$
19. $y = x + 8$
20. $(0,0), \left(\dfrac{4}{3}, -\dfrac{32}{27}\right)$
21. $16x^{15}$
22. $\dfrac{4}{5\sqrt[5]{x^4}}$
23. $\dfrac{22}{x^3}$
24. $\dfrac{3}{5}x^{-2/5}$, or $\dfrac{3}{5\sqrt[5]{x^2}}$
25. $-1.2x + 4.1$
26. $2x^2 - 8x + 10$
27. $-\dfrac{20}{(5-x)^2}$
28. $(x+1)^2(6-x)(16-5x)$
29. $-8x\left(6 + 5x^3 + 3x^4\right)\left(6x^2 + 2x^5 + x^6\right)^{-5}$
30. $2x\sqrt{x^3 - 2} + \dfrac{3x^4}{2\sqrt{x^3 - 2}}$, or $\dfrac{7x^4 - 8x}{2\sqrt{x^3 - 2}}$
31. $120x$
32. (a) $\dfrac{dS}{dr} = 4\pi r + 16\pi$;
    (b) 40π in^2;   (c) 24π in^2/in.
33. $x + 4\sqrt{x}$; $\sqrt{x^2 + 4x}$
34. $\dfrac{1}{2}\left[\dfrac{(2-3x)^4}{(5+x)^3}\right]^{1/6} - 2\left[\dfrac{(5+x)^3}{(2-3x)^2}\right]^{1/6}$
35. 27
36. $(0.37435356, 3.0910298)$,
    $(1.8975696, -7.910868)$

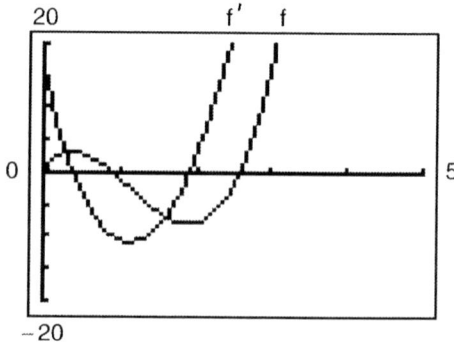

37. 0

CHAPTER 2, FORM C

1. Does not exist
2. -1
3. 2
4. 0
5. Does not exist
6. 4
7. Yes
8. No
9. Does not exist
10. 0
11. No
12. -1
13. -1
14. Yes
15. 14
16. -9
17. Does not exist
18. $4x + 5 + 2h$
19. $y = \dfrac{22}{5}x - 4$
20. $(0,0)$, $(1,-1)$
21. $15x^{14}$
22. $\dfrac{3}{\sqrt{x}}$
23. $\dfrac{49}{x^8}$
24. $\dfrac{2}{5}x^{-3/5}$, or $\dfrac{2}{5\sqrt[5]{x^3}}$
25. $-6x + 10.1$
26. $\dfrac{1}{2}x^4 + 12x^3 - 6$
27. $\dfrac{-9x - 4}{x^5}$
28. $(3-x)^3 (x+1)^2 (5-7x)$
29. $-8(6x-5)(6x^2 - 10x + 1)^{-5}$
30. $3x^2\sqrt{x^4 - 3} + \dfrac{2x^6}{\sqrt{x^4 - 3}}$, or $\dfrac{5x^6 - 9x^2}{\sqrt{x^4 - 3}}$
31. $360x^3 - 24$
32. (a) $\dfrac{dP}{dn} = 3n^2 - 6n + 2$; (b) 6840; (c) 1082 ways/club member
33. $-x^4 - 10x^2 - 22$; $x^4 - 6x^2 + 14$
34. $\dfrac{1}{2}\left[\dfrac{(2-x)^2}{81(10+x)^5}\right]^{1/6} - \left[\dfrac{(10+x)}{81(2-x)^4}\right]^{1/6}$
35. 12
36. $(0.1671, 0.4171)$, $(1.1986, -8.1351)$

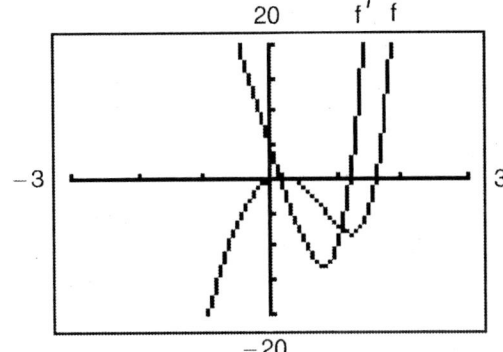

37. $\dfrac{1}{4}$

CHAPTER 2, FORM D

1. Does not exist
2. Does not exist
3. 1
4. -1
5. Does not exist
6. 1
7. Yes
8. No
9. -1
10. -1
11. Yes
12. Does not exist
13. 1
14. No
15. -9
16. $\dfrac{1}{18}$
17. Does not exist
18. $-2(2x+h)$
19. $y = -2x + 4$
20. $(0,0), (-1,3)$
21. $16x^{15}$
22. $\dfrac{13}{3\sqrt[3]{x^2}}$
23. $-\dfrac{32}{x^9}$
24. $\dfrac{3}{8}x^{-5/8}$, or $\dfrac{3}{8\sqrt[8]{x^5}}$
25. $-12x - 5$
26. $5x^7 - 24x^5 + 5$
27. $-\dfrac{2}{(2-x)^2}$
28. $2(x+2)^3(3-x)(4-3x)$
29. $-16x(3x-1)(4x^3 - 2x^2 + 5)^{-5}$
30. $3x^2\sqrt{x^2-1} + \dfrac{x^4}{\sqrt{x^2-1}}$, or $\dfrac{4x^4 - 3x^2}{\sqrt{x^2-1}}$
31. $240x^3$
32. (a) $\dfrac{dV}{dr} = \dfrac{8}{3}\pi r$; (b) $\dfrac{\pi}{3}$ in^3; (c) $\dfrac{4\pi}{3}$ in^3/in.
33. $x + 2\sqrt{x}$; $\sqrt{x^2 + 2x}$
34. $\dfrac{2}{5}(4-3x)\left[\dfrac{4-3x}{(1+x)^3}\right]^{1/5} - \dfrac{18}{5}\left[(1+x)^2(4-3x)\right]^{1/5}$
35. 75
36. $(-0.0999004, 2.04998)$, $(1.0313284, -2.015982)$

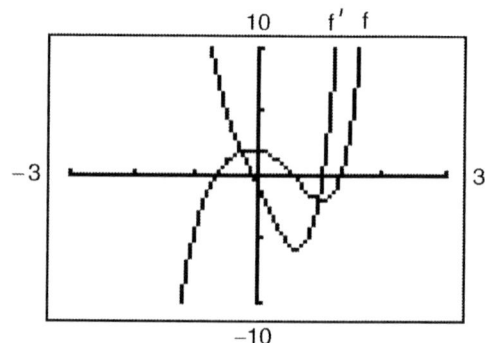

37. $\dfrac{1}{6}$

ANSWERS TO CHAPTER TEST FORMS

CHAPTER 2, FORM E

1. 5
2. Does not exist
3. 1
4. Does not exist
5. 3
6. 4
7. No
8. Yes
9. -3
10. 1
11. No
12. 4
13. 4
14. Yes
15. 4
16. $\dfrac{1}{10}$
17. Does not exist
18. $-2x - h + 5$
19. $y = -\dfrac{7}{2}x + 6$
20. $(0,0)$, $\left(\dfrac{4}{3}, -\dfrac{32}{27}\right)$
21. $106x^{105}$
22. $\dfrac{5}{3\sqrt[3]{x^2}}$
23. $-\dfrac{600}{x^6}$
24. $\dfrac{7}{3}x^{4/3}$
25. $-10x + \dfrac{1}{2}$
26. $3x^3 + 16x - 161$
27. $\dfrac{-3x - 16}{x^5}$
28. $(3-x)^3 (x+5)^2 (-7x - 11)$
29. $-3x(9x - 10)(3x^3 - 5x^2 + 8)^{-4}$
30. $3x^2\sqrt{x^4 - 3} + \dfrac{2x^6}{\sqrt{x^4 - 3}}$, or $\dfrac{5x^6 - 9x^2}{\sqrt{x^4 - 3}}$
31. $-180x^2$
32. (a) $\dfrac{dP}{da} = 0.8a - 40$; (b) 399; (c) -24 accidents/yr
33. $4x^2 - 16x^4$; $4x^4 - 8x^3 + 4x^2$
34. $\left[\dfrac{2(4-x)^3}{4+x}\right]^{1/2} - 3[2(4+x)(4-x)]^{1/2}$
35. 48
36. $(-1.1246, 9.3334)$, $(0.6508, -2.6279)$

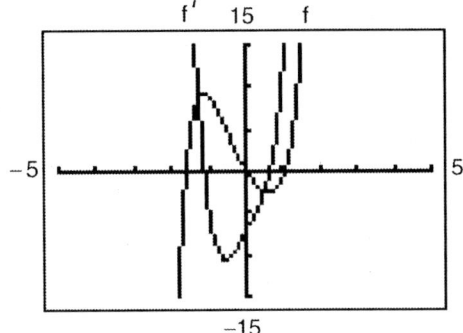

37. $\dfrac{3}{5}$

CHAPTER 2, FORM F

1. 0
2. Does not exist
3. -2
4. Does not exist
5. 3
6. -2
7. Yes
8. No
9. 2
10. 2
11. Yes
12. Does not exist
13. -2
14. No
15. -46
16. $-\dfrac{1}{11}$
17. Does not exist
18. $-8x - 4h$
19. $y = \dfrac{14}{3}x - 4$
20. $(-1, 6)$, $(1, -6)$
21. $50x^{49}$
22. $\dfrac{4}{\sqrt{x}}$
23. $-\dfrac{12}{x^5}$
24. $\dfrac{2}{7}x^{-5/7}$, or $\dfrac{2}{7\sqrt[7]{x^5}}$
25. $-8x - 3.1$
26. $-2x^2 + 32x + 4$
27. $\dfrac{-3}{(3+x)^2}$
28. $(x+2)^2 (2-x)(2-5x)$
29. $-4(6x^2 + 32x - 3)(2x^3 + 16x^2 - 3x)^{-5}$
30. $2x\sqrt{x^6+3} + \dfrac{3x^7}{\sqrt{x^6+3}}$, or $\dfrac{5x^7 + 6x}{\sqrt{x^6+3}}$
31. $480x^3$
32. (a) $\dfrac{dN}{dx} = \dfrac{1}{2}x^2 + x + \dfrac{1}{3}$;
    (b) 220; (c) $60\dfrac{1}{3}$ spheres/layer
33. $2x - 5$; $\sqrt{2x^2 - 5}$
34. $\dfrac{3}{4}\left[\dfrac{(5-2x)}{x+6}\right]^{1/4} - \dfrac{1}{2}\left[\dfrac{x+6}{5-2x}\right]^{3/4}$
35. 3
36. $(0.37260293, 1.2364915)$, $(1.734937, -1.511481)$

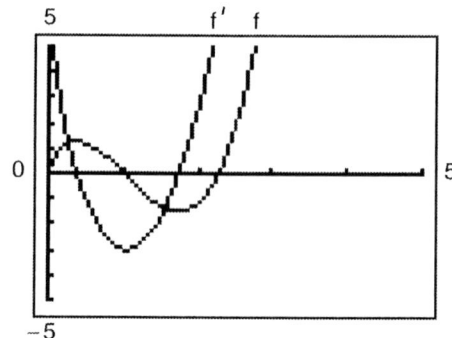

37. $-\dfrac{7}{8}$

ANSWERS TO CHAPTER TEST FORMS

CHAPTER 3, FORM A

1. Relative minimum at $(1, -9)$
   $f(x) = x^2 - 2x - 8$

   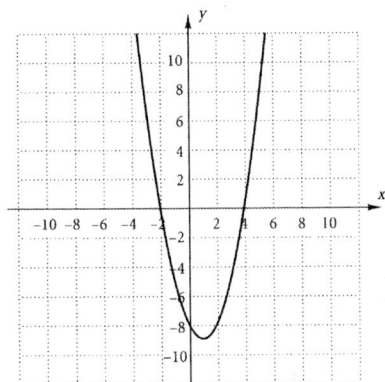

2. Relative minima at $\left(-\frac{\sqrt{6}}{2}, 0\right)$ and $\left(\frac{\sqrt{6}}{2}, 0\right)$; relative maximum at $(0, 9)$
   $f(x) = 4x^4 - 12x^2 + 9$

   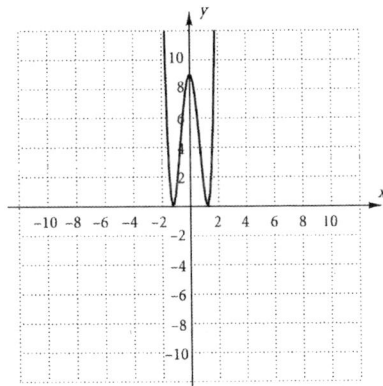

3. Relative minimum at $(2, 1)$
   $f(x) = (x-2)^{2/3} + 1$

4. Relative maximum at $(0, 8)$
   $f(x) = \frac{40}{x^2 + 5}$

   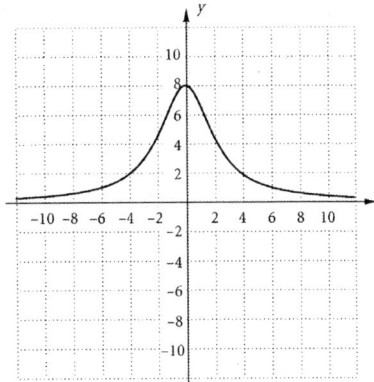

5. Relative maximum at $(-1, 0)$;
   relative minimum at $\left(\frac{1}{2}, \frac{-27}{4}\right)$
   $f(x) = 4x^3 + 3x^2 - 6x - 5$

   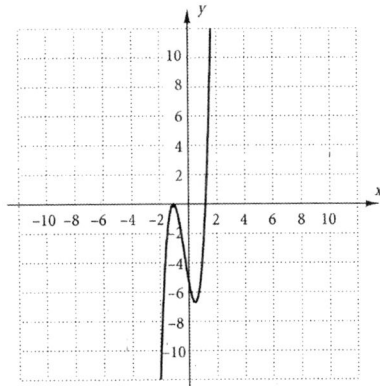

6. Relative maximum at $(-1, 5)$;
   relative minimum at $(1, -3)$
   $f(x) = 1 - 6x + 2x^3$

CHAPTER 3, FORM A (continued)

7. None
   $f(x) = (x+1)^3 + 3$

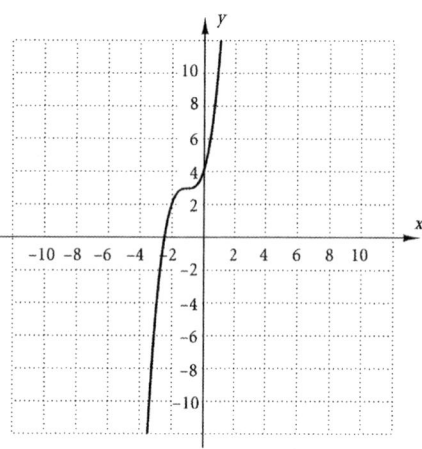

8. Relative minimum at $\left(-2\sqrt{2}, -8\right)$;
   relative maximum at $\left(2\sqrt{2}, 8\right)$
   $f(x) = x\sqrt{16 - x^2}$

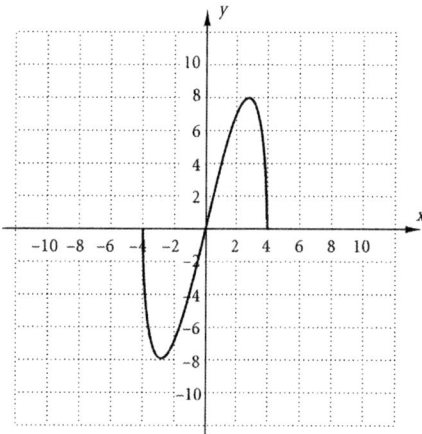

9. $f(x) = \dfrac{3}{x+2}$

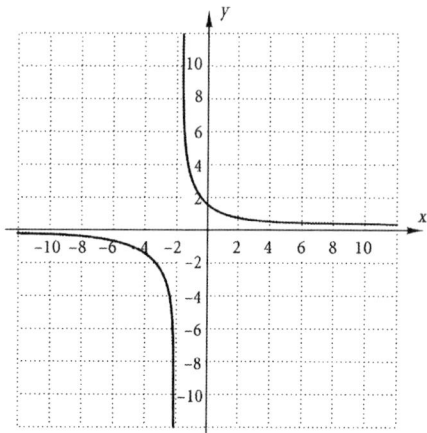

10. $f(x) = \dfrac{-4}{x^2 - 1}$

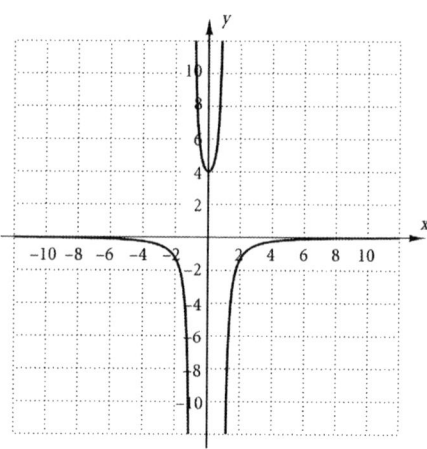

11. $f(x) = \dfrac{x^2 - 9}{x}$

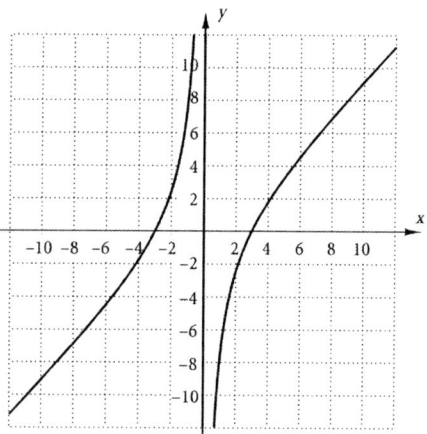

12. $f(x) = \dfrac{x+4}{x-5}$

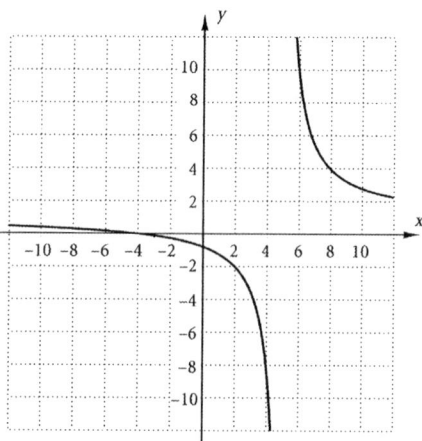

ANSWERS TO CHAPTER TEST FORMS

CHAPTER 3, FORM A (continued)

13. Maximum $= \frac{9}{4}$ at $x = \frac{3}{2}$
14. Minimum $= -13$ at $x = -2$; maximum $= 0$ at $x = -1$
15. Minimum $= -9.645$ at $x = -1.35$
16. Maximum $= 7$ at $x = -2$; minimum $= -1$ at $x = 2$
17. None
18. Maximum $= -\frac{11}{5}$ at $x = \frac{2}{5}$
19. Minimum $= 27$ at $x = 3$
20. 6 and -6
21. $x = 2$; $y = 2$; minimum $= 8$
22. Maximum profit $= 7970$; 200 units
23. 32 in. by 32 in. by 8 in.; 8192 in^3
24. 8 times at lot size 300
25. $\triangle y = 0.41$; $f'(x) \triangle x = 0.4$
26. 8.375
27. (a) $\dfrac{x\,dx}{\sqrt{x^2+6}}$; (b) 0.0063246
28. $-\dfrac{2x}{3y^2}$; $-\dfrac{8}{3}$
29. -0.4 ft/sec
30. Minimum $= 0$ at $x = 0$; maximum $= \frac{1}{3}$ at $x = 2$
31. (a) $A(x) = \dfrac{C(x)}{x} = 50 + \dfrac{50}{\sqrt{x}} + \dfrac{\sqrt{x}}{50}$; (b) minimum $= 52$ at $x = 2500$
32. Relative maximum at $(0.83, 17.55)$; relative minimum at $(4.19, -58.70)$

CHAPTER 3, FORM B

1. Relative minimum at $(-2, -9)$
   $f(x) = x^2 + 4x - 5$

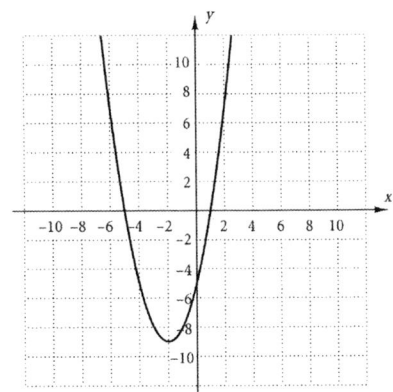

2. Relative minima at $\left(-\sqrt{2}, -8\right)$ and $\left(\sqrt{2}, -8\right)$; relative maximum at $(0, 4)$
   $f(x) = 3x^4 - 12x^2 + 4$

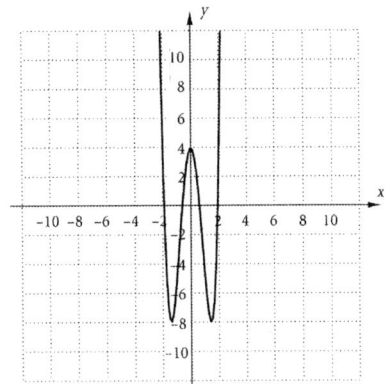

CHAPTER 3, FORM B (continued)

3. Relative minimum at $(2, 4)$
$f(x) = (x-2)^{2/3} + 4$

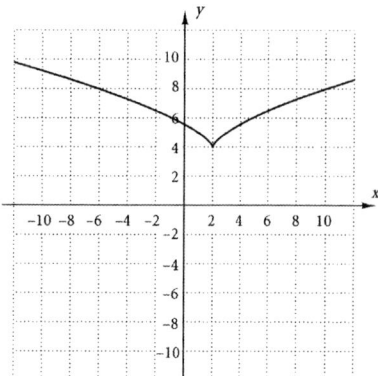

4. Relative minimum at $(0, -3)$
$f(x) = \frac{-3}{x^2 + 1}$

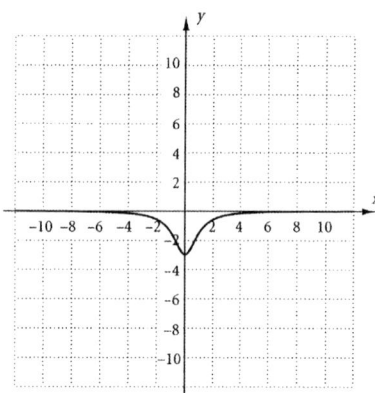

5. Relative minimum at $(2, -9)$;
relative maximum at $\left(-\frac{1}{3}, \frac{100}{27}\right)$
$f(x) = 2x^3 - 5x^2 - 4x + 3$

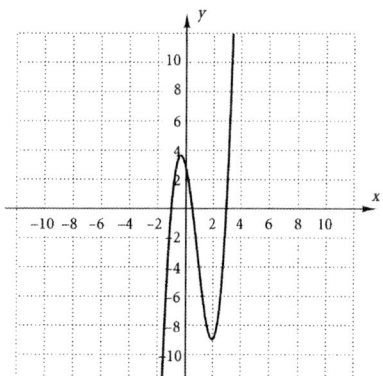

6. Relative maximum at $(-1, 8)$;
relative minimum at $(1, 4)$
$f(x) = 6 - 3x + x^3$

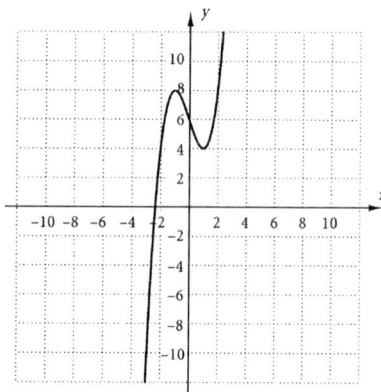

7. None
$f(x) = (x+3)^3 - 4$

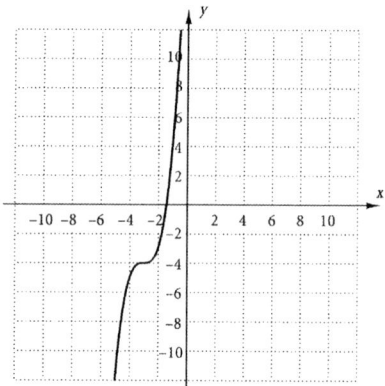

8. Relative minimum at $\left(-2\sqrt{2}, -8\right)$;
relative maximum at $\left(2\sqrt{2}, 8\right)$
$f(x) = x\sqrt{16 - x^2}$

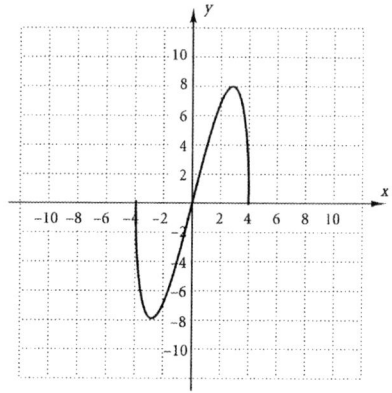

CHAPTER 3, FORM B (continued)

9. $f(x) = \frac{5}{x-3}$

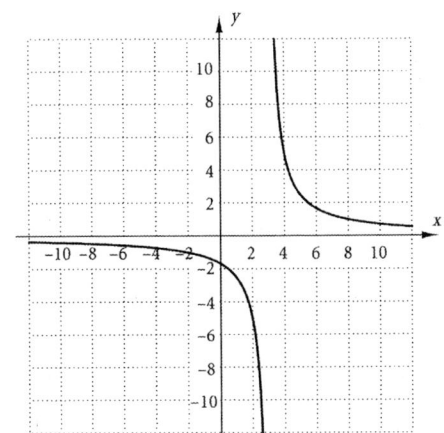

10. $f(x) = \frac{-2}{x^2-4}$

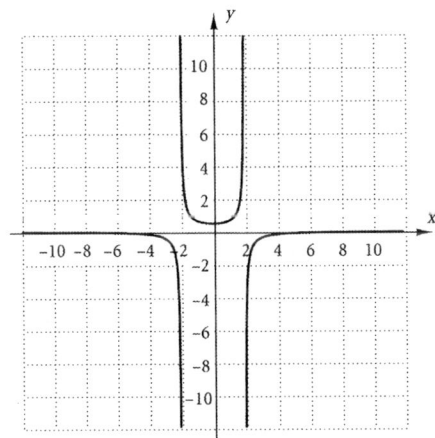

11. $f(x) = \frac{x^2-1}{x}$

12. $f(x) = \frac{x-1}{x+4}$

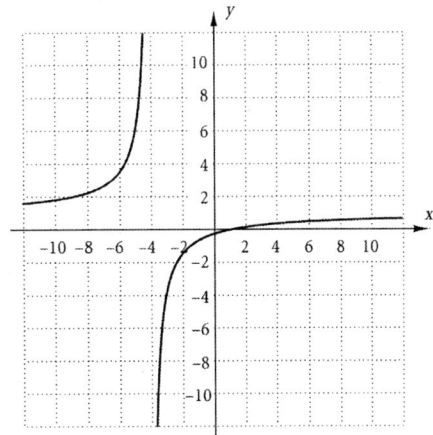

13. Maximum $= 9$ at $x = 3$

14. Maximum $= -25$ at $x = -2$; minimum $= \frac{100}{27}$ at $x = -\frac{1}{3}$

15. Maximum $= 7.63$ at $x = 1.1$

16. Minimum $= -7$ at $x = -2$; maximum $= 9$ at $x = 2$

17. None

18. Minimum $= \frac{71}{20}$ at $x = \frac{3}{10}$

19. Minimum $= 3$ at $x = 1$

20. $-\frac{9}{2}$ and $\frac{9}{2}$

21. $x = 4$; $y = 2$; minimum $= 24$

22. Maximum profit $= 16{,}800$; 410 units

23. 48 in. by 48 in. by 12 in.; 27,648 in^3

24. 15 times at lot size 36

25. $\triangle y = 1.61$; $f'(x)\, \triangle x = 1.6$

CHAPTER 3, FORM B (continued)

26. 3.5

27. (a) $\dfrac{x\,dx}{\sqrt{x^2+7}}$; (b) 0.0060302269

28. $-\dfrac{2x^2}{y^2}$; $-\dfrac{1}{2}$

29. -0.24 ft/sec

30. Maximum $= \dfrac{\sqrt[3]{4}}{3}$ at $x = 2\sqrt[3]{2}$; minimum $= 0$ at $x = 0$

31. (a) $A(x) = \dfrac{C(x)}{x} = 60 + \dfrac{60}{\sqrt{x}} + \dfrac{\sqrt{x}}{60}$; (b) minimum $= 62$ at $x = 3600$

32. Relative maximum at $(0.91, 18.14)$; relative minimum at $(3.77, -28.95)$

CHAPTER 3, FORM C

1. Relative minimum at $(1, -4)$
$f(x) = x^2 - 2x - 3$

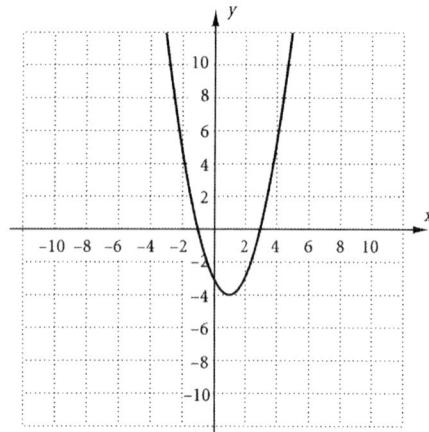

3. Relative minimum at $(4, 1)$
$f(x) = (x-4)^{2/3} + 1$

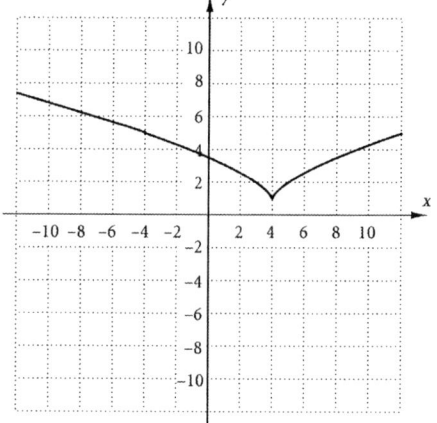

2. Relative minima at $\left(-\dfrac{\sqrt{2}}{2}, 0\right)$ and $\left(\dfrac{\sqrt{2}}{2}, 0\right)$; relative maximum at $(0, 1)$
$f(x) = 4x^4 - 4x^2 + 1$

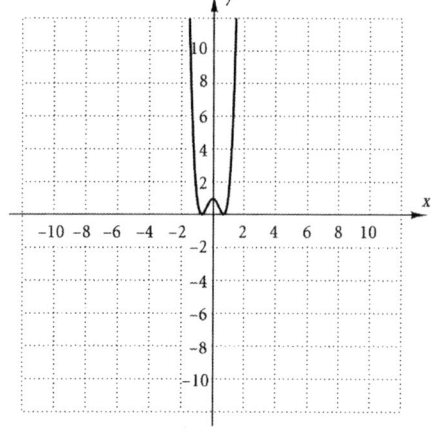

4. Relative maximum at $(0, 2)$
$f(x) = \dfrac{8}{x^2 + 4}$

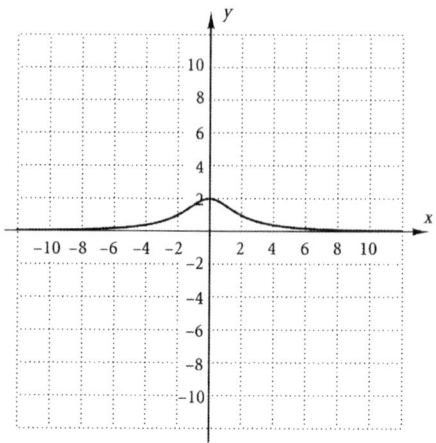

CHAPTER 3, FORM C (continued)

5. Relative maximum at $(-2, 8)$;
   relative minimum at $\left(1, -\frac{11}{2}\right)$
   $f(x) = x^3 + \frac{3}{2}x^2 - 6x - 2$

   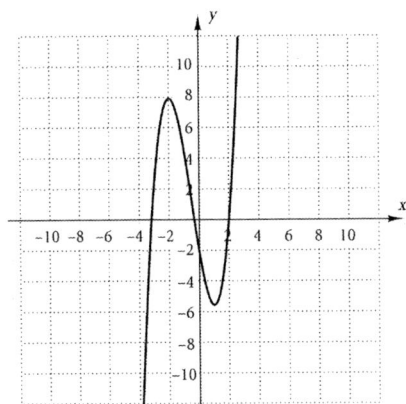

6. Relative minimum at $(1, 3)$;
   relative maximum at $(-1, 7)$
   $f(x) = 5 - 3x + x^3$

   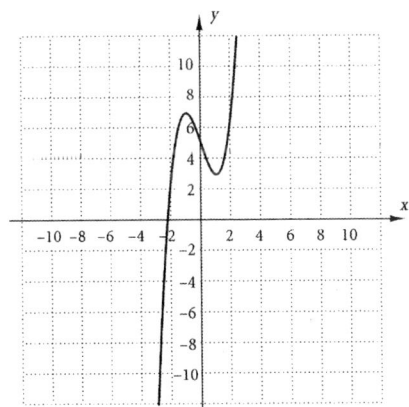

7. None
   $f(x) = (x - 1)^3$

   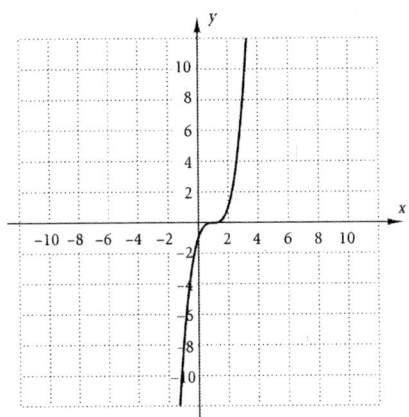

8. Relative minimum at $\left(-\frac{7}{3\sqrt{2}}, -\frac{49}{6}\right)$;
   relative maximum at $\left(\frac{7}{3\sqrt{2}}, \frac{49}{6}\right)$
   $f(x) = x\sqrt{49 - 9x^2}$

   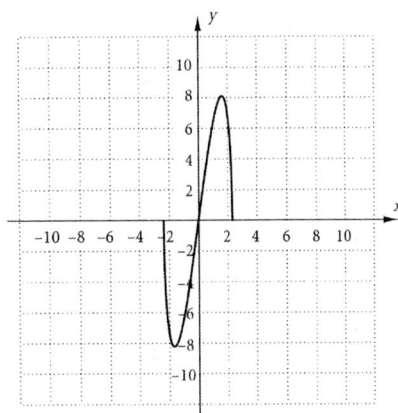

9. $f(x) = \frac{3}{x-6}$

   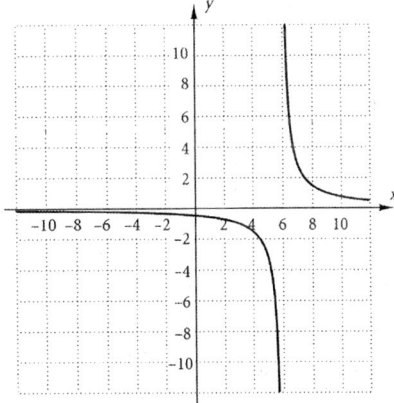

10. $f(x) = \frac{-6}{x^2 + 2x - 3}$

    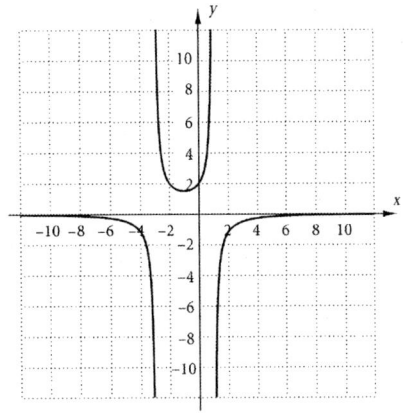

CHAPTER 3, FORM C (continued)

11. $f(x) = \frac{2x^2 - 1}{x}$

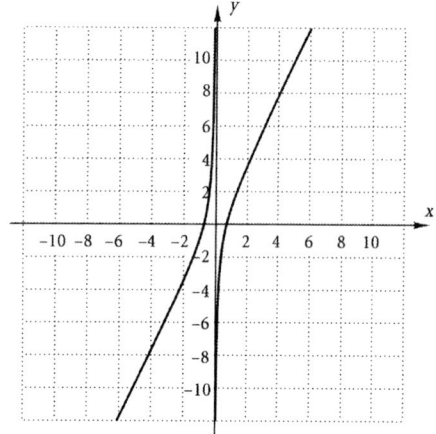

12. $f(x) = \frac{x-1}{x+3}$

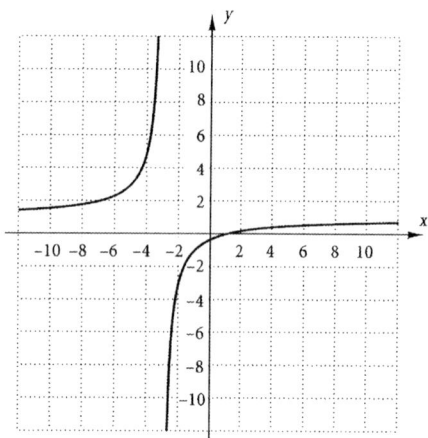

13. Maximum $= \frac{25}{4}$ at $x = \frac{5}{2}$

14. Maximum $= 8$ at $x = -2$; minimum $= -18$ at $x = -4$

15. Maximum $= 15.24$ at $x = 1.8$

16. Maximum $= 11$ at $x = -2$; minimum $= 3$ at $x = 2$

17. None

18. Minimum $= \frac{31}{16}$ at $x = \frac{1}{8}$

19. Minimum $= 75$ at $x = 5$

20. -4 and 4

21. $x = 1$; $y = 3$; minimum $= 12$

22. Maximum profit $= 8840$; 210 units

23. 56 in. by 56 in. by 14 in.; 43,904 in³

24. 10 times at lot size 25

25. $\triangle y = 1.41$; $f'(x) \triangle x = 1.4$

26. 7.75

27. (a) $\frac{x\,dx}{\sqrt{x^2 + 5}}$; (b) 0.0080178

28. $-\frac{2x^2}{y^2}$; -8

29. -0.6 ft/sec

30. Maximum $= 0$ at $x = 0$; minimum $= -\frac{2}{3}$ at $x = 2$

31. (a) $A(x) = \frac{C(x)}{x} = 80 + \frac{80}{\sqrt{x}} + \frac{\sqrt{x}}{80}$; (b) minimum $= 82$ at $x = 6400$

32. Relative maximum at $(0.70, 10.62)$; relative minimum at $(4.87, -98.11)$

ANSWERS TO CHAPTER TEST FORMS

CHAPTER 3, FORM D

1. Relative minimum at $\left(\frac{7}{2}, -\frac{25}{4}\right)$
$f(x) = x^2 - 7x + 6$

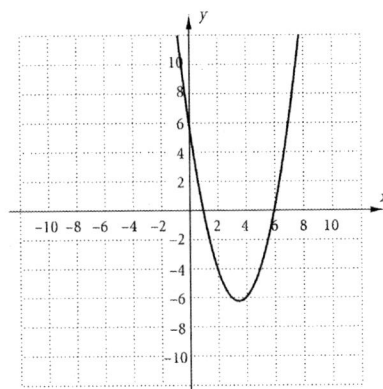

2. Relative minima at $\left(-\sqrt{2}, 0\right)$ and $\left(\sqrt{2}, 0\right)$; relative maximum at $(0, 4)$
$f(x) = x^4 - 4x^2 + 4$

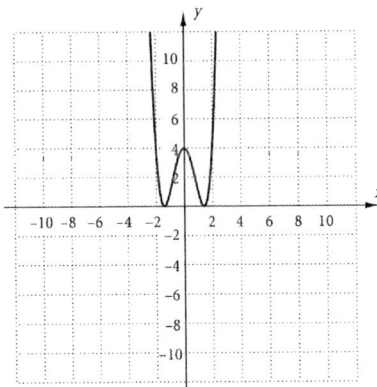

3. Relative minimum at $(4, -1)$
$f(x) = (x-4)^{2/3} - 1$

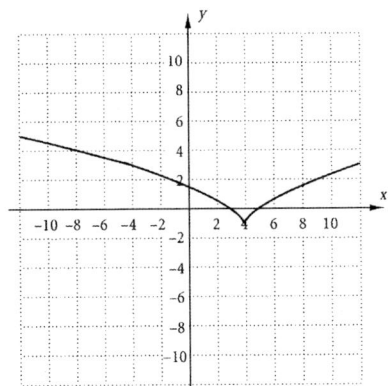

4. Relative maximum at $(0, 6)$
$f(x) = \frac{36}{x^2 + 6}$

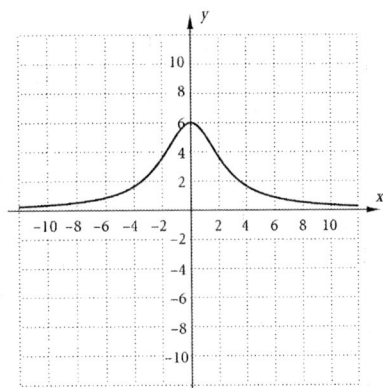

5. Relative maximum at $(1, 5)$; relative minimum at $\left(-\frac{1}{4}, -\frac{45}{16}\right)$
$f(x) = -8x^3 + 9x^2 + 6x - 2$

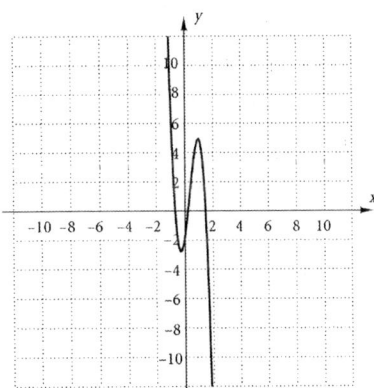

6. Relative minimum at $(-1, -9)$; relative maximum at $(1, -1)$
$f(x) = -5 + 6x - 2x^3$

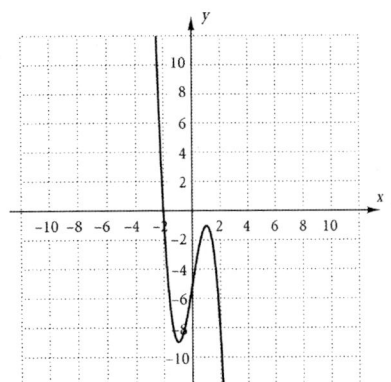

CHAPTER 3, FORM D (continued)

7. None
$f(x) = (x-4)^3$

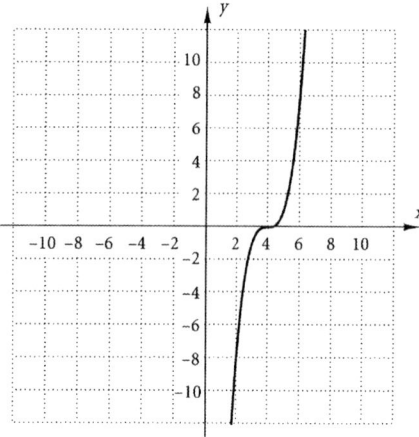

8. Relative minimum at $\left(\frac{5}{2\sqrt{2}}, \frac{25}{4}\right)$; relative maximum at $\left(-\frac{5}{2\sqrt{2}}, -\frac{25}{4}\right)$
$f(x) = x\sqrt{25 - 4x^2}$

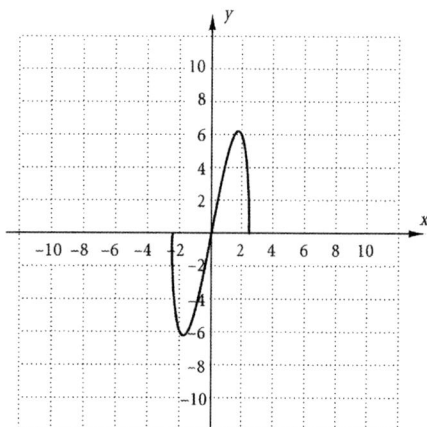

9. $f(x) = \frac{2}{x-3}$

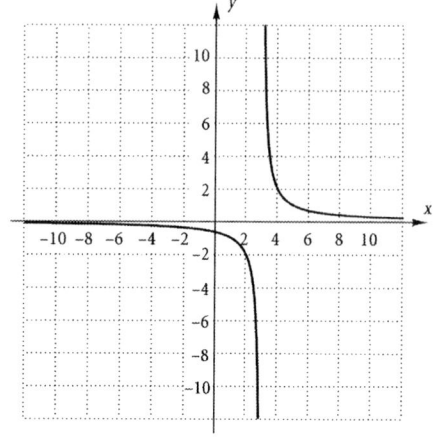

10. $f(x) = \frac{-2}{x^2+2x+1}$

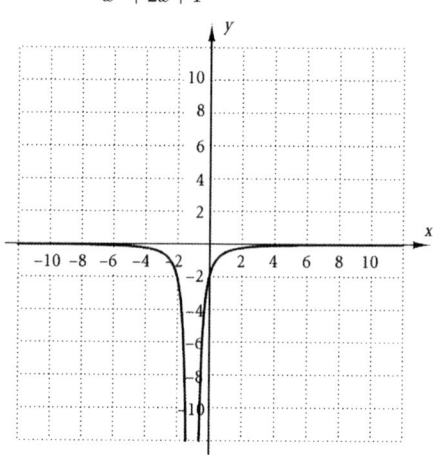

11. $f(x) = \frac{x^2-9}{x}$

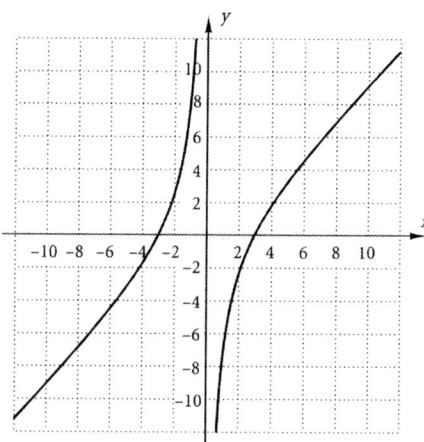

12. $f(x) = \frac{x-2}{x+3}$

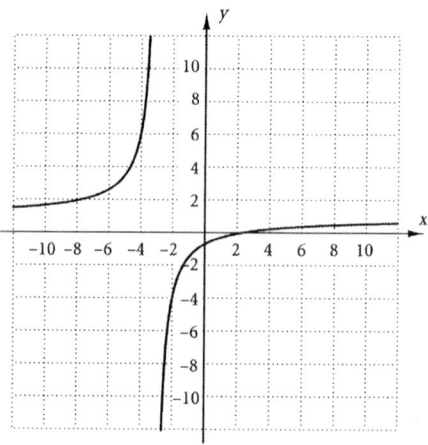

ANSWERS TO CHAPTER TEST FORMS

CHAPTER 3, FORM D (continued)

13. Maximum $= \frac{9}{4}$ at $x = \frac{3}{2}$

14. Minimum $= -\frac{45}{16}$ at $x = -\frac{1}{4}$; maximum $= 9$ at $x = -1$

15. Maximum $= 10.41$ at $x = 2.1$

16. Minimum $= -7$ at $x = -2$; maximum $= 1$ at $x = 2$

17. None

18. Minimum $= -\frac{1}{8}$ at $x = \frac{5}{4}$

19. Minimum $= 75$ at $x = -5$

20. -9 and 9

21. $x = \frac{8}{3}$; $y = -\frac{4}{3}$; minimum $= \frac{32}{3}$

22. Maximum profit $= 10{,}570$; 230 units

23. $66\frac{2}{3}$ in. by $66\frac{2}{3}$ in. by $16\frac{2}{3}$ in.; $74{,}074\frac{2}{27}$ in^3

24. 16 times at lot size 12

25. $\triangle y = -0.39$; $f'(x)\triangle x = -0.4$

26. $7.0\overline{714285}$

27. (a) $\dfrac{2x\,dx}{\sqrt{2x^2-1}}$; (b) 0.0142857143

28. $\dfrac{2x^2}{y^2}$; $\dfrac{8}{9}$

29. $-0.1\overline{6}$ ft/sec

30. Maximum $= 0$ at $x = 0$; minimum $= -\dfrac{5\sqrt[3]{4}}{3}$ at $x = \sqrt[3]{2}$

31. (a) $A(x) = \dfrac{C(x)}{x} = 180 + \dfrac{180}{\sqrt{x}} + \dfrac{\sqrt{x}}{180}$; (b) minimum $= 182$ at $x = 32{,}400$

32. Relative maximum at $(0.62, 8.78)$; relative minimum at $(4.55,\ 55.31)$

CHAPTER 3, FORM E

1. Relative minimum at $\left(-\frac{7}{2}, -\frac{1}{4}\right)$
   $f(x) = x^2 + 7x + 12$

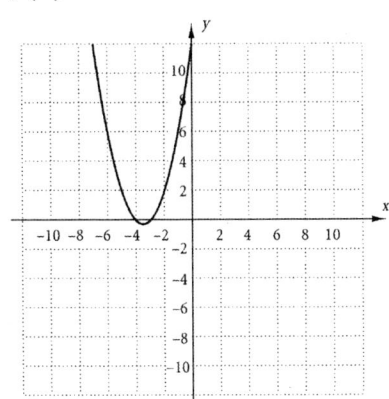

2. Relative minima at $\left(-\sqrt{3}, 0\right)$ and $\left(\sqrt{3}, 0\right)$; relative maximum at $(0, 9)$
   $f(x) = x^4 - 6x^2 + 9$

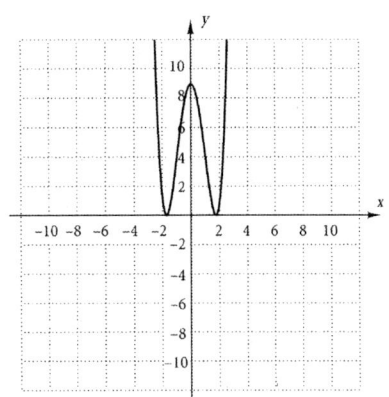

CHAPTER 3, FORM E (continued)

3. Relative minimum at $(1, 2)$
$f(x) = (x-1)^{2/3} + 2$

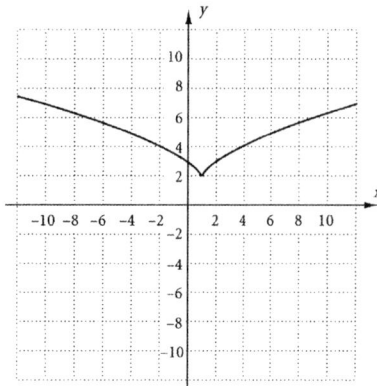

4. Relative minimum at $(0, -3)$
$f(x) = \frac{-6}{x^2 + 2}$

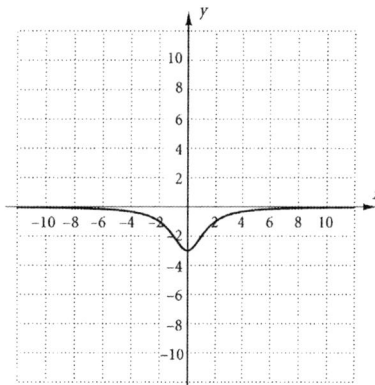

5. Relative maximum at $(-3, 10)$;
relative minimum at $\left(1, -\frac{2}{3}\right)$
$f(x) = \frac{x^3}{3} + x^2 - 3x + 1$

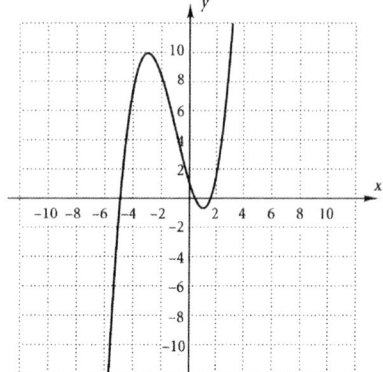

6. Relative minimum at $(-1, 0)$;
relative maximum at $(1, 4)$
$f(x) = 2 + 3x - x^3$

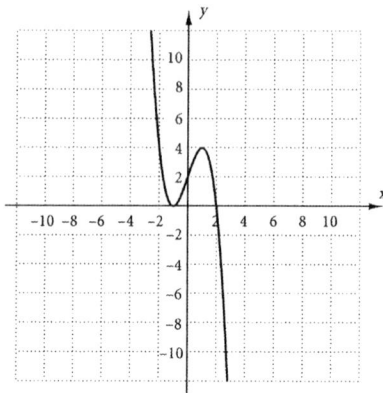

7. None
$f(x) = (x + 3)^3$

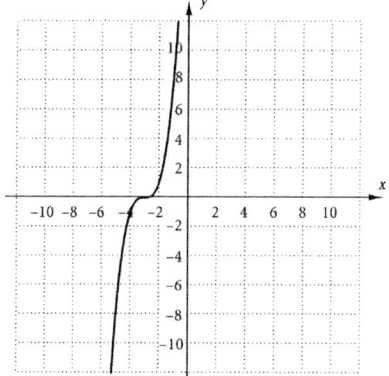

8. Relative minimum at $\left(-\frac{7}{4\sqrt{2}}, -\frac{49}{8}\right)$;
relative maximum at $\left(\frac{7}{4\sqrt{2}}, \frac{49}{8}\right)$
$f(x) = x\sqrt{49 - 16x^2}$

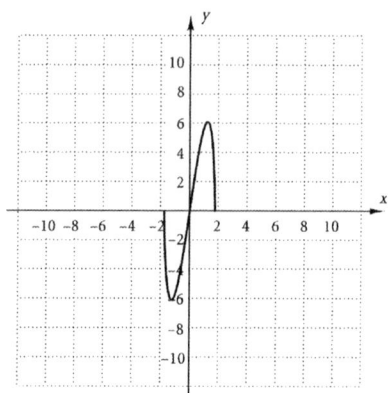

CHAPTER 3, FORM E (continued)

9. $f(x) = \frac{8}{x+5}$

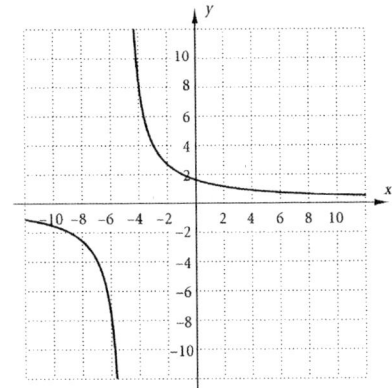

10. $f(x) = \frac{-8}{x^2 + 6x + 9}$

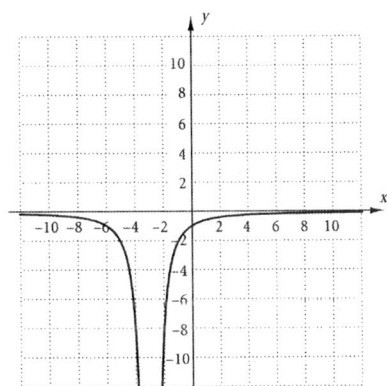

11. $f(x) = \frac{x^2 - 16}{x}$

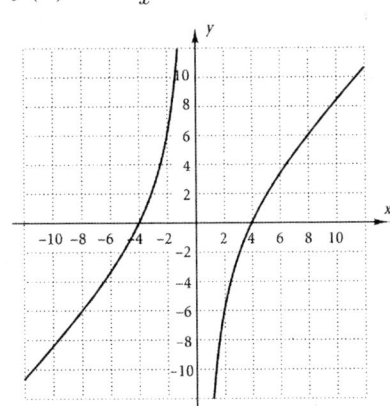

12. $f(x) = \frac{x+3}{x-2}$

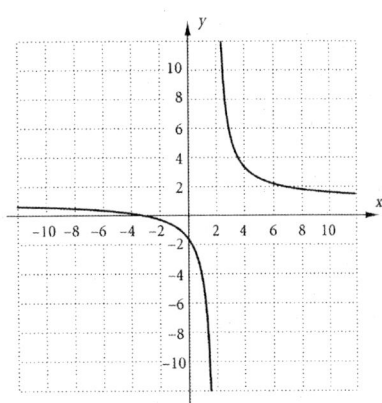

13. Maximum $= 25$ at $x = 5$

14. Minimum $= -\frac{2}{3}$ at $x = 1$; maximum $= \frac{79}{3}$ at $x = 4$

15. Maximum $= -3.31$ at $x = 1.3$

16. Maximum $= 7$ at $x = -2$; minimum $= 3$ at $x - 2$

17. None

18. Minimum $= -\frac{41}{8}$ at $x = \frac{3}{4}$

19. Minimum $= 192$ at $x = 8$

20. -4 and 4

21. $x = 5$; $y = 5$; minimum $= 100$

22. Maximum profit $= 20{,}240$; 450 units

23. 72 in. by 72 in. by 18 in.; $93{,}312$ in^3

24. 5 times at lot size 6

CHAPTER 3, FORM E (continued)

25. $\triangle y = 1.01$; $f'(x) \triangle x = 1.0$

26. $6.\overline{3}$

27. (a) $\dfrac{3x\,dx}{\sqrt{3x^2+2}}$; (b) 0.0169705627

28. $-\dfrac{3x^2}{y^2}$; $-\dfrac{4}{3}$

29. $-0.05\overline{3}$ ft/sec

30. Maximum $= \dfrac{4}{3} \cdot 2^{2/3}$ at $x = -\sqrt[3]{2}$; minimum $= 0$ at $x = 0$

31. (a) $A(x) = \dfrac{C(x)}{x} = 40 + \dfrac{40}{\sqrt{x}} + \dfrac{\sqrt{x}}{40}$; (b) minimum $= 42$ at $x = 1600$

32. Relative maximum at $(0.77, 9.64)$; relative minimum at $(3.68, -27.42)$

CHAPTER 3, FORM F

1. Relative minimum at $(-1, -9)$
$f(x) = x^2 + 2x - 8$

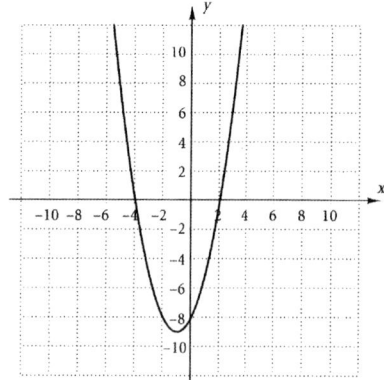

2. Relative minima at $(-1, 2)$ and $(1, 2)$; relative maximum at $(0, 3)$
$f(x) = x^4 - 2x^2 + 3$

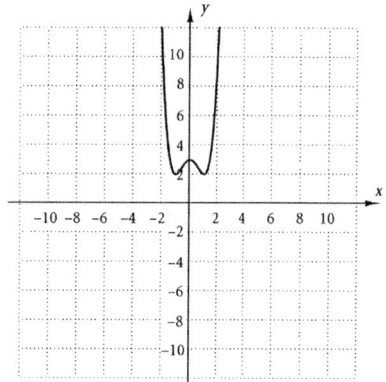

3. Relative minimum at $(3, -1)$
$f(x) = (x-3)^{2/3} - 1$

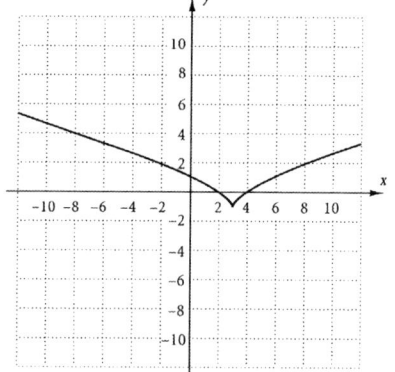

4. Relative maximum at $\left(0, \dfrac{5}{2}\right)$
$f(x) = \dfrac{10}{x^2 + 4}$

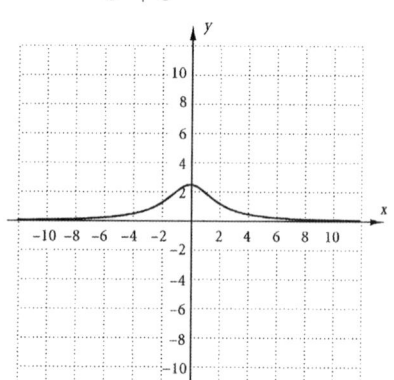

CHAPTER 3, FORM F (continued)

5. Relative minimum at $\left(-\frac{1}{4}, -\frac{93}{16}\right)$; relative maximum at $(1, 2)$
$f(x) = -8x^3 + 9x^2 + 6x - 5$

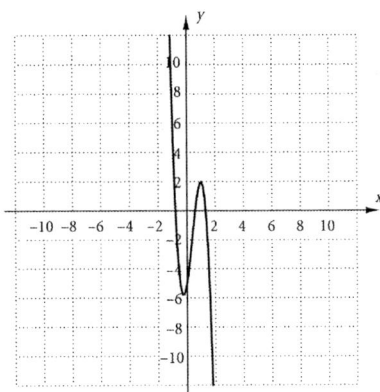

6. Relative maximum at $\left(1, \frac{19}{3}\right)$; relative minimum at $\left(-1, \frac{11}{3}\right)$
$f(x) = 5 + 2x - \frac{2}{3}x^3$

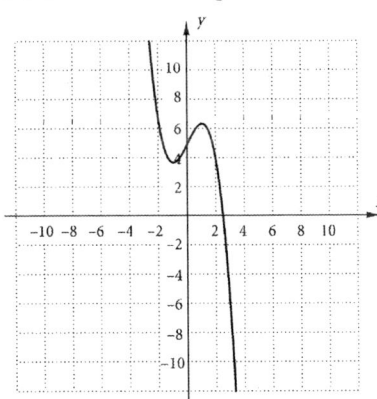

7. None
$f(x) = (x - 2)^3$

8. Relative minimum at $\left(-\frac{5}{4\sqrt{2}}, -\frac{25}{8}\right)$; relative maximum at $\left(\frac{5}{4\sqrt{2}}, \frac{25}{8}\right)$
$f(x) = x\sqrt{25 - 16x^2}$

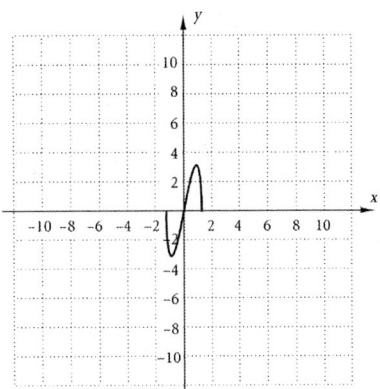

9. $f(x) = \frac{3}{x - 5}$

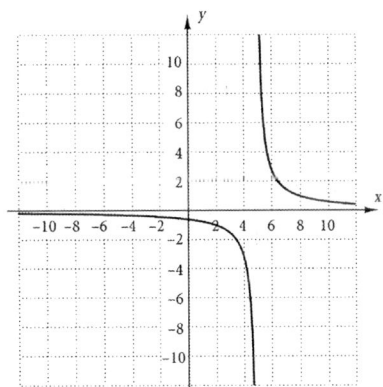

10. $f(x) = \frac{-6}{x^2 - 25}$

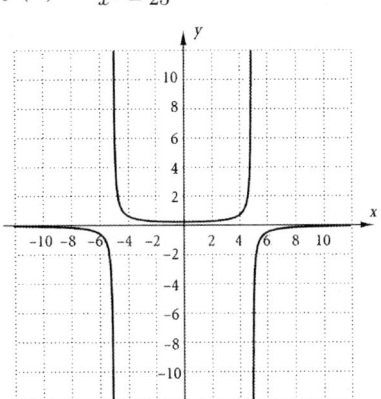

CHAPTER 3, FORM F (continued)

11. $f(x) = \frac{x^2 - 16}{x}$

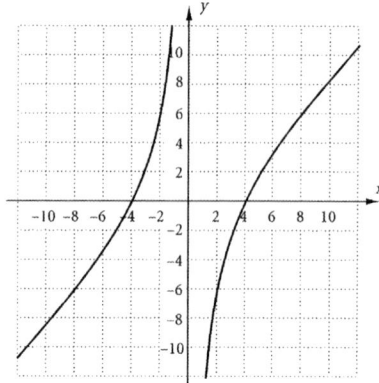

12. $f(x) = \frac{x+1}{x-2}$

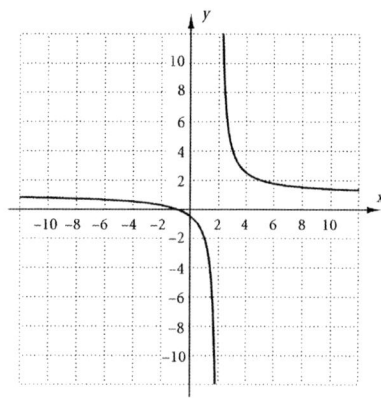

13. Maximum $= 1$ at $x = 1$

14. Maximum $= 83$ at $x = -2$; minimum $= \frac{-93}{16}$ at $x = -\frac{1}{4}$

15. Maximum $= 16.29$ at $x = 2.7$

16. Maximum $= 9$ at $x = -3$; minimum $= -3$ at $x = 3$

17. None

18. Minimum $= \frac{6}{5}$ at $x = \frac{2}{5}$

19. Minimum $= 12$ at $x = 2$

20. -8 and 8

21. $x = \frac{9}{2}$; $y = -\frac{3}{2}$; minimum $= 27$

22. Maximum profit $= 17{,}202.50$; 415 units

23. 54 in. by 54 in. by 13.5 in.; 39,366 in^3

24. 4 times at lot size 75

25. $\triangle y = -0.79$; $f'(x) \triangle x = -0.8$

26. $8.\overline{6}$

27. (a) $\dfrac{x\,dx}{\sqrt{x^2 + 7}}$; (b) 0.0075

28. $\dfrac{2x^2}{y^2}$; 8

29. -0.24 ft/sec

30. Maximum $= \frac{1}{3}$ at $x = -2$; minimum $= 0$ at $x = 0$

31. (a) $A(x) = \dfrac{C(x)}{x} = 120 + \dfrac{120}{\sqrt{x}} + \dfrac{\sqrt{x}}{120}$;
    (b) minimum $= 122$ at $x = 14{,}400$

32. Relative maximum at $(1, 16)$; relative minimum at $(2.70, 5.92)$

CHAPTER 4, FORM A

1. $3e^x$

2. $\dfrac{1}{x}$

3. $-4x^3 e^{-x^4}$

4. $\dfrac{1}{x}$

5. $e^x - 18x^2$

6. $2e^x \left(\dfrac{1}{x} + \ln x \right)$

7. $\dfrac{e^x - 2x}{e^x - x^2}$

8. $\dfrac{\frac{1}{x} - \ln x}{5e^x}$, or $\dfrac{1 - x \ln x}{5xe^x}$

9. 2.585

10. 0.5

11. 3.085

12. $E(t) = E_0 e^{kt}$

13. 8.7% per hour

14. 11 yr

15. (a) 0.0696682; $C(t) = \$0.05 e^{0.0696682 t}$; (b) \$1.63

16. (a) $N(t) = 100{,}000 e^{-0.12t}$; (b) 38,289; (c) 5.78 hr

17. 8.1 days

18. 0.043% per year

19. (a) 5.49%; 9.57%; 50.11%; (b) $P'(t) = \dfrac{3 e^{-0.15t}}{(1 + 20 e^{-0.15t})^2}$; (c)

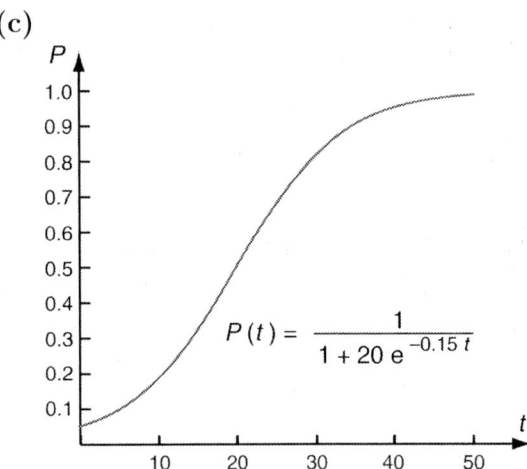

20. \$32,525.45

21. $(\ln 5) 5^x$

22. $\dfrac{1}{\ln 15} \cdot \dfrac{1}{x}$

23. (a) $E(p) = 0.24p$; (b) 1.2, elastic; (c) decrease; (d) $p = \$4.17$

24. $(\ln x)^3 + 3(\ln x)^2 - \ln x + 2$

25. Minimum $= 0$ at $x = 0$; maximum $= 4/e^2 \approx 0.54$ at $x = 2$

26.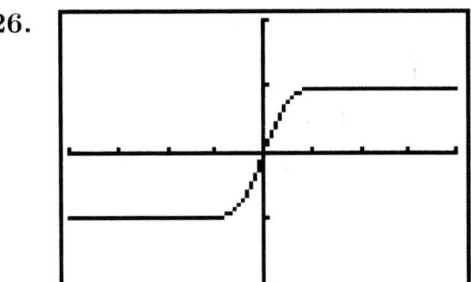

27. 0

CHAPTER 4, FORM B

1. e^x

2. $\dfrac{4}{x}$

3. $-3x^2 e^{-x^3}$

4. $\dfrac{1}{x}$

5. $e^x + 12x^3$

6. $10e^x \left(\dfrac{1}{x} + \ln x \right)$

7. $\dfrac{2x + e^x}{x^2 + e^x}$

8. $\dfrac{\frac{6}{x} - 6\ln x}{e^x}$, or $\dfrac{6 - 6x \ln x}{xe^x}$

9. -0.5283

10. 0.8617

11. 1.9183

12. $J(t) = J_0 e^{kt}$

13. 8.7% per hour

14. 9 yr

15. (a) 0.067701255; $C(t) = \$0.10 e^{0.067701255t}$; (b) $\$3.62$

16. (a) $N(t) = 80{,}000 e^{-0.25t}$; (b) 13,902; (c) 2.77 days

17. 11 yr

18. 4.85% per day

19. (a) 3.33%; 11.02%; 93.77%;
(b) $P'(t) = \dfrac{12.8 e^{-0.32t}}{(1 + 40 e^{-0.32t})^2}$;
(c)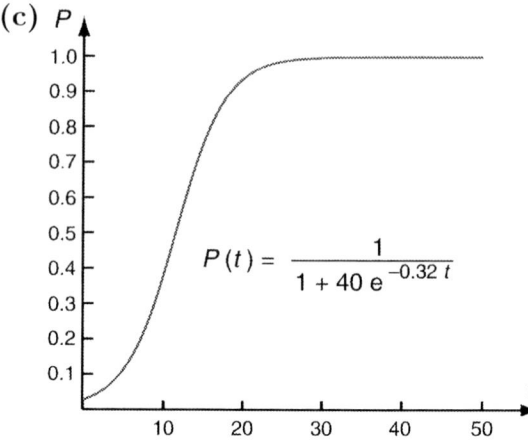

20. $\$46{,}875.17$

21. $(\ln 13)\, 13^x$

22. $\dfrac{1}{\ln 25} \cdot \dfrac{1}{x}$

23. (a) $E(p) = 0.15p$; (b) 0.9, inelastic; (c) increase; (d) $p = \$6.67$

24. $(\ln x)^3 + 3(\ln x)^2 - 4x \ln x$, or $\ln x \left[(\ln x)^2 + 3(\ln x) - 4x \right]$

25. Minimum $= 0$ at $x = 0$; maximum $= 3125/e^5 \approx 21.06$ at $x = 5$

26.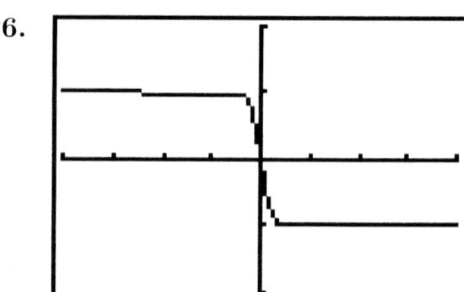

27. 0

CHAPTER 4, FORM C

1. e^x
2. $\dfrac{2}{x}$
3. $2e^{2x}$
4. $\dfrac{1}{x}$
5. $e^x - 8x$
6. $5e^x \left(\dfrac{1}{x} + \ln x \right)$
7. $\dfrac{e^x - 1}{e^x - x}$
8. $\dfrac{\frac{1}{x} - \ln x}{2e^x}$, or $\dfrac{1 - x \ln x}{2xe^x}$
9. 1.3023
10. 1.548
11. -0.5283
12. $Q(t) = Q_0 e^{kt}$
13. 11.6% per hour
14. 12 yr
15. (a) 0.0382205; $C(t) = 25 e^{0.0382205 t}$¢; (b) 99¢
16. (a) $N(t) = 80{,}000 e^{-0.1t}$; (b) 7257; (c) 6.93 hr
17. 14.29 days
18. 0.000000015% per year

19. (a) 6.03%; 14.86%; 88.12%;
    (b) $P'(t) = \dfrac{5 e^{-0.25 t}}{(1 + 20 e^{-0.25 t})^2}$;
    (c)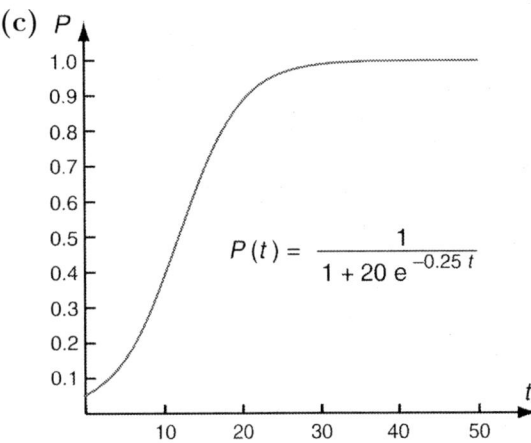

20. $26,763.07
21. $(\ln 20)\, 20^x$
22. $\dfrac{1}{\ln 15} \cdot \dfrac{1}{x}$
23. (a) $E(p) = \dfrac{p}{5}$; (b) 0.8, inelastic; (c) increase; (d) $p = \$5$
24. $\dfrac{1}{2\sqrt{\ln x}} + \sqrt{\ln x} + 3 \ln x$
25. Minimum $= 0$ at $x = 0$; maximum $= 1/e^3 \approx 0.0498$ at $x = 1$
26.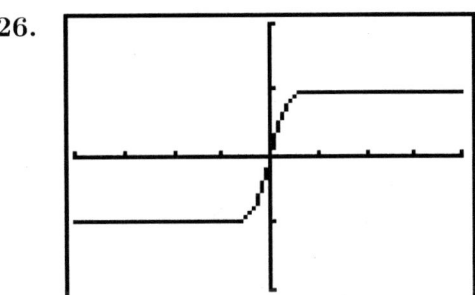
27. 0

CHAPTER 4, FORM D

1. $5e^x$

2. $\dfrac{1}{x}$

3. $6xe^{3x^2}$

4. $\dfrac{1}{x}$

5. $e^x + 20x^4$

6. $5e^x \left(\dfrac{1}{x} + \ln x \right)$

7. $\dfrac{5x^4 - e^x}{x^5 - e^x}$

8. $\dfrac{e^x \left(\ln x - \frac{1}{x} \right)}{(\ln x)^2}$, or $\dfrac{e^x (x \ln x - 1)}{x (\ln x)^2}$

9. -1.2925

10. 2.9535

11. 3.322

12. $F(t) = F_0 e^{kt}$

13. 5.8% per hour

14. 9 yr

15. (a) 0.049698133; $C(t) = \$0.25 e^{0.049698133t}$; (b) $\$4.93$

16. (a) $N(t) = 5{,}000{,}000 e^{-0.40t}$; (b) $1{,}505{,}971$; (c) 1.73 days

17. 12.4 years

18. 13.08% per year

19. (a) 5.49%; 9.57%; 50.11%;
(b) $P'(t) = \dfrac{3e^{-0.15t}}{\left(1 + 20e^{-0.15t}\right)^2}$;
(c)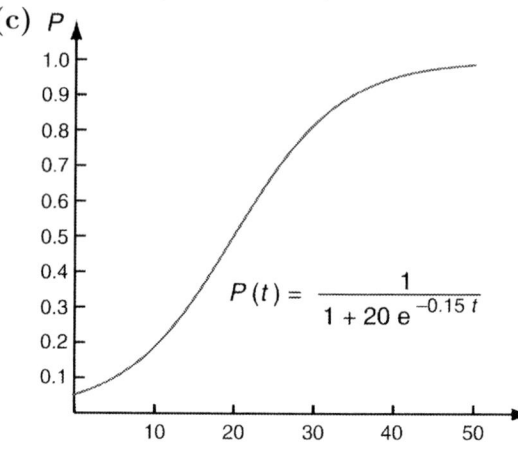

20. $\$18{,}421.07$

21. $(\ln 20)\, 20^x$

22. $\dfrac{1}{\ln 8} \cdot \dfrac{1}{x}$

23. (a) $E(p) = 0.02p$; (b) 0.5, inelastic; (c) increase; (d) $p = \$50$

24. $\dfrac{1}{2\sqrt{\ln x}} + \sqrt{\ln x} - 3 \ln x$

25. Minimum $= 0$ at $x = 0$; maximum $= 1/e^4 \approx 0.0183$ at $x = 1$

26.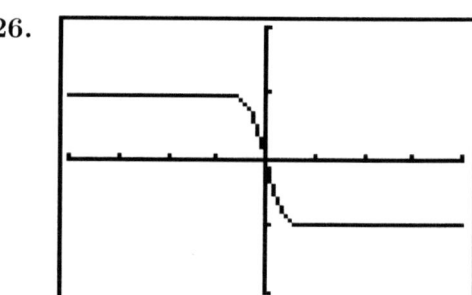

27. 0

ANSWERS TO CHAPTER TEST FORMS

CHAPTER 4, FORM E

1. e^x

2. $\dfrac{9}{x}$

3. $5x^4 e^{x^5}$

4. $\dfrac{1}{x}$

5. $-12x^3 + e^x$

6. $5e^x \left(\dfrac{1}{x} + \ln x \right)$

7. $\dfrac{2x - e^x}{x^2 - e^x}$

8. $\dfrac{\frac{4}{x} - 4\ln x}{e^x}$, or $\dfrac{4 - 4x \ln x}{xe^x}$

9. 2.7925

10. 0.5

11. -1.5

12. $N(t) = N_0 e^{kt}$

13. 3.9% per hour

14. 20 yr

15. (a) 0.0407734; $C(t) = \$3.75 e^{0.0407734 t}$; (b) $11.74

16. (a) $N(t) = 150{,}000 e^{-0.2t}$; (b) 45,179; (c) 3.5 hr

17. 5.3 yr

18. 0.043% per year

19. (a) 1.69%; 5.83%; 88.27%; (b) $P'(t) = \dfrac{25.6 e^{-0.32 t}}{(1 + 80 e^{-0.32 t})^2}$; (c)

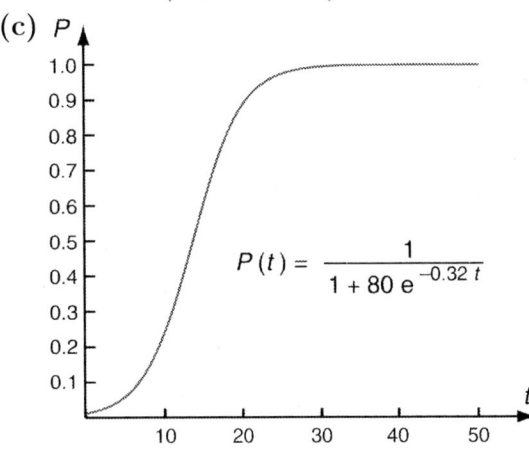

20. $37,157.67

21. $(\ln 15)\, 15^x$

22. $\dfrac{1}{\ln 5} \cdot \dfrac{1}{x}$

23. (a) $E(p) = 0.5p$; (b) 0.75, inelastic; (c) increase; (d) $p = \$2$

24. $3(\ln x)^4 + 12(\ln x)^3 - 4\ln x$, or $\ln x \left[3(\ln x)^3 + 12(\ln x)^2 - 4 \right]$

25. Minimum $= 0$ at $x = 0$; maximum $= 4/e^2 \approx 0.54$ at $x = 1$

26.

27. 0

CHAPTER 4, FORM F

1. $2e^x$

2. $\dfrac{1}{x}$

3. $-2xe^{-x^2}$

4. $\dfrac{1}{x}$

5. $e^x - 12x^2$

6. $8e^x \left(\dfrac{1}{x} + \ln x \right)$

7. $\dfrac{e^x + 12}{e^x + 12x}$

8. $\dfrac{5e^x \left(\ln x - \frac{1}{x} \right)}{(\ln x)^2}$, or $\dfrac{5e^x (x \ln x - 1)}{x (\ln x)^2}$

9. 0.5646

10. -1.2770

11. 1.0686

12. $S(t) = S_0 e^{kt}$

13. 17.3% per hour

14. 12 yr

15. (a) 0.0448667777; $C(t) = \$0.49 e^{0.0448667777t}$; (b) $2.46

16. (a) $N(t) = 250{,}000 e^{-0.6t}$; (b) 12,447; (c) 1.16 days

17. 1612 yr

18. 8.56% per day

19. (a) 2.32%; 4.47%; 37.47%;
    (b) $P'(t) = \dfrac{8.5 e^{-0.17t}}{(1 + 50 e^{-0.17t})^2}$;
    (c)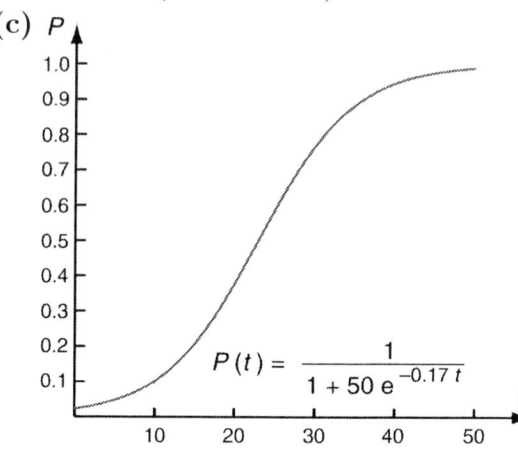

20. $52,880.08

21. $(\ln 5) 5^x$

22. $\dfrac{1}{\ln 4} \cdot \dfrac{1}{x}$

23. (a) $E(p) = \dfrac{p}{20}$; (b) 0.75, inelastic; (c) increase; (d) $p = \$20$

24. $2x (\ln x)^2 + 2x \ln x - \ln x + 1$

25. Minimum $= 0$ at $x = 0$; maximum $= 1/e^2 \approx 0.1353$ at $x = 1$

26.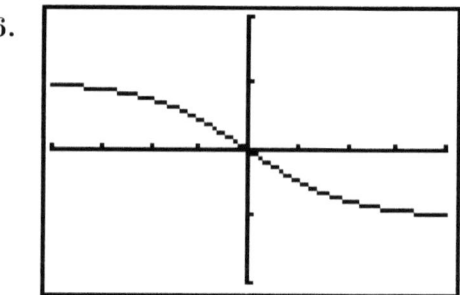

27. 0

ANSWERS TO CHAPTER TEST FORMS

CHAPTER 5, FORM A

1. $x + C$
2. $40x^5 + C$
3. $3\ln x + e^x + \dfrac{5}{8}x^{8/5} + C$
4. $\dfrac{9}{2}$
5. $2\ln 8$
6. An antiderivative, total water used in t hours.
7. 22.5
8. $-\dfrac{1}{4}\left(\dfrac{1}{e^{20}} - 1\right)$
9. $\ln 2$
10. Positive
11. $\ln|x+5| + C$
12. $-10e^{-0.1x} + C$
13. $\dfrac{1}{16}(t^2 - 5)^8 + C$
14. $\dfrac{x}{3}e^{3x} - \dfrac{e^{3x}}{9} + C$
15. $x^5 \ln x - \dfrac{x^5}{5} + C$
16. $\dfrac{3^x}{\ln 3} + C$
17. $\dfrac{1}{5}\ln\left(\dfrac{x}{5-x}\right) + C$
18. 67.5
19. 1
20. $22\dfrac{7}{9}$; answers may vary.
21. $32{,}000$
22. 80 m
23. $\dfrac{1}{6}\ln\left(\dfrac{x}{6-5x}\right) + C$
24. $5e^x(x^4 - 4x^3 + 12x^2 - 24x + 24) + C$
25. $\dfrac{1}{4}e^{x^4} + C$
26. $2x^{1/2}\ln x - 4x^{1/2} + C$
27. $\dfrac{1}{5}(x^2+6)^{3/2}(x^2 - 4) + C$
28. $\dfrac{1}{8}\ln\left(\dfrac{4+x}{4-x}\right) + C$
29. $-2e^{-0.5x}(x^4 + 8x^3 + 48x^2 + 192x + 384) + C$
30. $\dfrac{x^2}{2}\ln 12x - \dfrac{x^2}{4} + C$
31. $3\ln x + \dfrac{2}{3}(\ln x)^3 + \dfrac{(\ln x)^6}{6} + C$
32. $(x+9)\ln(x+9) - (x+2)\ln(x+2) + C$
33. 12

CHAPTER 5, FORM B

1. $x + C$
2. $3x^3 + C$
3. $\dfrac{2}{5}x^{5/2} + e^x + 4\ln x + C$
4. $\dfrac{243}{2}$
5. $3\ln 3$
6. An antiderivative, total sales in t days.
7. -675
8. $-\dfrac{1}{3}\left(\dfrac{1}{e^{15}} - 1\right)$
9. $\dfrac{2a\sqrt{a}}{3}$
10. Zero
11. $\ln(x-5) + C$
12. $-1.25e^{-0.8x} + C$
13. $\dfrac{1}{15}(t^5 - 2)^3 + C$

CHAPTER 5, FORM B (continued)

14. $\dfrac{x}{3}e^{3x} - \dfrac{e^{3x}}{9} + C$

15. $x^3 \ln x - \dfrac{x^3}{3} + C$

16. $\dfrac{5^x}{\ln 5} + C$

17. $\dfrac{1}{3} \ln\left(\dfrac{x}{x+3}\right) + C$

18. -43 19. 4

20. 22; answers may vary.

21. $37,000   22. 306 m

23. $\dfrac{1}{5} \ln\left(\dfrac{x}{2x+5}\right) + C$

24. $3e^x \left(x^3 - 3x^2 + 6x - 6\right) + C$

25. $\dfrac{1}{10}e^{x^{10}} + C$

26. $\dfrac{3}{4}x^{4/3}\left[\ln x - \dfrac{3}{4}\right] + C$

27. $\dfrac{2}{45}\left(x^3 + 2\right)^{3/2}\left(3x^3 - 4\right) + C$

28. $\dfrac{1}{10}\ln\left(\dfrac{5+x}{5-x}\right) + C$

29. $-\dfrac{5}{4}e^{-0.8x}\left(x^2 + \dfrac{5}{2}x + \dfrac{25}{8}\right) + C$

30. $\dfrac{x^2}{2}\ln 10x - \dfrac{x^2}{4} + C$

31. $(\ln x)^3 + \dfrac{5(\ln x)^2}{2} - 2\ln x + C$

32. $(x+4)\ln(x+4) - (x-3)\ln(x-3) + C$

33. $6\dfrac{2}{3}$

CHAPTER 5, FORM C

1. $x + C$ 2. $50x^5 + C$

3. $\dfrac{2}{3}x^{3/2} + 2e^x + \ln x + C$

4. $\dfrac{32}{3}$ 5. $6\ln 4$

6. An antiderivative, velocity.

7. -70

8. $-\dfrac{1}{6}\left(\dfrac{1}{e^{24}} - 1\right)$

9. $-\dfrac{3a\sqrt[3]{a}}{4}$

10. Negative

11. $\ln(x+6) + C$

12. $-5e^{-0.2x} + C$

13. $-\dfrac{(4-t^6)^3}{18} + C$

14. $\dfrac{x}{2}e^{2x} - \dfrac{e^{2x}}{4} + C$

15. $\dfrac{4x^9}{9}\ln x - \dfrac{4x^9}{81} + C$

16. $\dfrac{8^x}{\ln 8} + C$

17. $\ln\left(x + \sqrt{x^2 + 25}\right) + C$

18. $-\dfrac{43}{6}$ 19. $\dfrac{1}{3}$

20. 22; answers may vary.

21. $1150   22. $37.\overline{3}$ m

23. $\dfrac{1}{3}\ln\left(\dfrac{x}{3-x}\right) + C$

24. $10e^x(x^4 - 4x^3 + 12x^2 - 24x + 24) + C$

25. $\dfrac{1}{7}e^{x^7} + C$

CHAPTER 5, FORM C (continued)

26. $\dfrac{(\ln x)^2}{2} + C$

27. $\dfrac{2}{45}\left(x^3+1\right)^{3/2}\left(3x^3-2\right)$

28. $\dfrac{1}{14}\ln\left(\dfrac{7+x}{7-x}\right)+C$

29. $-\dfrac{e^{-1.1x}}{1.1}\left(x^4+\dfrac{4}{1.1}x^3+\dfrac{12}{1.1^2}x^2+\dfrac{24}{1.1^3}x+\dfrac{24}{1.1}\right)+C$

30. $\dfrac{x^2}{2}\ln 3x - \dfrac{x^2}{4}+C$

31. $10\ln x + \dfrac{5}{6}(\ln x)^6 - \dfrac{2}{11}(\ln x)^{11} + C$

32. $(x+8)\ln(x+8) - (x-2)\ln(x-2) + C$

33. 8

CHAPTER 5, FORM D

1. $x+C$ 2. $10x^6+C$

3. $3e^x + 5\ln x + \dfrac{6}{11}x^{11/6} + C$

4. $\dfrac{9}{2}$

5. $2\ln 4$, or $4\ln 2$

6. An antiderivative, total number returned in t days.

7. 90

8. $-\dfrac{1}{2}\left(\dfrac{1}{e^{10}} - \dfrac{1}{e^2}\right)$

9. $3\ln 3$

10. Positive

11. $\ln(x+2) + C$

12. $-4e^{-0.25x} + C$

13. $\dfrac{1}{6}(t^4+3)\sqrt{t^4+3} + C$

14. $\dfrac{x}{8}e^{8x} - \dfrac{e^{8x}}{64} + C$

15. $x^6\ln x - \dfrac{x^6}{6} + C$

16. $\dfrac{11^x}{\ln 11} + C$

17. $\dfrac{1}{6}\ln\left(\dfrac{x-3}{x+3}\right) + C$

18. 182 19. 20.25

20. 50; answers may vary.

21. \$76,400 22. 144 m

23. $\dfrac{1}{12}\ln\left(\dfrac{x}{12-x}\right) + C$

24. $e^x\left(x^4 - 4x^3 + 12x^2 - 24x + 24\right) + C$

25. $\dfrac{1}{9}e^{x^9} + C$

26. $-\dfrac{1}{x}\ln x - \dfrac{1}{x} + C$

27. $\dfrac{2}{15}\left(x^3+3\right)^{3/2}\left(x^3-2\right)$

28. $\dfrac{1}{18}\ln\left(\dfrac{9+x}{9-x}\right) + C$

29. $-\dfrac{5}{4}e^{-0.8x}\left(x^2 + \dfrac{5}{2}x + \dfrac{25}{8}\right) + C$

30. $\dfrac{x^2}{2}\ln 5x - \dfrac{x^2}{4} + C$

31. $\dfrac{(\ln x)^5}{5} - (\ln x)^3 + 4\ln x + C$

32. $(x-5)\ln(x-5) + (x+3)\ln(x+3) - 2x + C$

33. 2.8125, or $\dfrac{45}{16}$

CHAPTER 5, FORM E

1. $x + C$
2. $20x^9 + C$
3. $3\ln x + e^x + \dfrac{4}{7}x^{7/4} + C$
4. 36
5. $5\ln 10$
6. An antiderivative, total cost of n dozen units.
7. 129
8. $-\dfrac{1}{12}\left(\dfrac{1}{e^{72}} - 1\right)$
9. $1 - \dfrac{1}{a^2}$
10. Negative
11. $\ln(x - 21) + C$
12. $-5e^{-0.2x} + C$
13. $\dfrac{1}{20}\left(t^5 + 6\right)^4 + C$
14. $\dfrac{x}{3}e^{3x} - \dfrac{e^{3x}}{9} + C$
15. $\dfrac{7x^9}{9}\ln x - \dfrac{7x^9}{81} + C$
16. $\dfrac{11^x}{\ln 11} + C$
17. $\dfrac{x}{2} - \dfrac{5}{4}\ln(2x + 5) + C$
18. -14
19. $\dfrac{3}{10}$
20. $\dfrac{25}{3}$; answers may vary.
21. $\$4968.75$
22. 45 m
23. $\dfrac{1}{2}\ln\left(\dfrac{x}{2 - x}\right) + C$
24. $e^x\left(x^3 - 3x^2 + 6x - 6\right) + C$
25. $\dfrac{1}{5}e^{x^5} + C$
26. $2\sqrt{x}[\ln x - 2] + C$
27. $\dfrac{2}{45}\left(x^3 + 5\right)^{3/2}\left(3x^3 - 10\right) + C$
28. $\dfrac{1}{8}\ln\left(\dfrac{4 + x}{4 - x}\right) + C$
29. $4e^{0.25x}(x^3 - 12x^2 + 96x - 384) + C$
30. $\dfrac{x^2}{2}\ln 25x - \dfrac{x^2}{4} + C$, or $x^2\left(\dfrac{\ln x}{2} + \ln 5 - \dfrac{1}{4}\right) + C$
31. $2(\ln x)^4 - \dfrac{2}{3}(\ln x)^3 - 6\ln x + C$
32. $(x - 5)\ln(x - 5) - (x + 2)\ln(x + 2) + C$
33. 1

CHAPTER 5, FORM F

1. $x + C$
2. $20x^5 + C$
3. $3e^x + \ln x + \dfrac{2}{5}x^{5/2} + C$
4. $\dfrac{256}{3}$
5. $4\ln 10$
6. An antiderivative, total distance traveled in t units of time.
7. 75
8. $-\dfrac{1}{8}\left(\dfrac{1}{e^{32}} - 1\right)$
9. $\dfrac{2a\sqrt{a}}{3}$
10. Negative
11. $\ln(x - 1) + C$
12. $-2.5e^{-0.4x} + C$
13. $\dfrac{2}{15}\left(t^5 + 5\right)\sqrt{t^5 + 5} + C$

CHAPTER 5, FORM F (continued)

14. $\dfrac{x}{5}e^{5x} - \dfrac{e^{5x}}{25} + C$

15. $x^{10}\ln x - \dfrac{x^{10}}{10} + C$

16. $\dfrac{2^x}{\ln 2} + C$

17. $-\ln\left(\dfrac{1+\sqrt{1+x^2}}{x}\right) + C$

18. $-\dfrac{11}{2}$ 19. $\dfrac{1}{3}$

20. 54; answers may vary.

21. \$13,600   22. 18 m

23. $\dfrac{1}{4}\ln\left(\dfrac{x}{4-x}\right) + C$

24. $3e^x(x^5 - 5x^4 + 20x^3 - 60x^2 + 120x - 120) + C$

25. $\dfrac{1}{3}e^{x^3} + C$

26. $\dfrac{4}{5}x^{5/4}\left[\ln x - \dfrac{4}{5}\right] + C$

27. $\dfrac{(x^4+2)^{3/2}}{10}\left(x^4 - \dfrac{4}{3}\right) + C$

28. $\dfrac{1}{10}\ln\left(\dfrac{5+x}{5-x}\right) + C$

29. $-2e^{-0.5x}(x^2 + 4x + 8) + C$

30. $\dfrac{x^2}{2}\ln 6x - \dfrac{x^2}{4} + C$

31. $3\ln x - \dfrac{2}{3}(\ln x)^3 + (\ln x)^4 + C$

32. $(x-1)\ln(x-1) + (x+6)\ln(x+6) - 2x + C$

33. 20.25, or $\dfrac{81}{4}$

CHAPTER 6, FORM A

1. $(3, \$25)$
2. $54
3. $40.50
4. $24,591.23
5. $581.98
6. 78,101
7. 2056
8. $8489.89
9. $11,279.71
10. $25,000
11. Convergent, $\dfrac{2}{3}$
12. Divergent
13. $\dfrac{5}{1024}$; $f(x) = \dfrac{5x^4}{1024}$
14. 0.3935
15. $E(x) = \dfrac{28}{9}$
16. $E(x^2) = 10$
17. $\mu = \dfrac{28}{9}$
18. $\sigma^2 = \dfrac{26}{81}$
19. $\sigma = \dfrac{\sqrt{26}}{9}$
20. 0.4821
21. 0.1330
22. 0.9525
23. 0.0475
24. 8π
25. 20π
26. $y = C_1 e^{2x^3}$, where $C_1 = e^C$
27. $y = \sqrt{16x + C_1}$, $y = -\sqrt{16x + C_1}$, where $C_1 = 2C$
28. $y = 4e^{2t}$
29. $y = 6 - C_1 e^{-x^4/4}$, where $C_1 = e^{-C}$
30. $r = \pm\sqrt[4]{20t + C_1}$, where $C_1 = 4C$
31. $y = C_1 e^{5x + x^2/2}$, where $C_1 = e^C$
32. $x = \dfrac{C_1}{p^3}$, where $C_1 = e^C$
33. (a) $V(t) = 14\left(1 - e^{-kt}\right)$; (b) 0.04; (c) $V(t) = 14\left(1 - e^{-0.04t}\right)$; (d) $7.19; (e) 31.3
34. $b = \sqrt[7]{7}$
35. Convergent; $\dfrac{1}{5e}$
36. 6.28

CHAPTER 6, FORM B

1. $(2, \$16)$
2. $\$18.67$
3. $\$17.33$
4. $\$72{,}410.91$
5. $\$1145.59$
6. $155{,}693$ tons
7. 2084
8. $\$7107.83$
9. $\$67{,}678.25$
10. $\$150{,}000$
11. Convergent, $\dfrac{1}{24}$
12. Divergent
13. $\dfrac{2}{243}$; $f(x) = \dfrac{2}{243}x^5$
14. 0.8347
15. $E(x) = \dfrac{7}{9}$
16. $E(x^2) = \dfrac{5}{4}$
17. $\mu = \dfrac{7}{9}$
18. $\sigma^2 = \dfrac{209}{324}$
19. $\sigma = \dfrac{\sqrt{209}}{18}$
20. 0.4893
21. 0.2108
22. 0.8975
23. 0.0026
24. $\dfrac{128\pi}{9}$
25. 12π
26. $y = C_1 e^{2x^4}$, where $C_1 = e^C$
27. $y = \sqrt{8x + C_1}$, $y = -\sqrt{8x + C_1}$, where $C_1 = 2C$
28. $y = 5e^{8t}$
29. $y = 4 - C_1 e^{-x^5/5}$, where $C_1 = e^{-C}$
30. $r = \pm\sqrt[6]{36t + C_1}$, where $C_1 = 6C$
31. $y = C_1 e^{3x + x^2/2}$, where $C_1 = e^C$
32. $x = \dfrac{C_1}{p^{10}}$, where $C_1 = e^C$
33. (a) $V(t) = 40\left(1 - e^{-kt}\right)$; (b) 0.04; (c) $V(t) = 40\left(1 - e^{-0.04t}\right)$; (d) $\$20.53$; (e) 34.7
34. $b = \sqrt[5]{5}$
35. Convergent; $\dfrac{2}{3e}$
36. 9.07

CHAPTER 6, FORM C

1. $(2, \$9)$
2. $\$14.67$
3. $\$9.33$
4. $\$11{,}661.96$
5. $\$1027.68$
6. 274,040 tons
7. 2062
8. $\$9477.11$
9. $\$20{,}770.06$
10. $\$35{,}000$
11. Convergent, $\dfrac{5}{1024}$
12. Divergent
13. $\dfrac{1}{9}$; $f(x) = \dfrac{1}{9}x^2$
14. 0.8347
15. $E(x) = \dfrac{76}{15}$
16. $E(x^2) = 26$
17. $\mu = \dfrac{76}{15}$
18. $\sigma^2 = \dfrac{74}{225}$
19. $\sigma = \dfrac{\sqrt{74}}{15}$
20. 0.2881
21. 0.2563
22. 0.9502
23. 0.0179
24. $\dfrac{5\pi}{6}$
25. $\dfrac{\pi e^{-6}}{2}\left(e^{16} - 1\right)$
26. $y = C_1 e^{2x^3}$, where $C_1 = e^C$
27. $y = \sqrt{6x + C_1}$, $y = -\sqrt{6x + C_1}$, where $C_1 = 2C$
28. $y = 4e^{8t}$
29. $y = 5 - C_1 e^{-x^{11}/11}$, where $C_1 = e^{-C}$
30. $r = \pm\sqrt[4]{24t + C_1}$, where $C_1 = 4C$
31. $y = C_1 e^{12x - x^2/2}$, where $C_1 = e^C$
32. $x = \dfrac{C_1}{p^2}$, where $C_1 = e^C$
33. (a) $V(t) = 50\left(1 - e^{-kt}\right)$;
    (b) 0.11;
    (c) $V(t) = 50\left(1 - e^{-0.11t}\right)$;
    (d) $\$43.10$;
    (e) 14.6
34. $\sqrt[9]{9}$
35. Divergent
36. 8.89

ANSWERS TO CHAPTER TEST FORMS

CHAPTER 6, FORM D

1. $(2, \$4)$
2. $\$10.67$
3. $\$5.33$
4. $\$11{,}895.20$
5. $\$493.65$
6. $93{,}296$ tons
7. 2189
8. $\$8292.47$
9. $\$41{,}300.33$
10. $\$75{,}000$
11. Convergent, $\dfrac{3}{40{,}000}$
12. Divergent
13. $\dfrac{7}{128}$; $f(x) = \dfrac{7x^6}{128}$
14. 0.5507
15. $E(x) = \dfrac{21}{8}$
16. $E(x^2) = \dfrac{255}{32}$
17. $\mu = \dfrac{21}{8}$
18. $\sigma^2 = \dfrac{69}{64}$
19. $\sigma = \dfrac{\sqrt{69}}{8}$
20. 0.2995
21. 0.2464
22. 0.8647
23. 0.2119
24. 645π
25. 12π
26. $y = C_1 e^{2x^6}$, where $C_1 = e^C$
27. $y = \sqrt{10x + C_1}$, $y = -\sqrt{10x + C_1}$, where $C_1 = 2C$
28. $y = 8e^{6t}$
29. $y = 5 - C_1 e^{-x^6/6}$, where $C_1 = e^{-C}$
30. $r = \sqrt[11]{22t + C_1}$, where $C_1 = 11C$
31. $y = C_1 e^{10x - x^2/2}$, where $C_1 = e^C$
32. $x = \dfrac{C_1}{p^8}$, where $C_1 = e^C$
33. (a) $V(t) = 25\left(1 - e^{-kt}\right)$;
    (b) 0.12;
    (c) $V(t) = 25\left(1 - e^{-0.12t}\right)$;
    (d) $\$22.12$;
    (e) 4.3
34. $\sqrt[6]{6}$
35. Convergent; $\dfrac{2}{e}$
36. 11.11

CHAPTER 6, FORM E

1. $(4, \$25)$
2. $\$101.33$
3. $\$58.67$
4. $\$33{,}702.38$
5. $\$496.79$
6. 36,766 tons
7. 2100
8. $\$16{,}979.78$
9. $\$19{,}501.87$
10. $\$30{,}000$
11. Convergent, 1
12. Divergent
13. $\dfrac{5}{32}$; $f(x) = \dfrac{5}{32}x^4$
14. 0.3624
15. $E(x) = \dfrac{26}{15}$
16. $E(x^2) = 4$
17. $\mu = \dfrac{26}{15}$
18. $\sigma^2 = \dfrac{224}{225}$
19. $\sigma = \dfrac{\sqrt{224}}{15}$ or $\dfrac{4\sqrt{14}}{15}$
20. 0.4332
21. 0.3351
22. 0.9034
23. 0.1918
24. $\dfrac{175\pi}{3}$
25. $\dfrac{15\pi}{2}$
26. $y = C_1 e^{x^5}$, where $C_1 = e^C$
27. $y = \sqrt{12x + C_1}$, $y = -\sqrt{12x + C_1}$, where $C_1 = 2C$
28. $y = 6e^{4t}$
29. $y = 2 - C_1 e^{-x^4/4}$, where $C_1 = e^{-C}$
30. $r = \sqrt[5]{-10t + C_1}$, where $C_1 = 5C$
31. $y = C_1 e^{2x + x^2/2}$, where $C_1 = e^C$
32. $x = \dfrac{C_1}{p^6}$, where $C_1 = e^C$
33. (a) $V(t) = 8\left(1 - e^{-kt}\right)$;
    (b) 0.08;
    (c) $V(t) = 8\left(1 - e^{-0.08t}\right)$;
    (d) $\$6.10$;
    (e) 26
34. $\sqrt[8]{8}$
35. Convergent, $\dfrac{3}{5}$
36. 37.70

ANSWERS TO CHAPTER TEST FORMS

CHAPTER 6, FORM F

1. $(2, \$16)$
2. $\$18.67$
3. $\$5.33$
4. $\$56{,}236.44$
5. $\$1034.43$
6. $35{,}993$ tons
7. 2101
8. $\$24{,}394.18$
9. $\$37{,}599.03$
10. $\$83{,}333.33$
11. Convergent, $\dfrac{3}{2}$
12. Divergent
13. $\dfrac{4}{255}$; $f(x) = \dfrac{4}{255}x$
14. 0.6321
15. $E(x) = \dfrac{148}{21}$
16. $E(x^2) = 50$
17. $\mu = \dfrac{148}{21}$
18. $\sigma^2 = \dfrac{146}{441}$
19. $\sigma = \dfrac{\sqrt{146}}{21}$
20. 0.4641
21. 0.2481
22. 0.6466
23. 0.9918
24. $\dfrac{200\pi}{9}$
25. $\dfrac{133\pi}{2}$
26. $y = C_1 e^{2x^6}$, where $C_1 = e^C$
27. $y = \sqrt{6x + C_1}$, $y = -\sqrt{6x + C_1}$, where $C_1 = 2C$
28. $y = 4e^{8t}$
29. $y = 10 - C_1 e^{-x^5/5}$, where $C_1 = e^{-C}$
30. $r = \pm\sqrt[4]{-36t + C_1}$, where $C_1 = 4C$
31. $y = C_1 e^{2x - x^2/2}$, where $C_1 = e^C$
32. $x = \dfrac{C_1}{p^6}$, where $C_1 = e^C$
33. (a) $V(t) = 10\left(1 - e^{-kt}\right)$;
    (b) 0.12;
    (c) $V(t) = 10\left(1 - e^{-0.12t}\right)$;
    (d) $\$8.85$;
    (e) 5.8
34. 1
35. Convergent; $\dfrac{1}{10e}$
36. 12.70

CHAPTER 7, FORM A

1. $-130 - e^2$
2. $20x^4 y - e^x$
3. $2 + 4x^5$
4. $80x^3 y - e^x$
5. $20x^4$
6. $20x^4$
7. 0
8. Minimum $= -\dfrac{40}{27}$ at $\left(-\dfrac{4}{3}, \dfrac{2}{3}\right)$
9. None
10. (a) $y = \dfrac{7}{2}x + \dfrac{14}{3}$; (b) $18.67 million
11. Maximum $= 4560$ at $(-4, 32)$
12. $-\dfrac{29}{5}$
13. $300,000 for labor, $200,000 for capital
14. $f_x = \dfrac{-x^4 - 15x^2 t - 10xt}{(x^3 - 5t)^2}$; $f_t = \dfrac{5x^3 + 5x^2}{(x^3 - 5t)^2}$
15.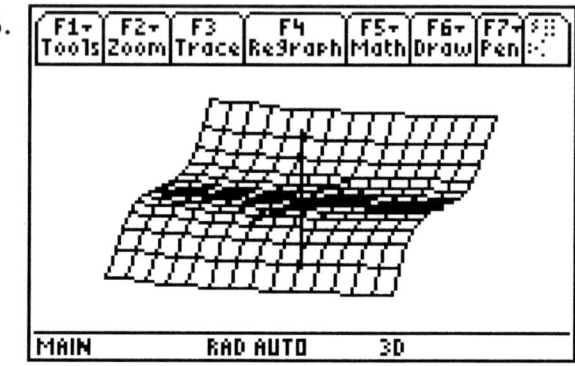

CHAPTER 7, FORM B

1. $e - 12$
2. $e^x + 8xy$
3. $4x^2 + 2$
4. $e^x + 8y$
5. $8x$
6. $8x$
7. 0
8. None
9. Minimum $= 0$ at $(0, 0)$
10. (a) $y = 2x + \dfrac{29}{3}$; (b) $17.67 million
11. Maximum $= -\dfrac{121}{10}$ at $\left(-\dfrac{11}{5}, -\dfrac{33}{10}\right)$
12. $\dfrac{28}{5}$
13. $156,250 for labor, $93,750 for capital
14. $f_x = \dfrac{x^6 + 20x^3 t + 15x^2 t}{(x^3 + 5t)^2}$;

   $f_t = \dfrac{-5x^3 - 5x^4}{(x^3 + 5t)^2}$
15.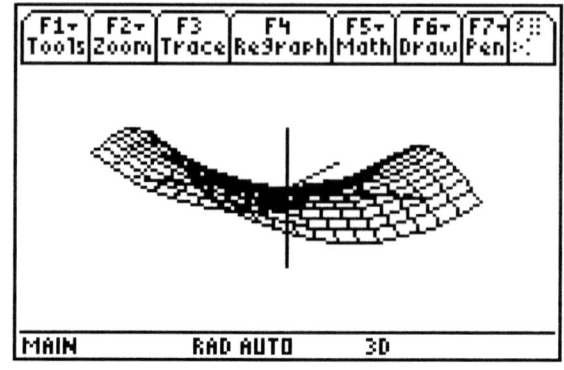

ANSWERS TO CHAPTER TEST FORMS

CHAPTER 7, FORM C

1. $\ln 2 + 51$
2. $\dfrac{1}{x} + 8xy$
3. $4x^2 + 1$
4. $-\dfrac{1}{x^2} + 8y$
5. $8x$
6. $8x$
7. 0
8. Minimum $= -448$ at $(24, 8)$
9. Minimum $= 0$ at $(0, 0)$
10. (a) $y = \dfrac{7}{2}x + 2$; (b) $16 million

11. Maximum $= -\dfrac{32}{7}$ at $\left(\dfrac{12}{7}, \dfrac{8}{7}\right)$
12. $\dfrac{11}{3}$
13. $120,000 for labor, $30,000 for capital
14. $f_x = \dfrac{-27x^2 t}{(2t + 3x^3)^2}$; $f_t = \dfrac{9x^3}{(2t + 3x^3)^2}$
15.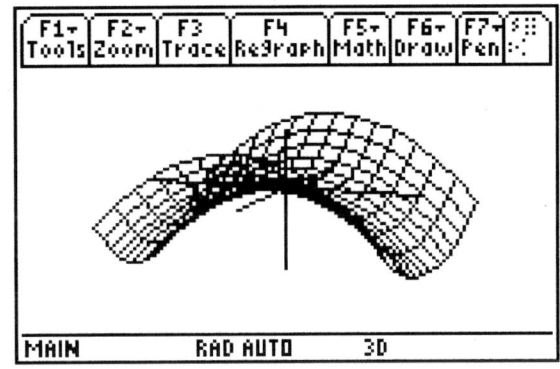

CHAPTER 7, FORM D

1. $-11 + 2e^2$
2. $6xy + 2e^x$
3. $3x^2 - 1$
4. $6y + 2e^x$
5. $6x$
6. $6x$
7. 0
8. Minimum $= -\dfrac{1}{11}$ at $\left(-\dfrac{2}{11}, -\dfrac{1}{11}\right)$
9. None
10. (a) $y = 3x + \dfrac{7}{3}$; (b) $14.3 million
11. Maximum $= -13$ at $(5, 3)$

12. $\dfrac{327}{20}$
13. $210,000 for labor, $90,000 for capital
14. $f_x = \dfrac{-16x^4 + 12x^2 t + 8xt}{(4x^3 + t)^2}$;

    $f_t = \dfrac{-4x^3 - 4x^2}{(4x^3 + t)^2}$
15.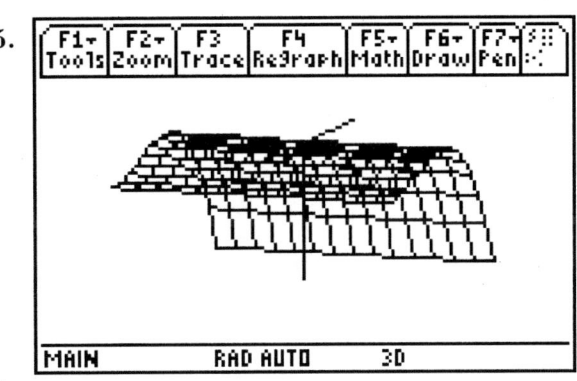

CHAPTER 7, FORM E

1. $2e - 6$
2. $2e^x + 8xy$
3. $4x^2 - 1$
4. $2e^x + 8y$
5. $8x$
6. $8x$
7. 0
8. Minimum $= -\dfrac{1}{8}$ at $\left(\dfrac{3}{4}, -1\right)$
9. None
10. (a) $y = 2x - \dfrac{4}{3}$; (b) $6.67 million
11. Minimum $= -600$ at $(13, 2)$
12. $\dfrac{2}{3}$
13. $150,000 for labor, $50,000 for capital
14. $f_x = \dfrac{9x^4 + 18x^2 t + 12xt}{(2t - 3x^3)^2}$;

    $f_t = \dfrac{-6x^3 - 6x^2}{(2t - 3x^3)^2}$
15.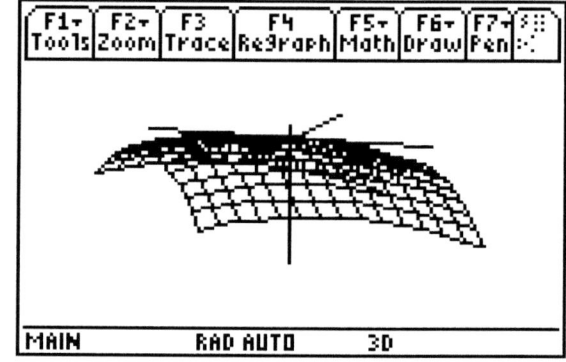

CHAPTER 7, FORM F

1. -17
2. $12xy + \dfrac{1}{x^2}$
3. $2 + 6x^2$
4. $12y - \dfrac{2}{x^3}$
5. $12x$
6. $12x$
7. 0
8. None
9. Minimum $= -\dfrac{5}{4}$ at $\left(1, -\dfrac{3}{2}\right)$
10. (a) $y = \dfrac{3}{2}x + \dfrac{20}{3}$; (b) $12.67 million
11. Maximum $= 33$ at $(1, 5)$
12. $-\dfrac{7}{30}$
13. $1,400,000 for labor, $2,800,000 for capital
14. $f_x = \dfrac{-x^6 + 4x^3 t + 3x^2 t}{(x^4 + t)^2}$; $f_t = \dfrac{-x^4 - x^3}{(x^4 + t)^2}$
15.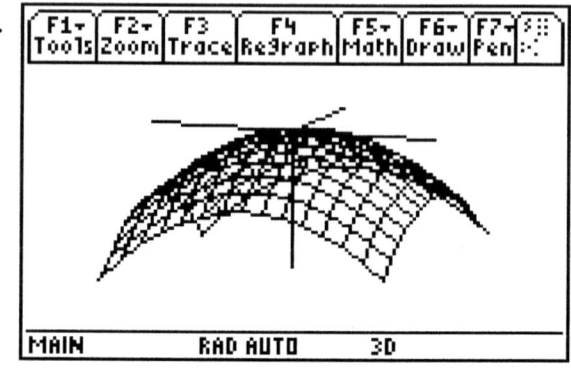

FINAL EXAM, FORM A

1. $y = -\dfrac{3}{4}x + \dfrac{7}{4}$

2. $5(2x+h)$

3. (a) $f(x) = \begin{cases} x^2+1, & \text{for } x \neq 0 \\ 0, & \text{for } x = 0 \end{cases}$

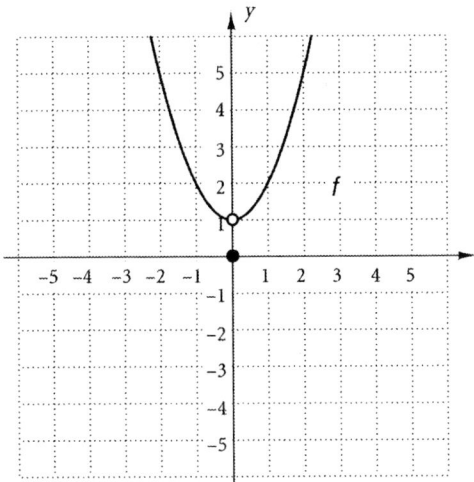

(b) 1; (c) 0; (d) No

4. -68

5. Does not exist

6. $\dfrac{1}{7}$

7. 0

8. $3x^2 - 4$

9. $\dfrac{5}{6}x^{-1/6}$

10. $-6x^{-7}$

11. $5(x+1)(x-3)^3$

12. $\dfrac{-3x^2 - 8x - 6}{(x^2 - 2)^2}$

13. $\dfrac{3x^2 + 1}{x^3 + x + 1}$

14. $5e^x$

15. $2(x+1)e^{x^2+2x}$

16. $40(2x+1)^4$

17. $60x$

18. $\dfrac{2x^2}{3y^2} - \dfrac{y}{3x}$

19. $y = 2x - 2$

20. Relative maximum at $(-3, 1)$; relative minimum at $(-1, -3)$

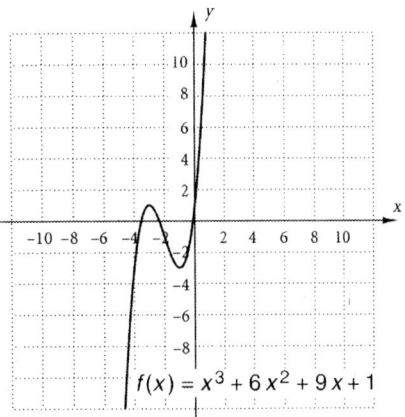

21. Relative minima at $\left(-\sqrt{\dfrac{5}{2}}, -\dfrac{9}{4}\right)$ and $\left(\sqrt{\dfrac{5}{2}}, -\dfrac{9}{4}\right)$; relative maximum at $(0, 4)$

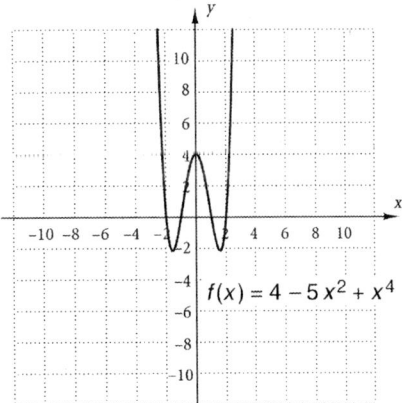

22. Relative maximum at $\left(-\sqrt{2}, \dfrac{3\sqrt{2}}{4}\right)$; relative minimum at $\left(\sqrt{2}, \dfrac{-3\sqrt{2}}{4}\right)$

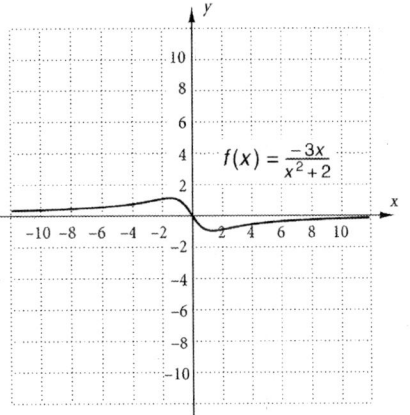

FINAL EXAM, FORM A (continued)

23. Relative maximum at $(0, -1)$

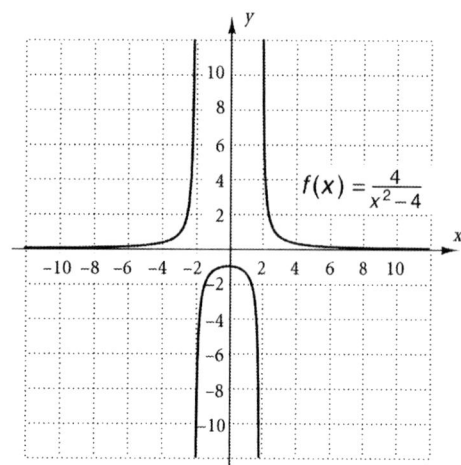

24. Minimum $= \dfrac{23}{16}$ at $x = \dfrac{3}{8}$

25. None

26. Maximum $= \dfrac{32}{3}$ at $x = -1$; minimum $= -\dfrac{22}{3}$ at $x = 2$

27. 25   **28.** 10 times; lot size 80

29. $\Delta y = 0.3003$; $f'(x)\Delta x = 0.3$

30. (a) $P(t) = 14{,}000 e^{0.03t}$; (b) 18,898; (c) 23.1 yr

31. (a) $E(p) = \dfrac{p}{24 - p}$;
(b) $E(10) = \dfrac{5}{7}$, inelastic;
(c) increase; (d) $p = \$12$

32. $\dfrac{x^8}{4} + C$ **33.** $3e^4 + 5$

34. $\ln\left(x + \sqrt{x^2 - 25}\right) + C$

35. $\dfrac{1}{3} e^{x^3 + 1} + C$

36. $\left(\dfrac{x^2}{2} - 2x\right)\ln x - \dfrac{x^2}{4} + 2x + C$

37. $18 \ln x + C$ **38.** 9   **39.** 28

40. \$19,918.37   **41.** \$55,646.74

42. Convergent, 1

43. Divergent

44. (a) $\dfrac{9}{4}$; (b) $\dfrac{3}{4}\sqrt{\dfrac{3}{5}}$, or $\dfrac{3\sqrt{15}}{20}$

45. 0.5925   **46.** $(3, \$49)$; \$40.50

47. $\pi\left(e^{10} - 1\right)$

48. $y = C_1 e^{x^4/2}$, where $C_1 = e^C$

49. $y = \pm\sqrt{16x + C_1}$, where $C_1 = 2C$

50. (a) $P(C) = 200(C + 4)^{-1/2} - 20$; (b) \$96

51. $6 - 2x^3$ **52.** $-6x^2$

53. Relative minimum $= 6$ at $x = (0, 0)$

54. Maximum $= \dfrac{4429}{9}$ at $\left(\dfrac{55}{3}, \dfrac{80}{9}\right)$

55. $e - \dfrac{1}{e} + 4$

56. $\ln(1 + e^x) + C$ **57.** 27

58.

FINAL EXAM, FORM B

1. $y = -\dfrac{2}{3}x + 1$ 2. $2(2x + h)$

3. (a) $f(x) = \begin{cases} x^2 - 5, & \text{for } x \neq 2 \\ 0, & \text{for } x = 2 \end{cases}$

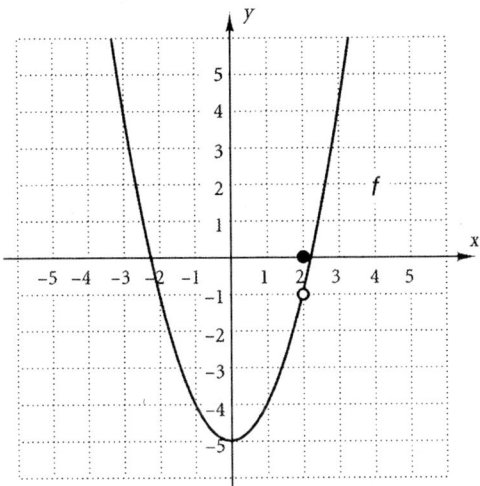

(b) -1; (c) 0; (d) No

4. -30 5. Does not exist

6. $-\dfrac{1}{6}$ 7. 0 8. $3x^2 + 2$

9. $\dfrac{4}{3}x^{1/3}$ 10. $-16x^{-17}$

11. $(5x + 18)(x - 2)^3$

12. $\dfrac{10x^3 - 24x^2 - 15}{(x^3 + 3)^2}$

13. $\dfrac{2x - 6}{x^2 - 6x + 4}$ 14. $5e^x$

15. $e^{4x^2 - 3x}(8x - 3)$

16. $160(5x - 2)^7$ 17. $48x - 40x^3$

18. $\dfrac{x^7}{y} - \dfrac{y}{x}$ 19. $y = -6x + 4$

20. Relative minimum at $(-1, -3)$; relative maximum at $(-3, 5)$

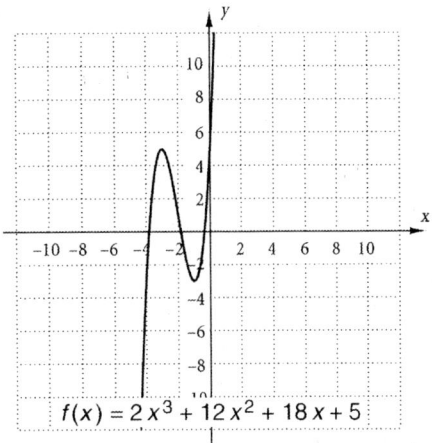

21. Relative minimum at $(0, -6)$

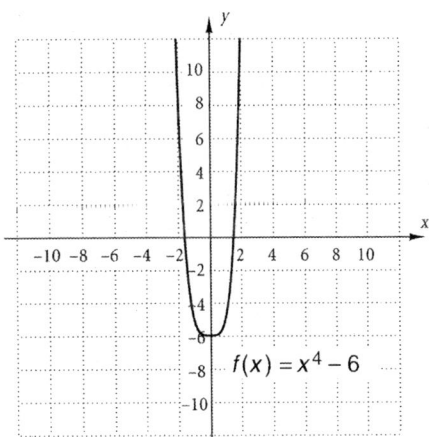

22. Relative maximum at $\left(-2, \dfrac{3}{2}\right)$; relative minimum at $\left(2, -\dfrac{3}{2}\right)$

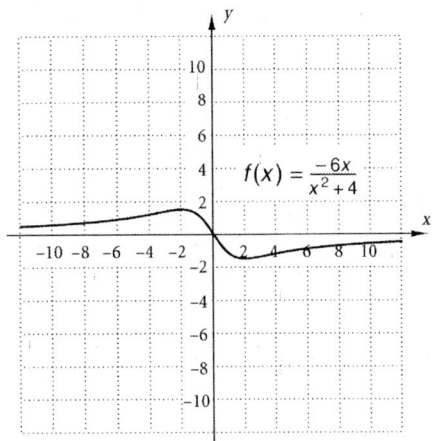

FINAL EXAM, FORM B (continued)

23. Relative minimum at $(0, -2)$

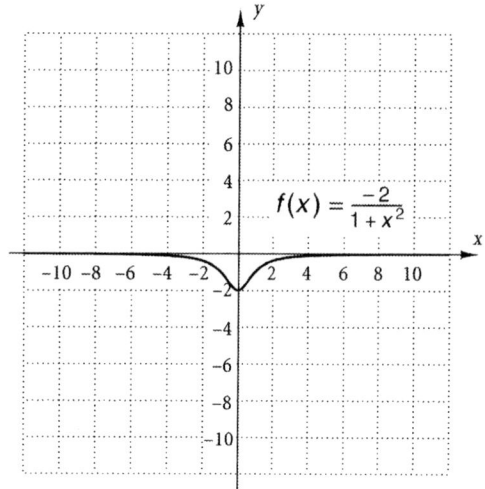

24. Minimum $= -\dfrac{65}{8}$ at $x = -\dfrac{3}{4}$

25. None

26. Minimum $= -\dfrac{16}{3}$ at $x = 2$; maximum $= \dfrac{28}{3}$ at $x = 4$

27. 15   **28.** 10 times; lot size 40

29. $\Delta y = -0.0597$; $f'(x)\,\Delta x = -0.06$

30. (a) $P(t) = 8500e^{0.06t}$;
(b) 20,907; (c) 11.6 yr

31. (a) $E(p) = \dfrac{2p}{25 - 2p}$;
(b) $E(8) = \dfrac{16}{9}$, elastic;
(c) decrease; (d) $p = \$6.25$

32. $\dfrac{2x^6}{3} + C$ **33.** $4e^2 + 2$

34. $\ln\left(\dfrac{x}{3x+1}\right) + C$ **35.** $6e^{x^2+4} + C$

36. $\left(\dfrac{x^2}{2} + 6x\right)\ln x - \dfrac{x^2}{4} - 6x + C$

37. $50\ln x + C$ **38.** $\dfrac{31}{5}$ **39.** $\dfrac{48}{5}$

40. \$32,627.98   **41.** \$27,983.54

42. Convergent, $\dfrac{1}{3}$

43. Divergent

44. (a) $\dfrac{15}{4}$; (b) $\dfrac{\sqrt{15}}{4}$

45. 0.4554   **46.** $(8, \$144)$; \$501.33

47. $\dfrac{\pi}{4}\left(e^{20} - 1\right)$

48. $y = C_1 e^{3x^4/2}$, where $C_1 = e^C$

49. $y = \pm\sqrt{C_1 - 20x}$, where $C_1 = 2C$

50. (a) $P(C) = 500(C+5)^{-1/2} - \dfrac{428}{9}$;
(b) \$105.54

51. $5xy^4 - 7e^y$ **52.** $5y^4$

53. None

54. Maximum $= 89$ at $(8, 5)$

55. $-\dfrac{2}{e} - 4$

56. $\dfrac{8x^3 t + 24x^2 + t^2}{(xt+2)^2}$

57. $\ln(e^x + 5) + C$

58.

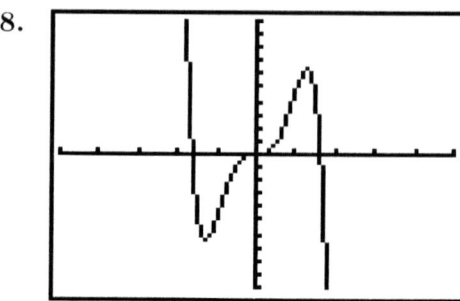

FINAL EXAM, FORM C

1. $y = -3x + 1$ **2.** $8x + 4h - 1$

3. (a) $f(x) = \begin{cases} -x^2 + 2, & \text{for } x \neq -1 \\ -1, & \text{for } x = -1 \end{cases}$

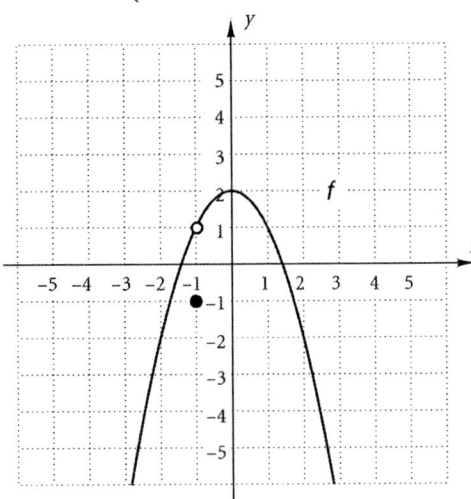

(b) 1; (c) -1; (d) No

4. -7 **5.** Does not exist

6. $-\dfrac{1}{4}$ **7.** 0   **8.** $4x^3 - 6x$

9. $\dfrac{3}{4}x^{-1/4}$ **10.** $-8x^{-9}$

11. $6(x+3)(x-2)^4$

12. $\dfrac{-3x^2 + 8x + 15}{(x^2 + 5)^2}$

13. $\dfrac{2x - 3}{x^2 - 3x + 1}$ **14.** $10e^x$

15. $e^{x^2 - 4x}(2x - 4)$

16. $-36(3x - 8)^5$ **17.** $60x^4 - 24x$

18. $\dfrac{1}{y^3} - \dfrac{y}{4x}$ **19.** $y = 3x - 2$

20. Relative minimum at $(-1, -2)$; relative maximum at $(-3, 2)$

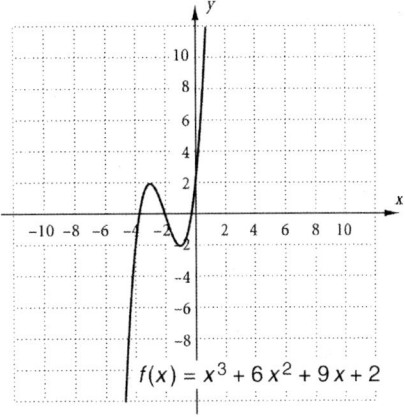

21. Relative minima at $(-1, -9)$ and $(1, -9)$; relative maximum at $(0, -8)$

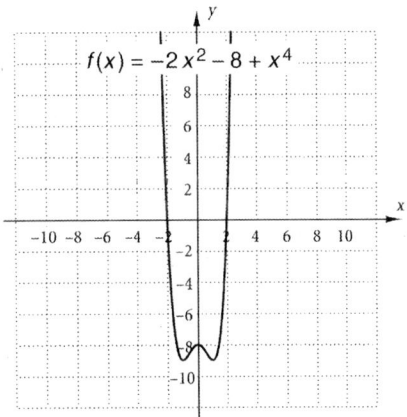

22. Relative minimum at $(-2, -3)$; relative maximum at $(2, 3)$

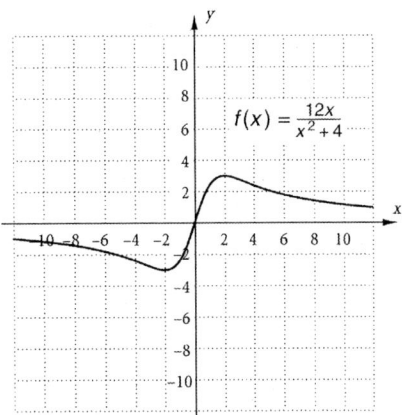

FINAL EXAM, FORM C (continued)

23. Relative maximum at $(0, -3)$

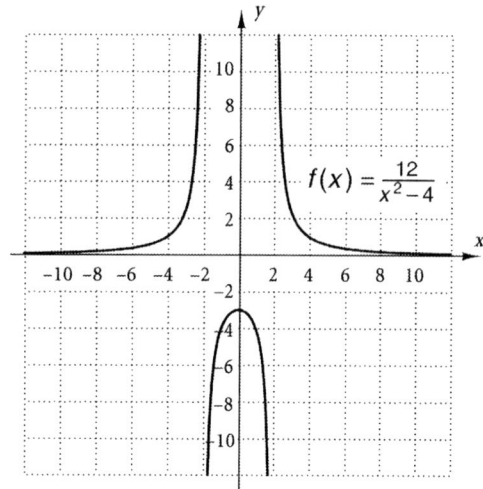

24. Maximum $= 2$ at $x = 1$

25. None

26. Minimum $= -9$ at $x = -3$; maximum $= \frac{25}{3}$ at $x = -1$

27. 5   **28.** 5 times; lot size 60

29. $\Delta y = 0.2404$; $f'(x)\Delta x = 0.24$

30. (a) $P(t) = 400e^{0.08t}$; (b) 890; (c) 8.7 yr

31. (a) $E(p) = \dfrac{p}{24 - p}$; (b) $E(18) = 3$, elastic; (c) decrease; (d) $p = \$12$

32. $\dfrac{6}{5}x^5 + C$ **33.** $4e^2 + 2$

34. $\dfrac{3}{4(2x+3)} + \dfrac{1}{4}\ln(2x+3) + C$

35. $\dfrac{1}{2}e^{x^4+2} + C$

36. $\left(\dfrac{x^2}{2} + 6x\right)\ln x - \dfrac{x^2}{4} - 6x + C$

37. $-2\ln x + C$ **38.** 72   **39.** 4

40. \$32,525.45   **41.** \$18,799.52

42. Divergent

43. Convergent, $\dfrac{1}{3}$

44. (a) $\dfrac{8}{5}$; (b) $\dfrac{2}{5}\sqrt{\dfrac{2}{3}}$, or $\dfrac{2\sqrt{6}}{15}$

45. 0.6326   **46.** $(3, \$49)$; \$63

47. $5\pi e^{2/5}\left(e^{3/5} - 1\right)$

48. $y = C_1 e^{2x^6/3}$, where $C_1 = e^C$

49. $y = \pm\sqrt{C_1 + 3x}$, where $C_1 = 2C$

50. (a) $P(C) = 400(C+8)^{-1/2} - 16$; (b) \$617

51. $4y + 3xy^2$ **52.** $3y^2$

53. Relative minimum $= 2$ at $(0, 0)$

54. Minimum $= -67.75$ at $(-6, 8.5)$

55. $2e^4 - 2e^2 + 2$

56. $3x^2 + x + 2x\ln x$

57. $2\ln(1 + e^x) + C$

58. 6.66

FINAL EXAM, FORM D

1. $y = -3x + \dfrac{19}{2}$

2. $6x + 3h - 1$

3. (a) $f(x) = \begin{cases} 4 - x^2, & \text{for } x \neq 1 \\ 1, & \text{for } x = 1 \end{cases}$

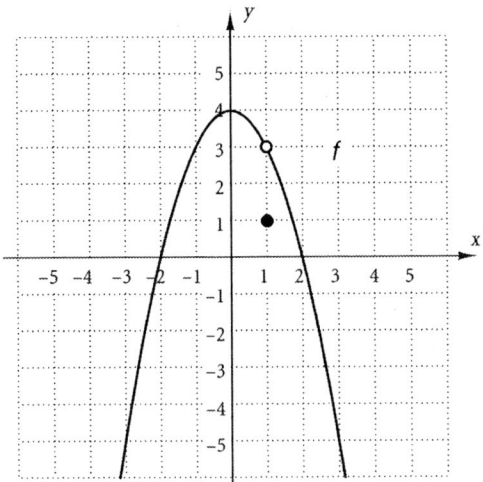

(b) 3; (c) 1; (d) No

4. 2

5. Does not exist

6. $-\dfrac{1}{7}$

7. 0

8. $8x^7 - 6x$

9. $\dfrac{3}{4}x^{-1/4}$

10. $-x^{-2}$

11. $(6x + 11)(x - 4)^4$

12. $\dfrac{5x^2 - 4x + 15}{(3 - x^2)^2}$

13. $\dfrac{2x - 7}{x^2 - 7x + 1}$

14. $8e^x$

15. $e^{x^2 - x - 3}(2x - 1)$

16. $-72(6x + 5)^2$

17. $60x^2$

18. $\dfrac{1}{8xy\sqrt{x}} - \dfrac{y}{2x}$

19. $y = 3x + 2$

20. Relative minimum at $(3, -8)$; relative maximum at $(1, -4)$

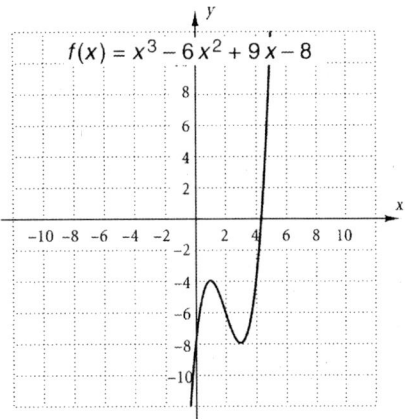

21. Relative maxima at $(-1, 6)$ and $(1, 6)$; relative minimum at $(0, 5)$

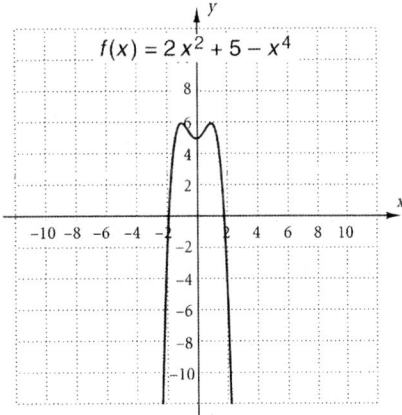

22. Relative maximum at $(-3, 2)$; Relative minimum at $(3, -2)$

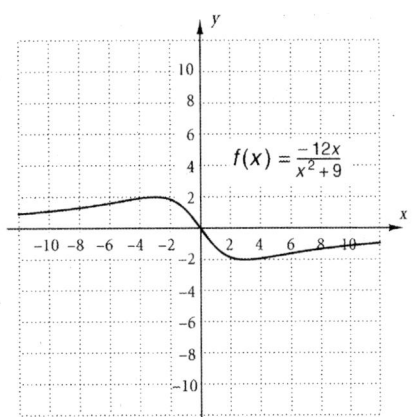

FINAL EXAM, FORM D (continued)

23. Relative minimum at $\left(0, \frac{5}{2}\right)$

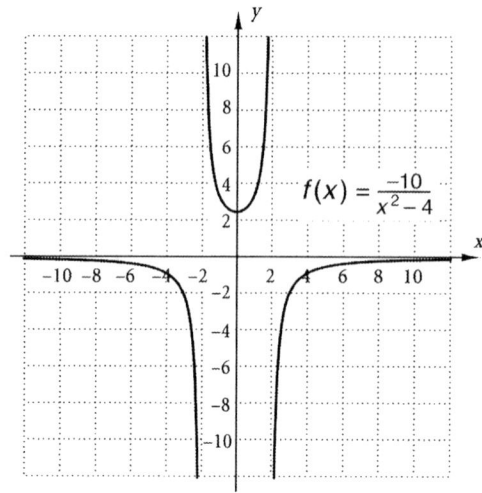

24. Maximum $= -\dfrac{23}{16}$ at $x = \dfrac{5}{8}$

25. None

26. Minimum $= -59\frac{2}{3}$ at $x = 5$; maximum $= 173\frac{2}{3}$ at $x = 10$

27. 45   **28.** 4 times; lot size 400

29. $\Delta y = 0.3003$; $f'(x)\Delta x = 0.3$

30. (a) $P(t) = 8000 e^{0.15 t}$; (b) 16,936; (c) 4.6 hr

31. (a) $E(p) = \dfrac{p}{40 - p}$;
(b) $E(25) = \dfrac{5}{3}$, elastic;
(c) decrease; (d) $p = \$20$

32. $x^4 + C$ **33.** $4e^2 - 4e + 9$

34. $\ln\left(x + \sqrt{x^2 + 9}\right) + C$

35. $\dfrac{1}{4} e^{x^4 + 6} + C$

36. $\left(\dfrac{x^2}{2} + 2x\right)\ln x - \dfrac{x^2}{4} - 2x + C$

37. $-8\ln x + C$ **38.** 48   **39.** 8

40. \$17,552.04   **41.** \$38,877.27

42. Divergent

43. Convergent, $\dfrac{1}{7}$

44. (a) 4; (b) $\sqrt{\dfrac{2}{3}}$, or $\dfrac{\sqrt{6}}{3}$

45. 0.3218   **46.** (3, \$49); \$36

47. $2\pi(e^4 - 1)$

48. $y = C_1 e^{-x^4/4}$, where $C_1 = e^C$

49. $y = \pm\sqrt{10x + C_1}$, where $C_1 = 2C$

50. (a) $P(C) = 200(C + 3)^{-1/2} - 38$; (b) \$24.70

51. $2x + 2y$ **52.** 2

53. Relative maximum $= 6$ at $(0, 0)$

54. Maximum $= 4$ at $(3, 1)$

55. $e - \dfrac{1}{e} - 2$

56. $e^x\left(x + \dfrac{1}{x} + \ln x + 1\right)$

57. $5\ln(5 + e^x) + C$

58. 8.89

FINAL EXAM, FORM E

1. $y = -\frac{1}{2}x - \frac{5}{2}$

2. $3(2x + h)$

3. (a) $f(x) = \begin{cases} x^2 - 4, & \text{for } x \neq 2 \\ 4, & \text{for } x = 2 \end{cases}$

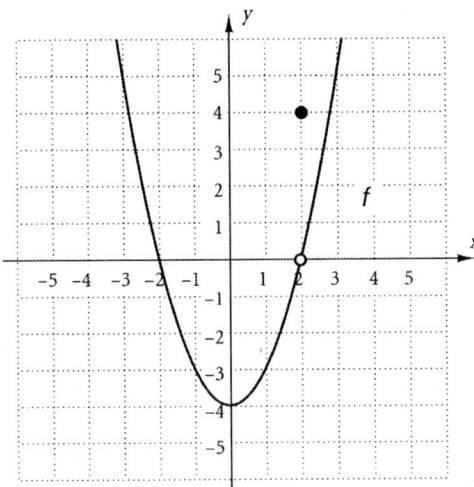

(b) 0; (c) 4; (d) No

4. 11 5. Does not exist

6. 1 7. 0 8. $9x^2 + 1$

9. $\frac{3}{5}x^{-2/5}$ 10. $-3x^{-4}$

11. $5(x-1)(x+3)^3$

12. $\dfrac{-3x^2 + 2x + 18}{(x^2 + 6)^2}$

13. $\dfrac{4x^3 - 2}{x^4 - 2x}$ 14. $5e^x$

15. $4x^3 e^{x^4 - 1}$

16. $24(2x+3)^2$ 17. $78x$

18. $2 + \dfrac{4}{x} - \dfrac{y}{x}$ 19. $y = 6x - 7$

20. Relative minimum at $(-3, 5)$; relative maximum at $(-1, 9)$

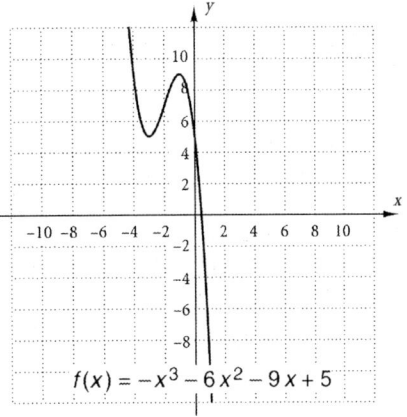

21. Relative minima at $(-2, -8)$ and $(2, -8)$; relative maximum at $(0, 8)$

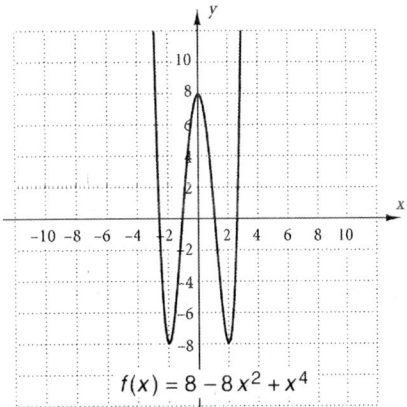

22. Relative minimum at $\left(-\sqrt{3}, -\sqrt{3}\right)$; relative maximum at $\left(\sqrt{3}, \sqrt{3}\right)$

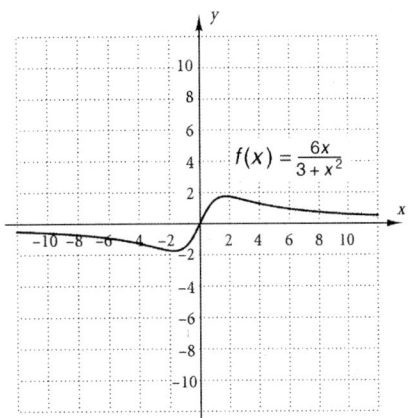

FINAL EXAM, FORM E (continued)

23. Relative maximum at $(0, -2)$

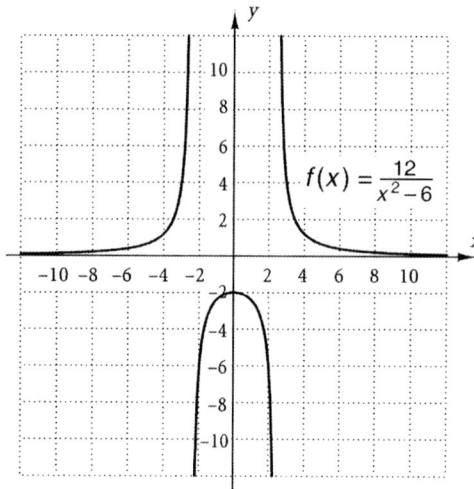

24. Minimum $= -18$ at $x = -2$

25. None

26. Minimum $= -\frac{44}{3}$ at $x = -2$; maximum $= \frac{10}{3}$ at $x = 1$

27. 85   **28.** 8 times; lot size 30

29. $\Delta y = 0.4808$; $f'(x)\Delta x = 0.48$

30. (a) $P(t) = 10{,}000e^{0.1t}$; (b) 13,499; (c) 6.9 hr

31. (a) $E(p) = \dfrac{3p}{50 - 3p}$; (b) $E(6) = \dfrac{9}{16}$, inelastic; (c) increase; (d) $p = \$8.33$

32. $\dfrac{3}{5}x^5 + C$ **33.** $8e^2 - 8e + 7$

34. $\dfrac{12^x}{\ln 12} + C$

35. $\dfrac{1}{4}e^{x^4+3} + C$

36. $\left(\dfrac{x^2}{2} - 8x\right)\ln x - \dfrac{x^2}{4} + 8x + C$

37. $2\ln x + C$ **38.** 16   **39.** 6

40. $25{,}237.16   **41.** $17{,}979.10

42. Convergent, $\dfrac{1}{18}$

43. Divergent

44. (a) $\dfrac{5}{3}$; (b) $\dfrac{1}{3}\sqrt{\dfrac{5}{7}}$, or $\dfrac{\sqrt{35}}{21}$

45. 0.4435   **46.** $(4, \$25)$; $58.67

47. $\dfrac{\pi}{2}\left(e^{-2} - e^{-6}\right)$

48. $y = C_1 e^{-x^8}$, where $C_1 = e^C$

49. $y = \pm\sqrt{8x + C_1}$, where $C_1 = 2C$

50. (a) $P(C) = 600(C+5)^{-1/2} - 23.5$; (b) $646.88

51. $\dfrac{1}{x} - 12y$ **52.** $-\dfrac{1}{x^2}$

53. None

54. Maximum $= \dfrac{21}{2}$ at $\left(\dfrac{5}{4}, \dfrac{5}{4}\right)$

55. $6e - 6$

56. $\dfrac{6x^2 + 2x}{(x + 2t)^2}$

57. 48   **58.** $\dfrac{64}{3}$

FINAL EXAM, FORM F

1. $y = -4x - 7$ 2. $4x + 2h + 3$

3. (a) $f(x) = \begin{cases} -x^2 + 2, & \text{for } x \neq 1 \\ 0, & \text{for } x = 1 \end{cases}$

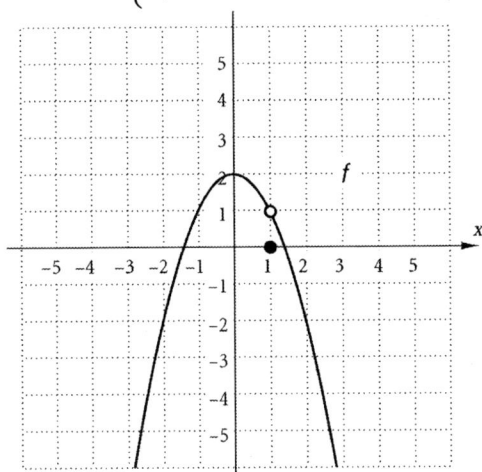

(b) 1; (c) 0; (d) No

4. -38 5. Does not exist

6. $-\dfrac{1}{7}$ 7. 0   8. $20x^3 - 6x^2$

9. $\dfrac{4}{3}x^{1/3}$ 10. $-6x^{-7}$

11. $(5x - 19)(x + 1)^3$

12. $\dfrac{-3x^2 - 4x - 24}{(x^2 - 8)^2}$

13. $\dfrac{3x^2 - 5}{x^3 - 5x}$ 14. $-4e^x$

15. $e^{x^5 - x}(5x^4 - 1)$

16. $200(5x - 2)^4$ 17. $48x^2 + 6$

18. $-\dfrac{1}{4x\sqrt{x}} - \dfrac{y}{x}$ 19. $y = 3x - 1$

20. Relative minimum at $(1, -2)$; relative maximum at $(3, 6)$

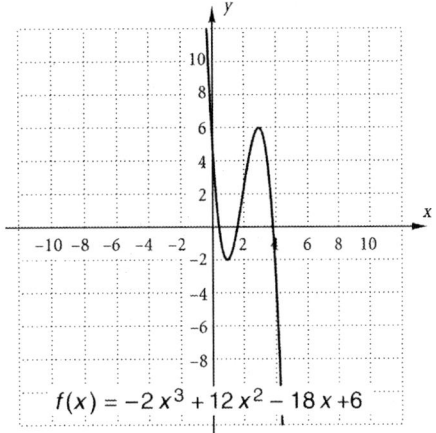

21. Relative maxima at $(-2, 8)$ and $(2, 8)$; relative minimum at $(0, -8)$

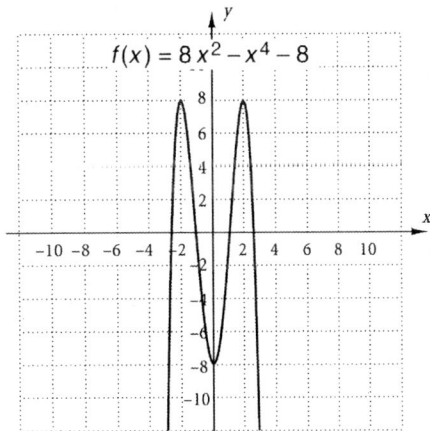

22. Relative minimum at $\left(-\sqrt{5}, -2\sqrt{5}\right)$; relative maximum at $\left(\sqrt{5}, 2\sqrt{5}\right)$

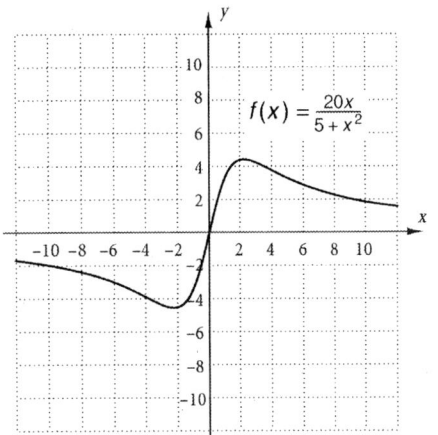

FINAL EXAM, FORM F (continued)

23. Relative minimum at $(0, 2)$

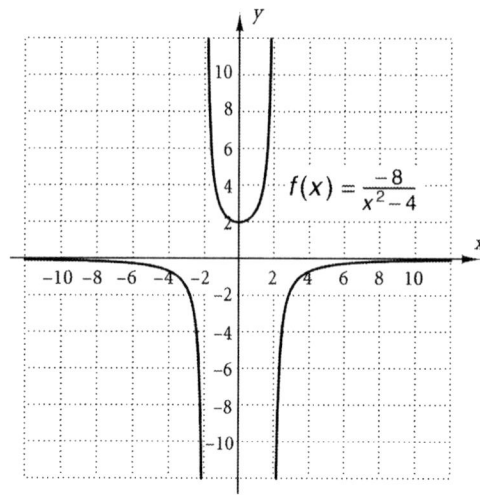

24. Minimum $= -\dfrac{19}{8}$ at $x = -\dfrac{1}{4}$

25. None

26. Minimum $= -10$ at $x = -3$; maximum $= \dfrac{22}{3}$ at $x = -1$

27. 20   **28.** 10 times; lot size 60

29. $\Delta y = 0.2406$; $f'(x)\Delta x = 0.24$

30. (a) $P(t) = 40{,}000 e^{0.12t}$; (b) 82,177; (c) 5.8 hr

31. (a) $E(p) = \dfrac{2p}{25 - 2p}$; (b) $E(7) = \dfrac{14}{11}$, elastic; (c) decrease; (d) $p = \$6.25$

32. $\dfrac{8x^5}{5} + C$ **33.** $2e^2 + 2$

34. $\dfrac{1}{4}\ln\left(\dfrac{x}{3x+4}\right) + C$

35. $3e^{x^5+2} + C$

36. $\left(\dfrac{x^2}{2} - 6x\right)\ln x - \dfrac{x^2}{4} + 6x + C$

37. $10 \ln x + C$ **38.** $\dfrac{76}{3}$ **39.** $\dfrac{37}{6}$

40. \$34,408.97   **41.** \$22,374.37

42. Divergent

43. Convergent, $\dfrac{1}{2}$

44. (a) $\dfrac{8}{5}$; (b) $\dfrac{2}{5}\sqrt{\dfrac{2}{3}}$, or $\dfrac{2\sqrt{6}}{15}$

45. 0.3721

46. $(15, \$1225)$; \$4500

47. $\dfrac{\pi}{6}\left(e^{36} - 1\right)$

48. $y = C_1 e^{3x^5/5}$, where $C_1 = e^C$

49. $y = \pm\sqrt{220x + C_1}$, where $C_1 = 2C$

50. (a) $P(C) = 400(C+8)^{-1/2} - \dfrac{110}{9}$; (b) \$1063.07

51. $-\dfrac{3x}{y^2} + 2y$ **52.** $-\dfrac{3}{y^2}$

53. Relative maximum $= 8$ at $(0, 0)$

54. Maximum $= 34$ at $(-5, 4)$

55. $\ln 2 + \dfrac{1}{3}$ **56.** -78

57. $\dfrac{1 + \dfrac{1}{x} - x - \ln x}{e^x}$ **58.** $\dfrac{63}{4}$

ANSWERS TO EVEN-NUMBERED EXERCISES IN THE EXERCISE SETS

CHAPTER 1

Exercise Set 1.1

2.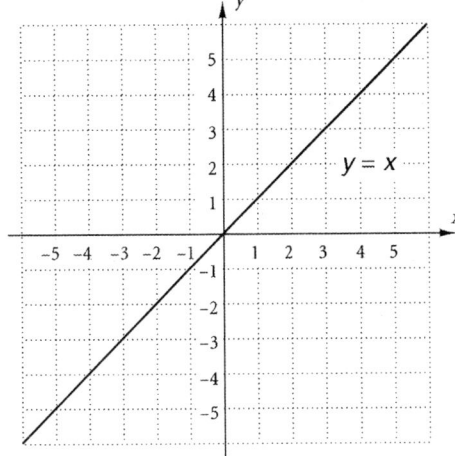

4.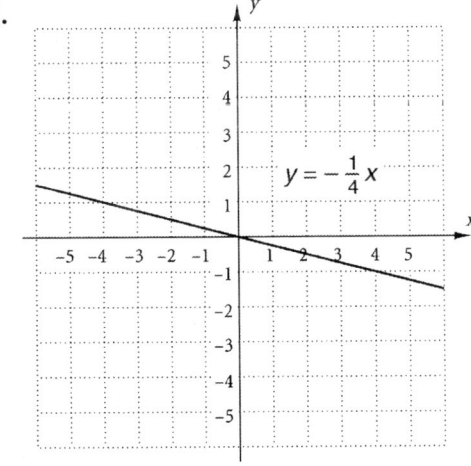

6.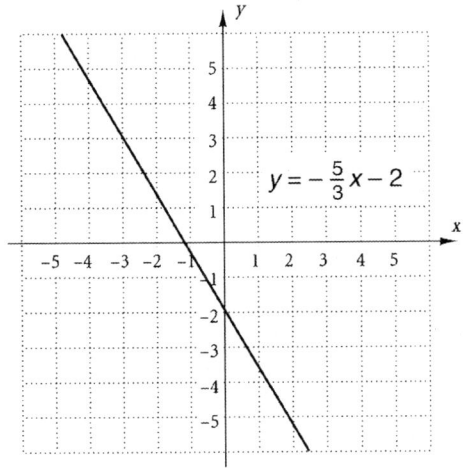

8.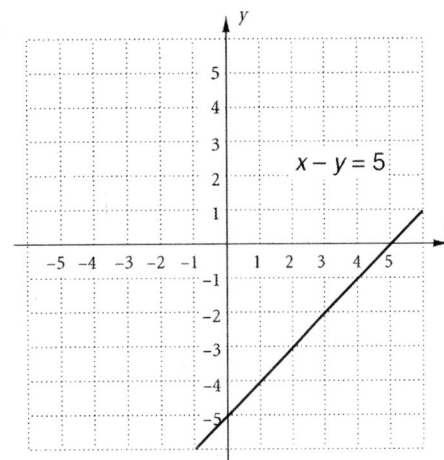

10.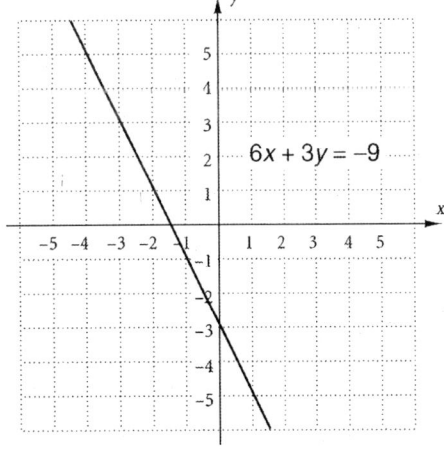

12.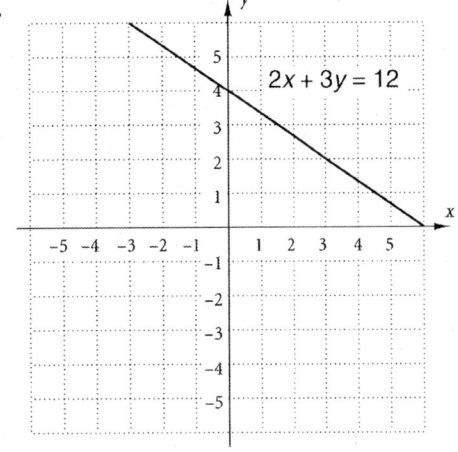

283

14.

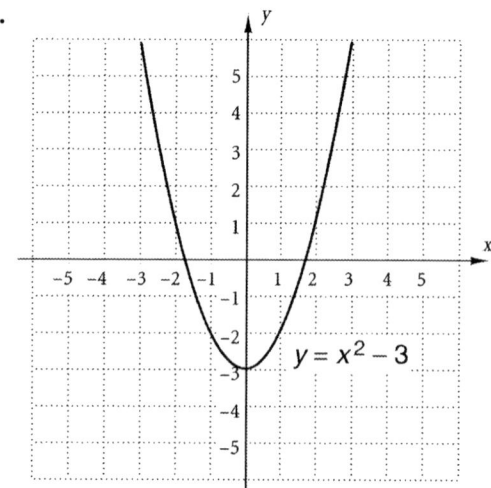

16.

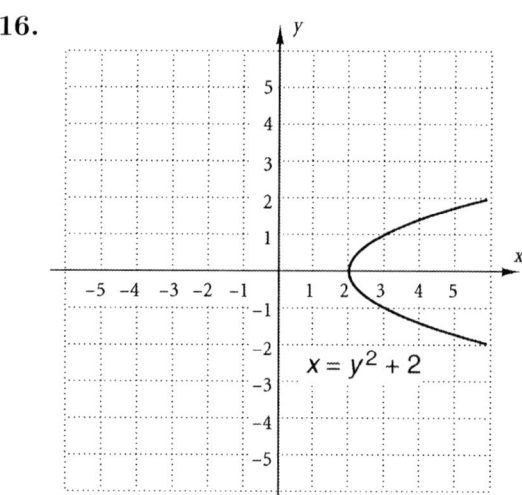

18.

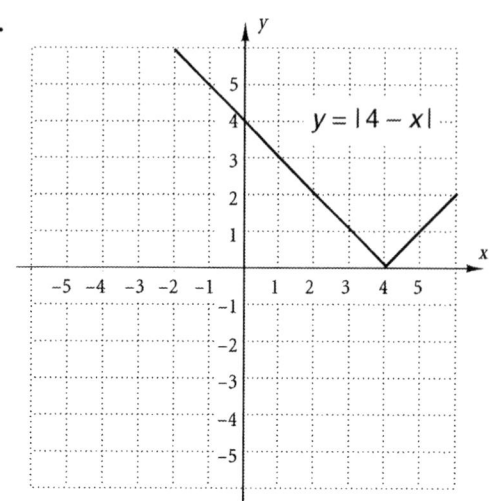

20. (a) $1085; (b) $1086.81; (c) $1087.75; (d) $1088.71; (e) $1088.72

22. $648.60

24. $1186.74

26. (a) 1.8 million, 3.7 million, 4.4 million, 4.5 million; (b) 44 and 67; (c) About 58; (d) tw

28. (a) 1962; (b) 1890; (c) tw

30. $y = 9 - abs(x)$

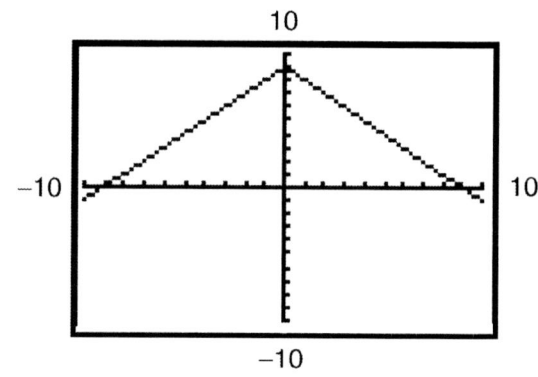

32. $y = \sqrt{23 - 7x}$

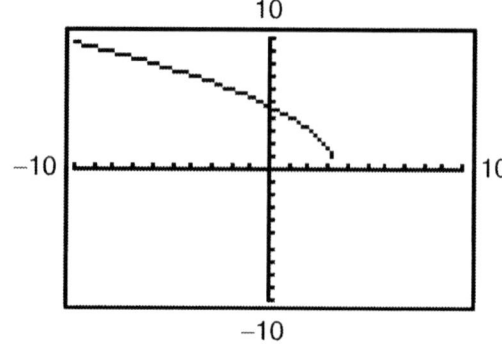

34. $y = -2.3x^2 + 4.8x - 9$

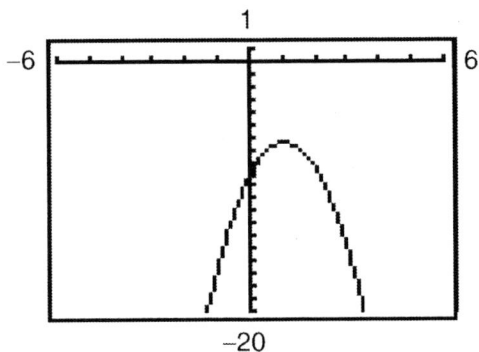

36. $x = 8 - y^2$

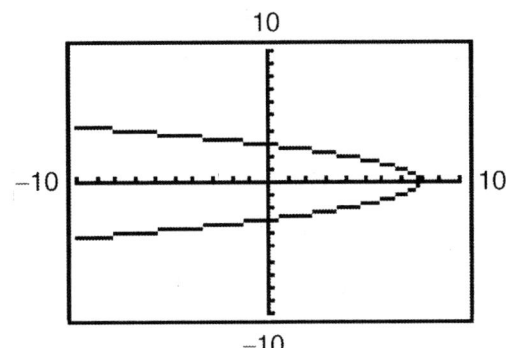

Exercise Set 1.2

2. Yes **4.** Yes **6.** No

8. Yes **10.** Yes **12.** Yes **14.** No

16. (a)

x	$f(x)$
5.1	1.9048
5.01	1.99
5.001	1.999
5	2

(b) 4, does not exist, $-\dfrac{4}{5}$, $\dfrac{4}{k-3}$, $\dfrac{4}{t-2}$, $\dfrac{4}{x+h-3}$

18. 13, 4, 5, 53, $v^2 + 4$, $a^2 + 2ah + h^2 + 4$, $5 - 2t + t^2$

20. (a) 49, 4, 16, $k^2 + 8k + 16$, $t^2 + 6t + 9$, t^2, $x^2 + 2xh + h^2 + 8x + 8h + 16$
(b) This function takes an input, squares it, adds 8 times the input, then adds 16.

22.

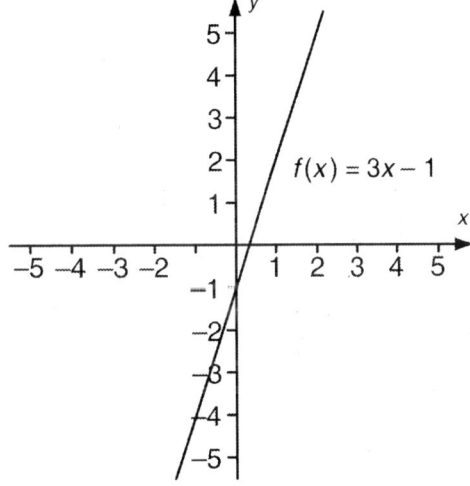

24.

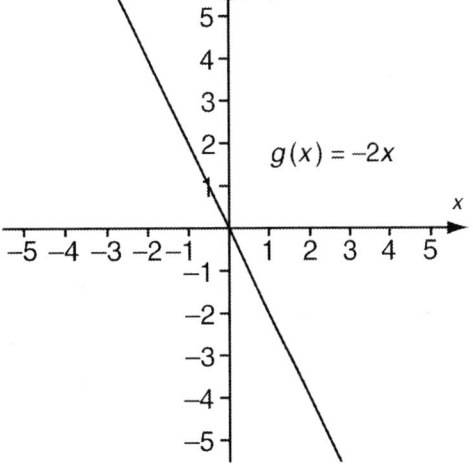

26.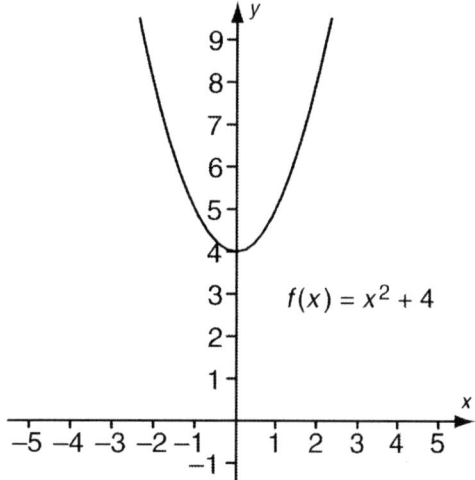

28.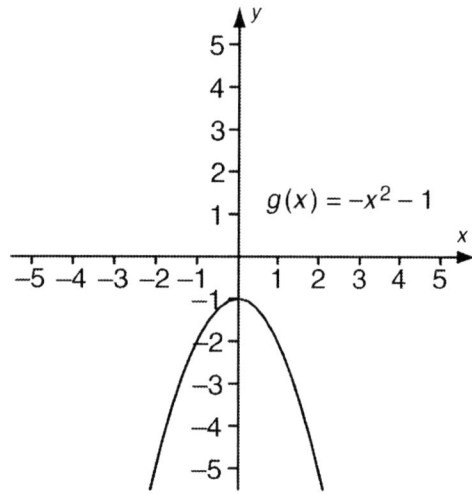

30.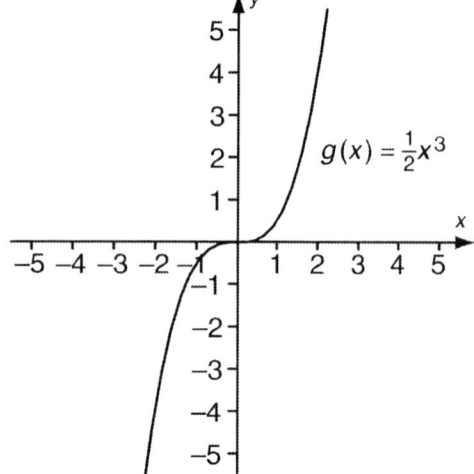

32. Yes **34.** No **36.** Yes

38. Yes **40.** Yes **42.** Yes

44. (a)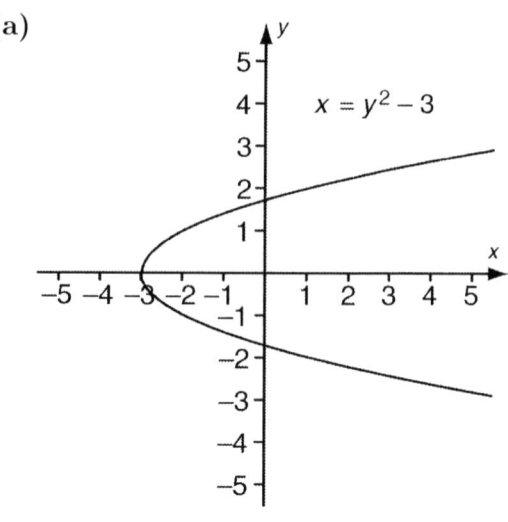

(b) No

46. $x^2 + 2xh + h^2 + 4x + 4h$

48.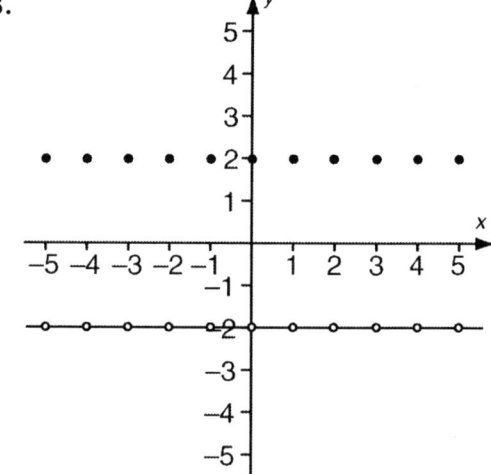

50.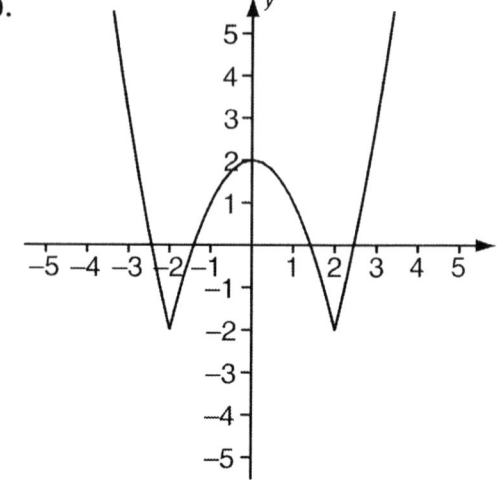

ANSWERS TO EVEN-NUMBERED EXERCISES

52. $106; $1060

54. (a) Yes; (b) $900 billion; (c) 1983; (d) 1997; (e) 1975; (f) 2000

56. (a) 1.708 m²; (b) 1.871 m²

58. $y = 5$; function

60. $y = \pm\sqrt{x}$; not a function

62. tw

64.

X	Y₁
-3	-10
-1	-8
1	-14
3	20
5	142
7	400
9	842

X = -3

66. 2, 0, −2, 2

68. Answers may vary. Some ordered pairs are $(-2.978723, 1.0617)$, $(-1.382979, 2.8438301)$, $(0, 3.1622777)$, $(1.7021277, 2.6651006)$, and $(2.6595745, 1.7107494)$

Exercise Set 1.3

2. $[-1, 2]$ **4.** $(-9, -5]$ **6.** $(x, x + h]$

8. $(-\infty, q]$ **10.** $(-4, 4)$ **12.** $(6, 20]$

14. $(5, \infty)$ **16.** $[-10, 4)$ **18.** $[12.5, \infty)$

20. (a) 3; (b) $\{-4, -3, -2, -1, 0, 1, 2\}$; (c) $-2, 0$; (d) $\{1, 2, 3, 4\}$

22. (a) 2; (b) $\{-6, -4, -2, 0, 1, 3, 4\}$; (c) 1, 3; (d) $\{-5, -2, 0, 2, 5\}$

24. (a) $2\frac{1}{2}$; (b) $\{x| -3 \leq x \leq 5\}$, or $[-3, 5]$; (c) $2\frac{1}{4}$; (d) $\{y|1 \leq y \leq 4\}$, or $[1, 4]$

26. (a) $2\frac{1}{4}$; (b) $\{x| -4 \leq x \leq 3\}$, or $[-4, 3]$; (c) 0; (d) $\{y| -5 \leq y \leq 4\}$, or $[-5, 4]$

28. (a) 2; (b) $\{x| -5 \leq x \leq 4\}$, or $[-5, 4]$; (c) $\{x|1 \leq x \leq 4\}$, or $[1, 4]$; (d) $\{y| -3 \leq y \leq 2\}$, or $[-3, 2]$

30. (a) 2; (b) $\{x| -4 \leq x \leq 4\}$, or $[-4, 4]$; (c) $\{x|0 < x \leq 2\}$, or $(0, 2]$; (d) $\{1, 2, 3, 4\}$

32. $\{x|x \text{ is a real number } and \ x \neq -3\}$, or $(-\infty, -3) \cup (-3, \infty)$

34. All real numbers

36. All real numbers

38. $\left\{x|x \text{ is a real number } and \ x \neq \frac{14}{5}\right\}$, or $\left(-\infty, \frac{14}{5}\right) \cup \left(\frac{14}{5}, \infty\right)$

40. All real numbers

42. $\left\{x|x \text{ is a real number } and \ x \neq \frac{3}{2}\right\}$, or $\left(-\infty, \frac{3}{2}\right) \cup \left(\frac{3}{2}, \infty\right)$

44. $\left\{x|x \leq \frac{2}{3}\right\}$, or $\left(-\infty, \frac{2}{3}\right]$

46. All real numbers

48. All real numbers

50. $\left\{x \mid x \text{ is a real number } and\ x \neq \dfrac{5}{2}\right\}$, or $\left(-\infty, \dfrac{5}{2}\right) \cup \left(\dfrac{5}{2}, \infty\right)$

52. All real numbers

54. $\left\{x \mid x \text{ is a real number } and\ x \neq -\dfrac{5}{4}\right\}$, or $\left(-\infty, -\dfrac{5}{4}\right) \cup \left(-\dfrac{5}{4}, \infty\right)$

56. $\{x \mid x \text{ is an integer}\}$

58. (a) $\{x \mid 0 \leq x \leq 84.7\}$, or $[0, 84.7]$; (b) $\boxed{\text{tw}}$; (c) $\boxed{\text{tw}}$

60. (a) 35; (b) 49, 67; (c) About 72.321 per 100,000 males

62. $\boxed{\text{tw}}$

64. Exercise 32: $(-\infty, 0) \cup (0, \infty)$, Exercise 35: $[2, \infty)$, Exercise 39: $[0, \infty)$, Exercise 40: $[-4, \infty)$, Exercise 47: all real numbers

Exercise Set 1.4

2.

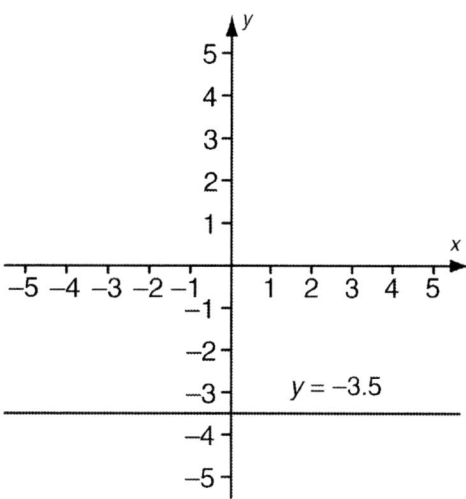

4.

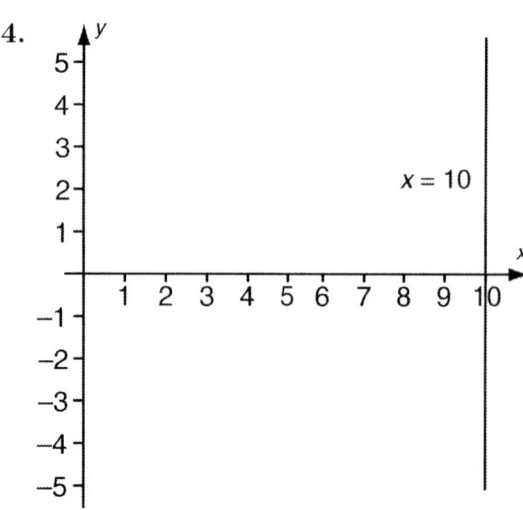

6. Slope: -0.5
y-intercept: $(0, 0)$

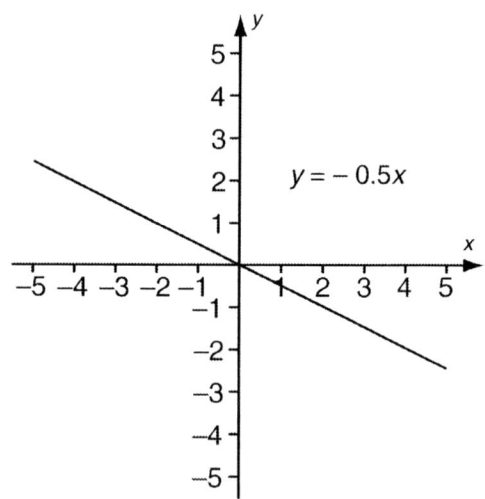

8. Slope: 3
y-intercept: $(0, 0)$

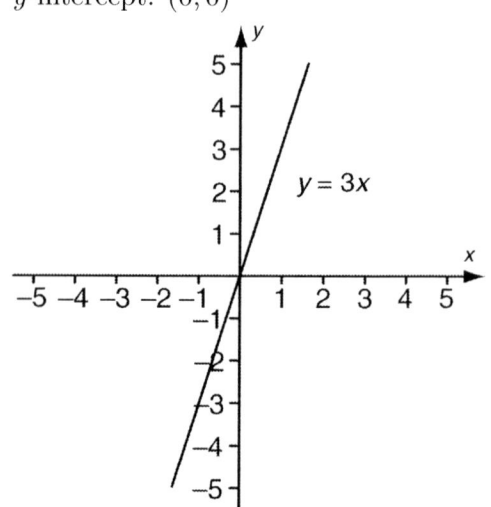

ANSWERS TO EVEN-NUMBERED EXERCISES

10. Slope: -1
y-intercept: $(0, 4)$

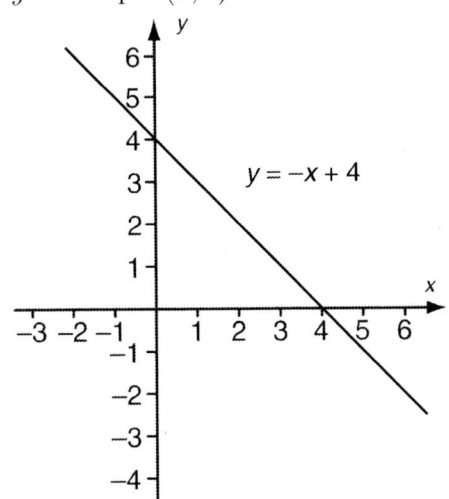

12. Slope: -3
y-intercept: $(0, 2)$

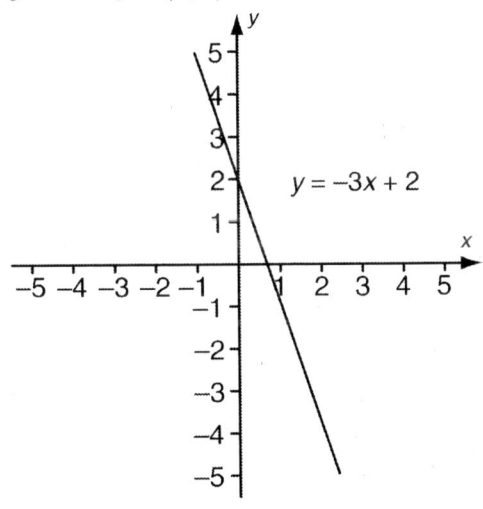

14. $m = 2$, y-int.: $(0, 3)$

16. $m = 1$, y-int.: $(0, 2)$

18. $m = -\dfrac{1}{4}$, y-int.: $\left(0, \dfrac{3}{4}\right)$

20. $y = 7x$

22. $y = -3x + 13$

24. $y = -5x + 25$

26. $y = \dfrac{4}{3}x + 7$ **28.** $y = 8$

30. $\dfrac{1}{4}$ **32.** $-\dfrac{13}{4}$ **34.** No slope

36. 0 **38.** 4 **40.** 3

42. $y = \dfrac{1}{4}x + \dfrac{3}{2}$

44. $y = -\dfrac{13}{4}x - \dfrac{29}{16}$

46. $x = -4$ **48.** $y = \dfrac{1}{2}$

50. $y = 4x$ **52.** $y = 3x - 1$

54. About 31.7%

56. About 91.7%

58. $733 per year

60. (a) $C(x) = 40x + 22{,}500$; (b) $R(x) = 85x$; (c) $P(x) = 45x - 22{,}500$; (d) A profit of $112,500; (e) 500 pairs

62. (a) $V(t) = \$5200 - \$512.50t$; (b) $5200, $4687.50, $4175, $3662.50, $3150, $1612.50, $1100

64. $743,590

66. (a) The number -700 signifies that the value of the photocopier decreases $700 per year; 3500 signifies that the original value of the photocopier was $3500; (b) 5 yr; (c) [tw]

68. About 0.02 sec

70. (a) $M = 0.4W$; (b) 40%W. The weight of the muscles is 40% of the body weight.; (c) 48 lb

72. (a) 0.5 ft, 11.5 ft, 22.5 ft, 55.5 ft, 72 ft;
(b)

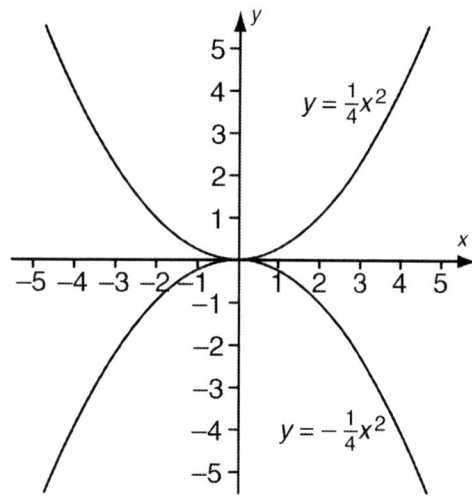

(c) tw

74. (a) $y = 35.5x - 70{,}749.5$; (b) 250 or 251;
(c) tw

76. (a) $N = P + 2\%P = P + 0.02P = 1.02P$;
(b) 204,000; (c) 360,000

78. tw

80. (a) III; (b) IV; (c) I; (d) II

Exercise Set 1.5

2.

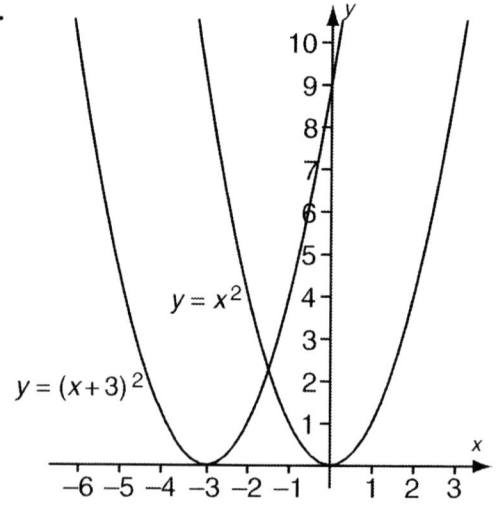

6.

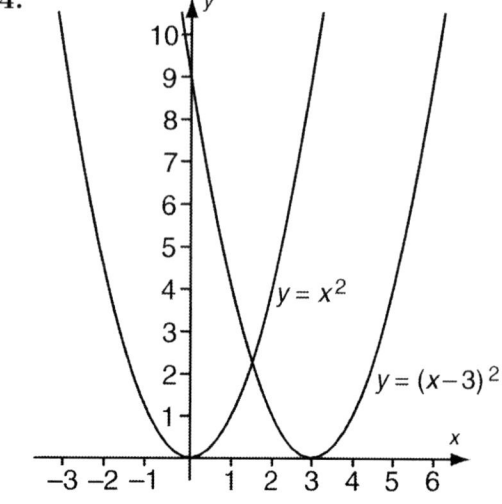

4.

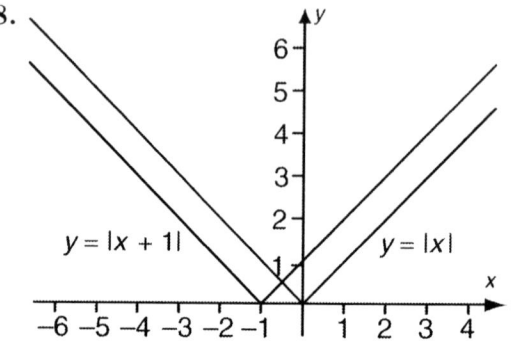

8.

ANSWERS TO EVEN-NUMBERED EXERCISES

10.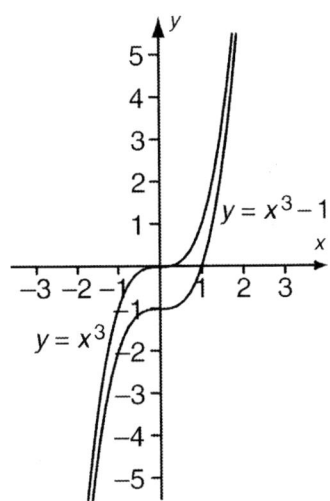

12.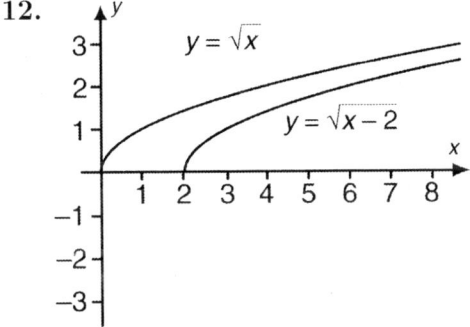

14. No   **16.** Yes; $(1, -3)$

18.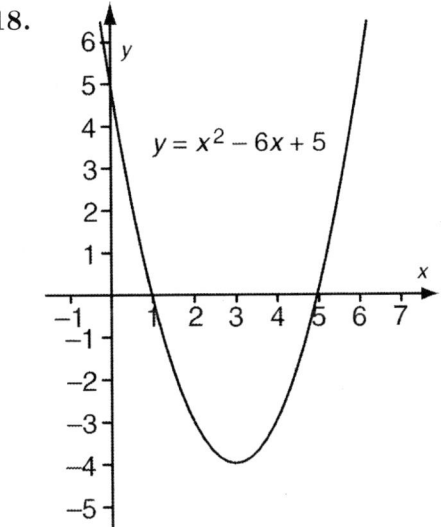

20.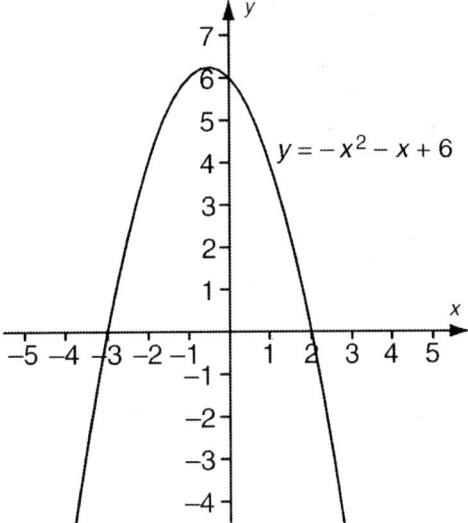

22.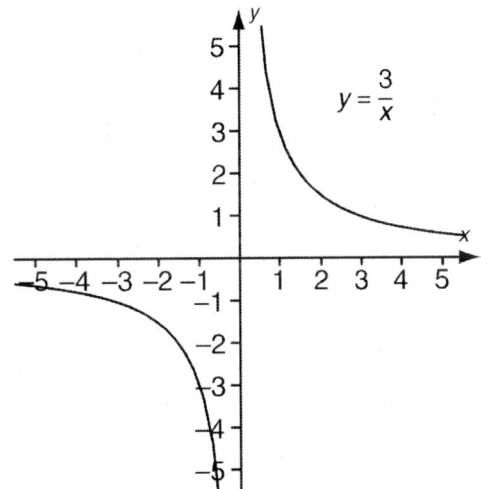

24.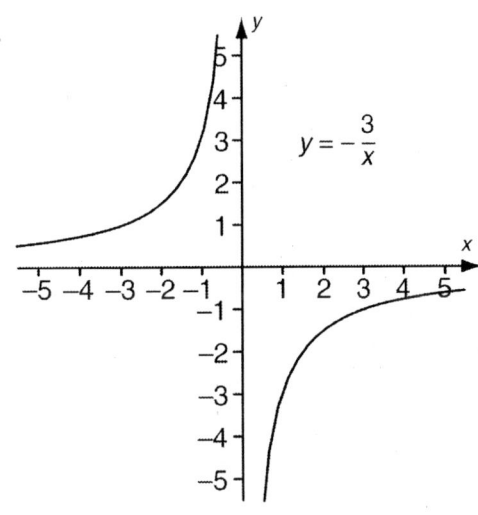

26.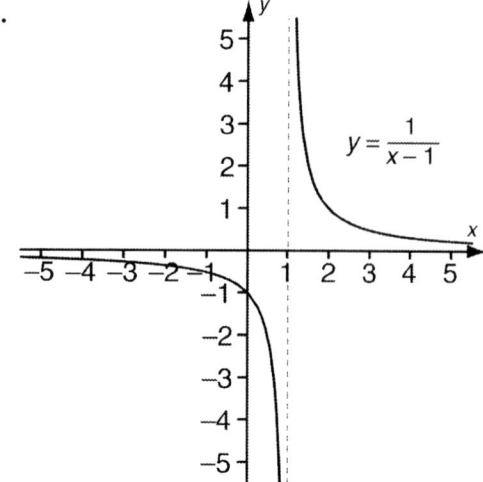

28.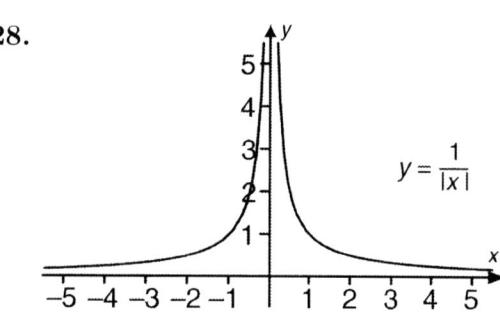

30.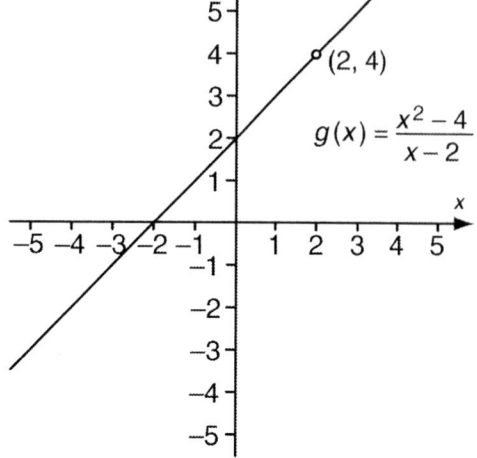

32.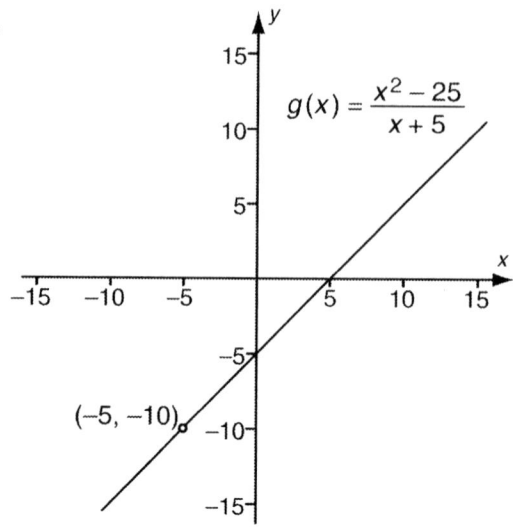

34. $1 \pm \sqrt{5}$ **36.** $-2 \pm \sqrt{7}$ **38.** $\dfrac{-1 \pm \sqrt{2}}{2}$

40. $\dfrac{5 \pm \sqrt{33}}{4}$ **42.** $\dfrac{1 \pm \sqrt{5}}{2}$ **44.** $x^{5/2}$

46. $b^{1/2}$ **48.** $c^{1/8}$ **50.** $b^{-6/5}$

52. $m^{-1/2}$ **54.** $(x^3+4)^{1/2}$ **56.** $\sqrt[7]{t}$

58. $\sqrt[5]{t^2}$ **60.** $\dfrac{1}{\sqrt[3]{y^2}}$ **62.** $\dfrac{1}{\sqrt[5]{b}}$

64. $\dfrac{1}{\sqrt[6]{m^{19}}}$ **66.** $\dfrac{1}{\sqrt[4]{y^2+7}}$ **68.** $\sqrt[5]{w^4}$

70. 1024   **72.** 4   **74.** 3125

76. All real numbers except -2

78. All real numbers except $-5, -1$

80. $[3, \infty)$ **82.** $(\$40, 7600)$

84. $(\$4, 1)$ **86.** $(\$1, 9)$

88. $(\$3, 4)$ **90.** 160,000 fax machines

ANSWERS TO EVEN-NUMBERED EXERCISES

92.

W	0	10	20	30	40	50	100	150
T	0	20	51	86	126	168	417	709

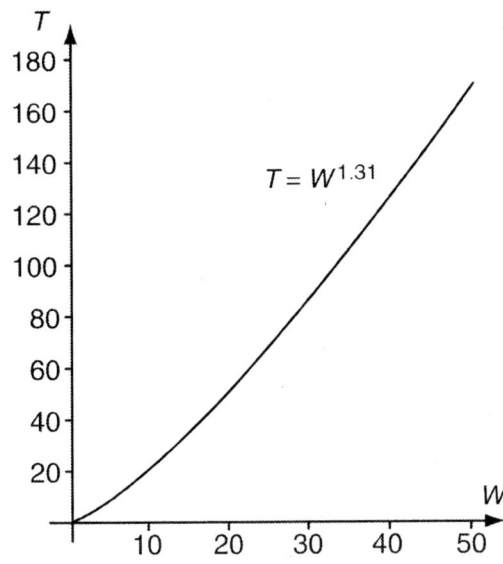

94. (a) 1.932 m^2; (b) 1.8778 m^2;
(c) f(h)

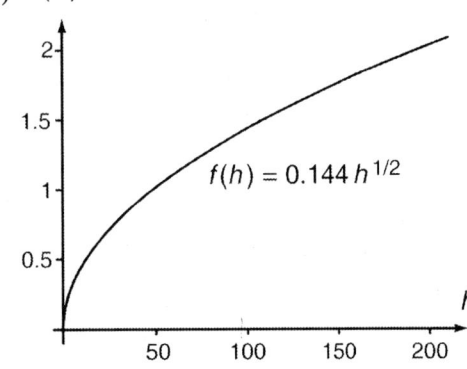

96. tw   **98.** 0, −1, 1

100. −1.831, −0.856, 3.188

102. 1.489, 5.673   **104.** −2, 3

106. All real numbers in the interval $[-1, 2]$

108. ($7.89, 75)

Exercise Set 1.6

2. Quadratic, $a > 0$

4. Polynomial, neither linear nor quadratic

6. Polynomial, neither linear nor quadratic

8. Quadratic, $a < 0$

10. (a) $y = -0.41x + 9.9$;
(b) 7.44%

12. (a) $y = 158.25x^2 - 802x + 2546.75$;
(b) $3431.75 billion; (c) tw

14. (a) $y = 0.05x^2 - 5.5x + 250$;
(b) 100 accidents per 200,000,000 km driven

16. Using the points $(30, 8.6)$ and $(70, 57.3)$
(a) $y = 1.2175x - 27.925$;
(b)

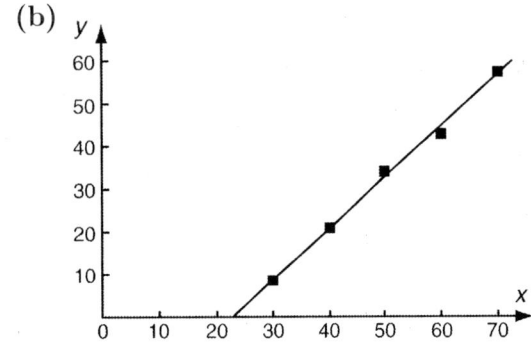

(c) 39.0375%

18. tw

20. (a) $y = 2.5246x + 33.1775$, where x is the number of years after 1995;
(b) $58.42, $71.05; (c) tw

CHAPTER 2

Exercise Set 2.1

2. No **4.** No

6. (a) $-2, -2, -2$; (b) -2; (c) yes; (d) does not exist; (e) -3; (f) no

8. (a) $\frac{1}{3}, \frac{1}{3}, \frac{1}{3}$ (b) $\frac{1}{3}$; (c) yes; (d) does not exist; (e) does not exist; (f) no

10. (a) 1; (b) -1; (c) does not exist; (d) 1; (e) no; (f) yes

12. (a) F; (b) F; (c) F; (d) F; (e) T; (f) F; (g) F; (h) T; (i) T; (j) F

14. (a) T; (b) T; (c) T; (d) T; (e) T

16. Yes, no, yes, no

18. $0.60, $0.83, does not exist

20. Does not exist

22. Yes, no, yes, yes

24. $2.60, $2.60, $2.60

26. $2.90, $3.20, does not exist

28. Does not exist

30. Birth at $t = 0.5, 0.75, 1.5, 1.75$; death at $t = 1.25$

32. 11 deer

34. Birth at $t = 0.1$ (twins), $0.4, 0.5, 0.6$ (twins); death at $t = 0.3, 0.8$

36. 33 bears

38. (a) 4; (b) 4; (c) 4; (d) 4; (e) 4; (f) no; (g) yes

40. $\boxed{\text{tw}}$

Exercise Set 2.2

2. 5 **4.** Does not exist **6.** -7

8. -8 **10.** $\frac{2}{5}$ **12.** -9 **14.** 3

16. 19 **18.** ∞ **20.** $-\frac{4}{3}$ **21.** $\frac{8}{7}$

24. $10x$ **26.** $\frac{-5}{x^2}$

28. (a) 11; (b) 11; (c) 11; (d) 9; (e) no, yes

30. 0 **32.** -2 **34.** 1

36. -3 **38.** 0.5 **40.** 0.5

42. 0.378 or $\frac{1}{\sqrt{7}}$ **44.** 0

Exercise Set 2.3

2. (a) $5(2x + h)$; (b) 50, 45, 40.5, 40.05

4. (a) $-5(2x + h)$; (b) $-50, -45, -40.5, -40.05$

6. (a) $5(3x^2 + 3xh + h^2)$; (b) 380, 305, 246.05, 240.6005

8. (a) $\dfrac{-4}{x(x+h)}$; (b) $-0.1667, -0.2, -0.2439, -0.2494$

ANSWERS TO EVEN-NUMBERED EXERCISES

10. (a) 2; (b) All 2

12. (a) $2x + h + 1$; (b) 11, 10, 9.1, 9.01

14. (a) $2x + h$; (b) 10.9, 8.1, 8.01

16. (a) 300 units/thousand dollars, 180 units/thousand dollars, 120 units/thousand dollars, 100 units/thousand dollars; (b) tw

18. (a) $10.09; (b) $14.99; (c) $4.90; (d) $0.98 per year

20. (a) $10,519.95; (b) $10,500; (c) $19.95; (d) $19.95; (e) tw

22. (a) $1.\overline{3}$ lb/mo; (b) 1.5 lb/mo; (c) $1.\overline{6}$ lb/mo; (d) tw

24. (a) 40 miles, 250 miles; (b) 210 miles, this represents the distance traveled in the 3 hr period from $t = 2$ to $t = 5$; (c) 70 miles/hr

26. (a) 14.5 miles/gallon; (b) 0.069 gal/mi

28. tw   **30.** m

32. $3ax^2 + 3axh + ah^2 + 2bx + bh$

34. $4x^3 + 6x^2h + 4xh^2 + h^3$

36. $\dfrac{1}{(1-x)\left[1-(x+h)\right]}$

38. $\dfrac{-2}{\sqrt{3-2x-2h} + \sqrt{3-2x}}$

Exercise Set 2.4

2. (a), (b)

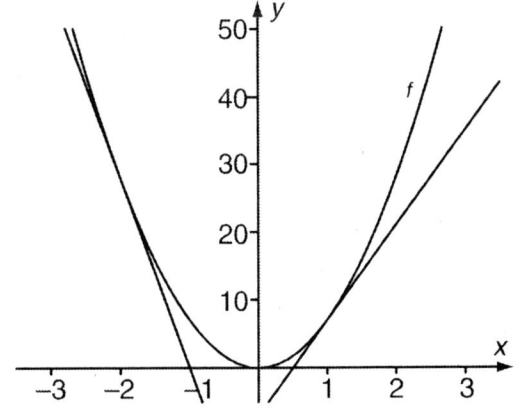

(c) $f'(x) = 14x$
(d) -28, 0, 14

4. (a), (b)

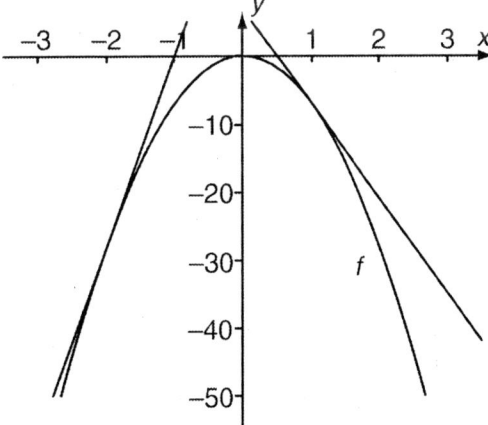

(c) $f'(x) = -14x$
(d) 28, 0, -14

296 ANSWERS TO EVEN-NUMBERED EXERCISES

6. (a), (b)

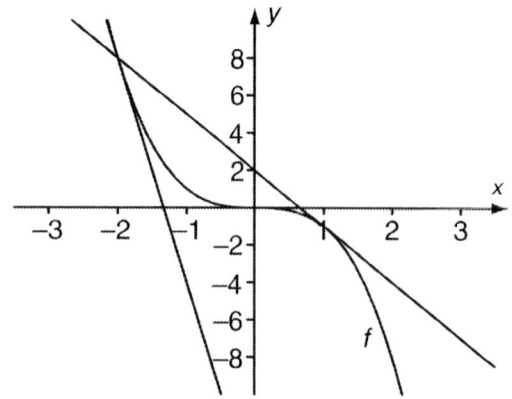

(c) $f'(x) = -3x^2$; (d) $-12, 0, -3$

8. (a), (b) All the tangent lines are identical to the graph of the original function.

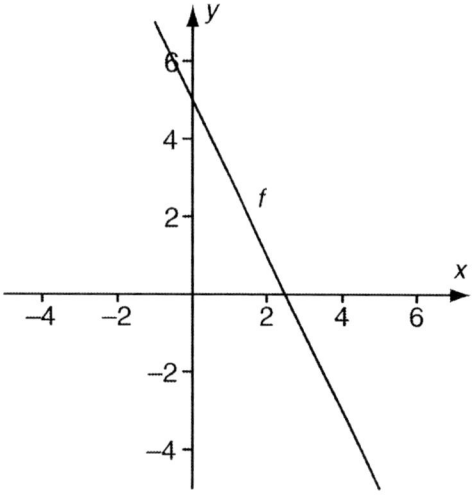

(c) $f'(x) = -2$; (d) $-2, -2, -2$

10. (a), (b) All the tangent lines are identical to the graph of the original function.

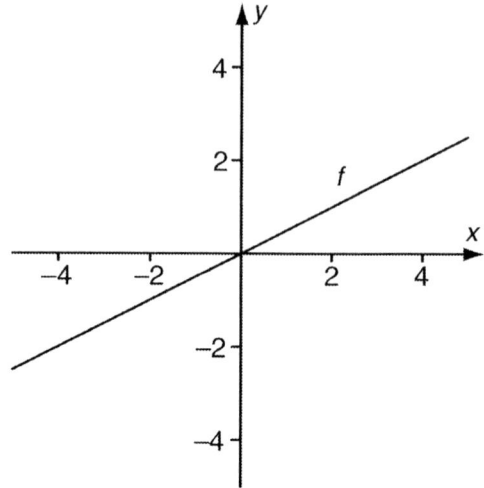

(c) $f'(x) = \dfrac{1}{2}$; (d) $\dfrac{1}{2}, \dfrac{1}{2}, \dfrac{1}{2}$

12. (a), (b)

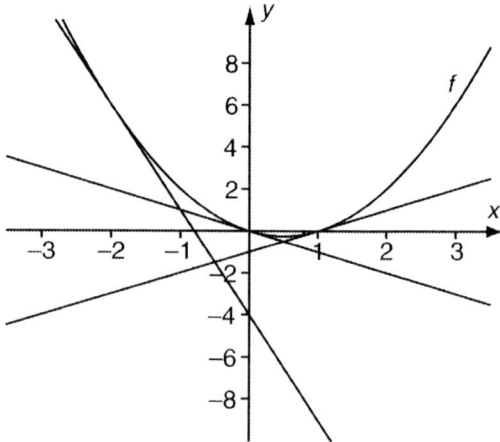

(c) $f'(x) = 2x - 1$; (d) $-5, -1, 1$

ANSWERS TO EVEN-NUMBERED EXERCISES

14. (a), (b)

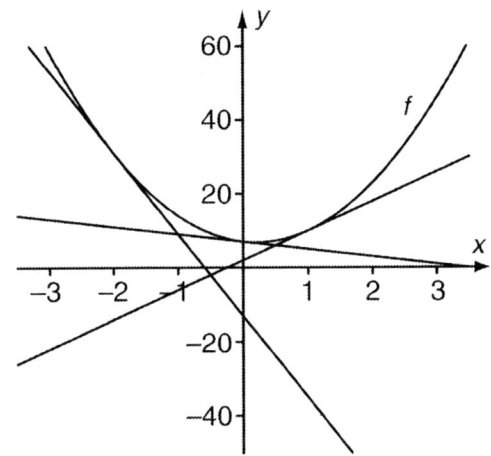

(c) $f'(x) = 10x - 2$; (d) $-22, -2, 8$

16. (a), (b)

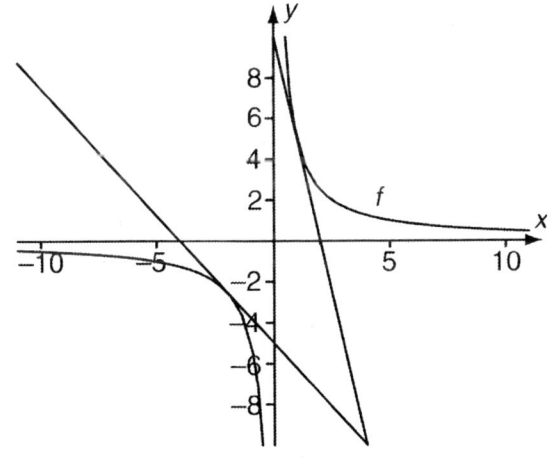

(c) $f'(x) = \dfrac{-5}{x^2}$
(d) -1.25, does not exist, -5

18. $f'(x) = 2ax + b$

20. $y = 12x + 16$; $y = 0$, $y = 48x - 128$

22. $y = -2x - 4$; $y = -\dfrac{1}{2}x + 2$; $y = -\dfrac{1}{50}x + \dfrac{2}{5}$

24. $y = -6x - 4$; $y = -1$; $y = 6x - 16$

26. $x_2, x_4, x_5, x_6, x_7, x_8$

28. $0, 0.2, 0.4, 0.6$, and so on   **30.** tw

32. $4x^3$ **34.** $\dfrac{1}{(1-x)^2}$

36. $\dfrac{1}{2\sqrt{x}}$ **38.** $\dfrac{-2}{(x-1)^2}$

40.–44. Left to the student.

46. (a)

(b) $V = 21.545t + 3.455$;
(c) \$21.545 million per year;
(d) $V = 1.666667t + 23.33333$,
about \$1.67 million per year;
$V = -8.169873t + 33.169873$,
about $-\$8.17$ million per year;
$V = -4.002551t + 29.002551$,
about $-\$4$ million per year; (e) 0

Exercise Set 2.5

2. $8x^7$ **4.** 0 **6.** $1400x^{199}$

8. $4x^3 - 7$ **10.** $\dfrac{2}{x^{1/2}}$, or $\dfrac{2}{\sqrt{x}}$

12. $\dfrac{0.78}{x^{0.22}}$ **14.** $\dfrac{-1.6}{\sqrt[3]{x^2}}$ **16.** $\dfrac{-4}{x^5}$

18. $8x - 7$ **20.** $\dfrac{1}{5\sqrt[5]{x^4}} + \dfrac{2}{x^2}$

22. $\dfrac{4\sqrt[3]{x}}{3}$ **24.** $4x^{11.5}$

26. $\dfrac{-4}{x^2} - 1$ **28.** 7 **30.** 7

32. x^2 **34.** $-0.02x + 0.4$

36. $-\dfrac{3}{4}x^{-7/4} - 2x^{-1/3} + \dfrac{5}{4}x^{1/4} - 8x^{-5}$

38. $\dfrac{1}{5} - \dfrac{5}{x^2}$

40. $-100x^{-6} - 3x^{-4} + 2x^{-2}$

42. $3x^2 - 6x + 2 + 8x^{-2} - 8x^{-3}$

44. $\dfrac{1}{2\sqrt{x}}, \dfrac{1}{4}$ **46.** $1 - \dfrac{1}{x^2}, 0$

48. $y = \dfrac{3}{2}x - \dfrac{3}{2};\ y = \dfrac{31}{4}x - 17;$
$y = \dfrac{107}{6}x - \dfrac{165}{2}$

50. $(0,0)$ **52.** $(0,0)$

54. $\left(\dfrac{3}{10}, \dfrac{151}{20}\right)$, or $(0.3, 7.55)$

56. $(20, 54)$ **58.** There are none.

60. The tangent is horizontal at all points on graph.

62. $\left(1, -\dfrac{166}{3}\right), \left(11, \dfrac{334}{3}\right)$

64. $\left(\sqrt{2}, 1 - 4\sqrt{2}\right), \left(-\sqrt{2}, 1 + 4\sqrt{2}\right)$

66. $(3, 0)$ **68.** $\left(\dfrac{5}{2}, \dfrac{35}{4}\right)$ **70.** $(50, 75)$

72. $\left(1 + \sqrt{6}, -\dfrac{11}{3} - 3\sqrt{6}\right),$
$\left(1 - \sqrt{6}, -\dfrac{11}{3} + 3\sqrt{6}\right)$

74. $(0, -2), \left(\sqrt{\dfrac{1}{3}}, -\dfrac{55}{27}\right), \left(-\sqrt{\dfrac{1}{3}}, -\dfrac{55}{27}\right)$

76. $2x$ **78.** $\dfrac{5}{4}x - 1$ **80.** $50x$

82. $\dfrac{\sqrt{7}}{2\sqrt{x}}$ **84.** $2x - 6$

86. $1 + \dfrac{5}{3\sqrt[6]{x}} + \dfrac{2}{3\sqrt[3]{x}}$

88. $4x^3 - 15x^2 + 12x + 4$ **90.** $\boxed{\text{tw}}$

92. $(-1.225, -1.25), (0, 1), (1.225, -1.25)$

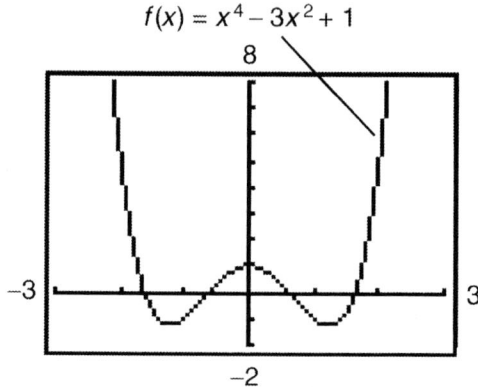

94. $(0, 0), (0.507, -0.225)$

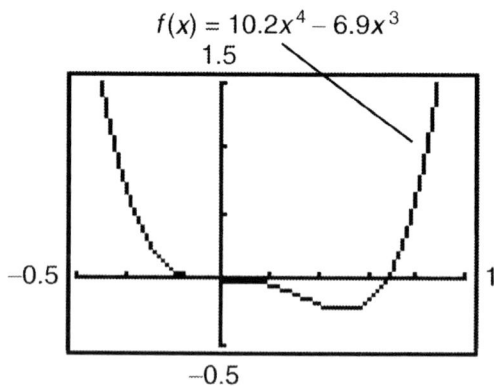

96.

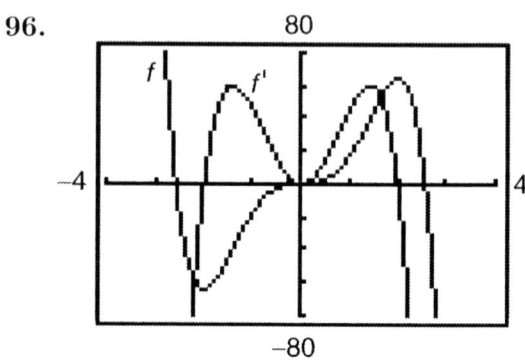

98.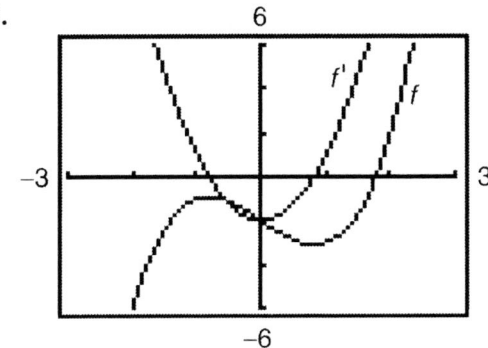

100.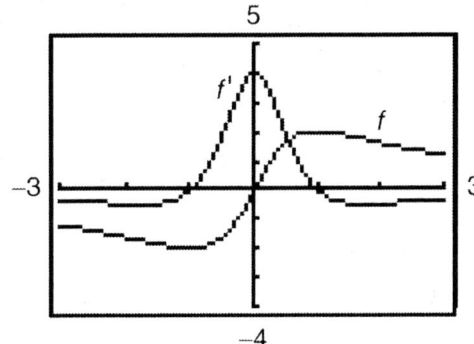

Exercise Set 2.6

2. (a) $v(t) = 3$; (b) $a(t) = 0$;
(c) $v(2) = 3$ mi/hr; $a(2) = 0$ mi/hr^2;
(d) tw

4. (a) $v(t) = 2t - \frac{1}{2}$; (b) $a(t) = 2$;
(c) $v(1) = \frac{3}{2}$ m/sec, $a(1) = 2$ m/sec^2

6. 497,500 computers

8. (a) $P(x) = -0.5x^2 + 46x - 10$;
(b) $R(20) = \$800$, $C(20) = \$90$,
$P(20) = \$710$; (c) $R'(x) = 50 - x$,
$C'(x) = 4$, $P'(x) = 46 - x$;
(d) $R'(20) = \$30$ per unit,
$C'(20) = \$4$ per unit,
$P'(20) = \$26$ per unit

10. (a) \$342,000, \$528,000, \$472,000, \$358,000, \$4,560,000;
(b) $S'(t) = 6t^2 - 80t + 220$;
(c) \$146,000 per month, $-\$4000$ per month, $-\$44,000$ per month, $-\$14,000$ per month, \$1,020,000 per month; (d) tw

12. (a) $\frac{dS}{dp} = 0.24p^2 + 4p + 10$;
(b) about 61 units; (c) 24.16 units per dollar; (d) tw

14. (a) $p'(x) = 0.18x - 0.19$;
(b) \$16.51; (c) \$1.61 per year; (d) positive; ticket prices increase every year

16. (a) \$50,784;
(b) \$812.48; (c) \$811.20; (d) \$53,217.60

18. (a) $D'(F) = 2$; (b) tw

20. (a) $A'(r) = 2\pi r$; (b) tw

22. (a) $B'(x) = 0.1x - 0.9x^2$; (b) tw

24. (a) $\frac{dH}{dW} = 1.41W^{0.41}$; (b) tw

26. (a) $P'(t) = 4000t$; (b) 300,000 people;
(c) 40,000 people/yr; (d) tw

28. (a) $\frac{dV}{dh} = 0.61/\sqrt{h}$; (b) 244 mi;
(c) 0.00305 mi/ft; (d) tw

30. tw

32. $y_1 = 5x$, $y_2 = 0.001x^2 + 1.2x + 60$,
$y_3 = 5$, $y_4 = 0.002x + 1.2$

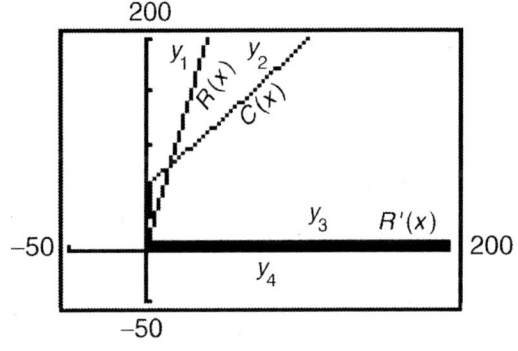

34. $y_1 = 50x - 0.5x^2$, $y_2 = 10x + 3$, $y_3 = 50 - x$, $y_4 = 10$

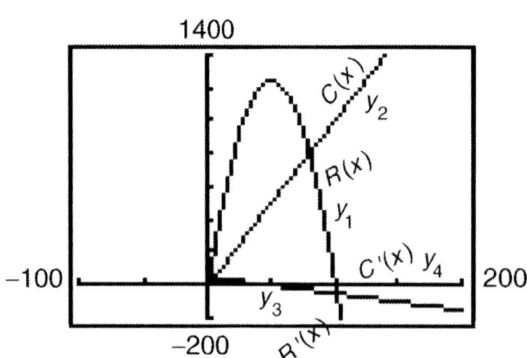

Exercise Set 2.7

2. $13x^{12}$ **4.** $\dfrac{-1}{x^2}$ **6.** $4x^3$

8. $(7x^6 + 4x^3 - 50)\left(90x^9 - \dfrac{7}{2\sqrt{x}}\right) + (42x^5 + 12x^2)(9x^{10} - 7\sqrt{x})$ **10.** $400 - 2x$ **12.** 1

14. $(\sqrt[3]{x} - 5x^2 + 4)(8x + 11) + \left(\dfrac{1}{3}x^{-2/3} - 10x\right)(4x^2 + 11x - 5)$ **16.** $50x - 40$

18. $2(3x^2 - 4x + 5)(6x - 4)$

20. $-36x^{-2} - 120x^{-3} + 144x^{-4} - 72x^{-5}$

22. $16x^3 - 3x^2 - 40x + 5$

24. $\dfrac{400}{(400 - x)^2}$ **26.** $\dfrac{-13}{(x - 5)^2}$

28. $\dfrac{x^4 + 3x^2 + 2x}{(x^2 + 1)^2}$ **30.** $\dfrac{3}{(3 - x)^2}$

32. $\dfrac{-4}{(x - 2)^2}$ **34.** $\dfrac{-2x}{(x^2 + 1)^2}$

36. $\dfrac{-1}{(x + 2)^2}$ **38.** $\dfrac{5x^2 - 6x + 5}{(x^2 - 1)^2}$

40. $\dfrac{-9x - 8}{x^5}$, or $-9x^{-4} - 8x^{-5}$

42. $\dfrac{6x - 1}{2\sqrt{x}}$

44.–84. Left to the student.

86. (a) $\dfrac{-2x}{(x^2 - 1)^2}$; (b) $\dfrac{-2x}{(x^2 + 1)^2}$; (c) $\boxed{\text{tw}}$

88. $y = \dfrac{1}{2}$, $y = \dfrac{12}{25}x + \dfrac{7}{25}$

90. $y = 4x$; $y = -2$

92. (a) $R'(x) = \dfrac{240{,}000x^{-1/2} - 120{,}000x}{(4 + x^{3/2})^2}$;
(b) -2500

94. (a) $R(p) = 100p - p^{3/2}$;
(b) $R'(p) = 100 - \dfrac{3}{2}\sqrt{p}$

96. (a) $R(p) = 500p - p^2$;
(b) $R'(p) = 500 - 2p$

ANSWERS TO EVEN-NUMBERED EXERCISES

98. (a) $R(p) = 3000 + 5p$;
(b) $R'(p) = 5$

100. $D'(p) = \dfrac{pR'(p) - R(p)}{p^2}$

102. (a) $P'(t) = \dfrac{4500 - 1000t^2}{(2t^2 + 9)^2}$;
(b) 20,202; (c) -1581 people/mo

104. $\dfrac{3 - 10t^4\sqrt{t} - 9t^5}{2\sqrt{t}\,(t^5 + 3)^2}$

106. $\dfrac{2(1 - z^2)}{(1 - z + z^2)^2}$

108. $\dfrac{-10x^2 - 18x + 6}{3x^4\left(\dfrac{3}{x^2} + 5\right)^2}$

110. $30t^2 + 10t - 15$

112. $\dfrac{3x^2(x^2 + 1)}{x^2 - 1} + \dfrac{(-4x)(x^3 - 8)}{(x^2 - 1)^2}$

114. $\dfrac{(x^4 - 3x^3 - 5)[(2x + 3)(x^5 - 7x^2 - 3) + (x^2 + 3x)(5x^4 - 14x)] - (4x^3 - 9x^2)(x^2 + 3x)(x^5 - 7x^2 - 3)}{(x^4 - 3x^3 - 5)^2}$

116. $\dfrac{(t^5 - 3)(t^4 + 7)(40t^7 - 6t^2) - [5t^4(t^4 + 7) + 4t^3(t^5 - 3)](5t^8 - 2t^3)}{[(t^5 - 3)(t^4 + 7)]^2}$

118. $(-1.41421, -4)$, $(0,0)$, $(1.41421, -4)$;

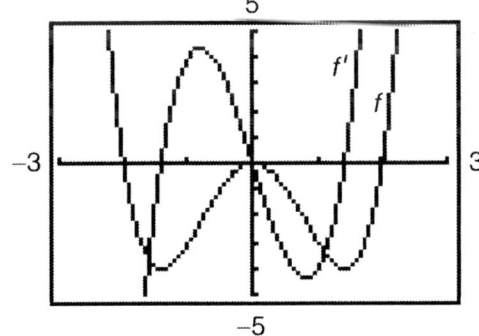

122. $(0, 0)$, $(-0.4, 0.03125)$, $(0.4, 0.03125)$

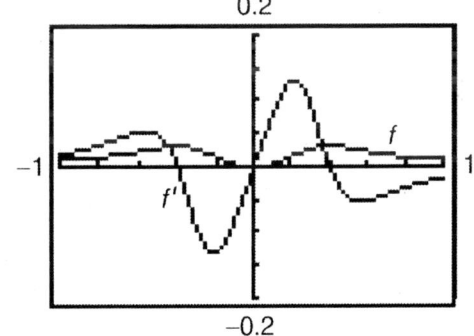

120. $(0, -1)$, $(-0.59607, -0.89411)$;

124. y_3

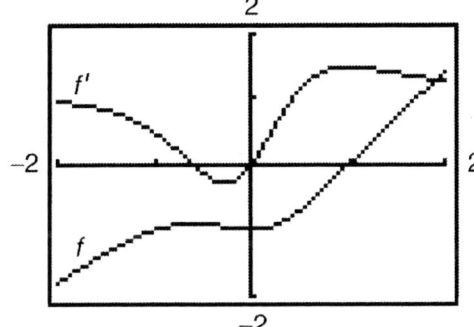

Exercise Set 2.8

2. $8x - 12$ **4.** $-100(1-x)^{99}$

6. $\dfrac{-1}{2\sqrt{1-x}}$ **8.** $\dfrac{4x}{\sqrt{4x^2+1}}$

10. $-400x(4x^2+1)^{-51}$ **12.** $\dfrac{6x-7}{\sqrt{4x-7}}$

14. $\dfrac{7x^3+6x^2}{2\sqrt{x+1}}$ **16.** $\dfrac{-8}{(4x+5)^3}$

18. $3x^2(1+x^3)^3(1+5x^3)$

20. $4x - 200$ **22.** $\dfrac{2}{\sqrt[3]{(2x-1)^2}} + 2x - 8$

24. $(x-4)^7(x+3)^8(17x-12)$

26. $(x+6)^9(5-x)^8(-19x-4)$

28. $-5(3x+5)^5(21x+5)$

30. $\dfrac{1}{2}\left(\dfrac{3+x}{2-x}\right)^{-1/2} \cdot \dfrac{5}{(2-x)^2}$

32. $\left(\dfrac{x}{x^2+1}\right)^2 \cdot \dfrac{3(1-x^2)}{(x^2+1)^2}$

34. $\dfrac{1}{2}(x^2+5x)^{-1/2}(2x+5)$

36. $204(7x^4+6x^3-x)^{203}(28x^3+18x^2-1)$

38. $\left(\dfrac{1-3x}{2-7x}\right)^{-6} \cdot \dfrac{-5}{(2-7x)^2}$

40. $\dfrac{1}{3}\left(\dfrac{4-x^3}{1-x^2}\right)^{-2/3} \cdot \dfrac{x^4-3x^2+8x}{(1-x^2)^2}$

42. $\dfrac{(5x-4)^6(120x+107)}{(6x+1)^4}$

44. $2(x^2+x)^{-2/3}(2x+1)(x^4-6x)^3 + 18\sqrt[3]{x^2+x}(x^4-6x)^2(4x^3-6)$

46. $-45u^{-4}$, 2, $\dfrac{-90}{(2x+1)^4}$

48. $\dfrac{-2}{(u-1)^2}$, $\dfrac{1}{2}x^{-1/2}$, $-x^{-3/2}$

50. $2u$, $3x^2$, $6x^2(x^3+1)$

52. $y = 0$ **54.** $y = -735x + 1813$

56. (a) $6x^5 + 40x^3 + 50x$; (b) $6x^5 + 40x^3 + 50x$; (c) same

58. $8x^2 - 17$, $32x^2 + 48x + 13$

60. $\dfrac{3}{2x^2+3}$, $\dfrac{18}{x^2} + 3$

62. $\dfrac{1}{(x+2)^2}$, $\dfrac{1}{x^2} + 2$

64. $f(x) = \dfrac{1}{\sqrt{x}}$, $g(x) = 7x + 2$

66. $f(x) = x^4$, $g(x) = \sqrt{x} + 5$

68. $-1/2$

70. $\dfrac{-70 \cdot 13^4}{6^6} \approx -42.8513$

72. (a) $R'(x) = \dfrac{2000x}{\sqrt{x^2+3}}$, $R'(20) = \dfrac{40{,}000}{\sqrt{403}} \approx 1992.54$; (b) tw

74. (a) $\dfrac{5x}{\sqrt{60+5x^2}}$; (b) $\dfrac{100(20t+40)}{\sqrt{5(20t+40)^2+60}}$; (c) tw ; (d) \$467.10 per month

76. (a) $u'(x) = \left(\dfrac{2x+1}{3x+4}\right)^{-1/2} \cdot \dfrac{200}{(3x+4)^2}$; (b) tw

ANSWERS TO EVEN-NUMBERED EXERCISES

78. (a) $\dfrac{dA}{dt} = \$5000\left(1+\dfrac{i}{4}\right)^{19}$; (b) [tw]

80. (a) $P(t) = 2t^2 + 400.8t + 340.08$;
(b) $P'(t) = 4t + 400.8$;
(c) $\$592.80$ per month

82. $\dfrac{4x-6}{\sqrt{(2x-3)^2+1}}$

84. $\dfrac{2t^3+3t}{2\sqrt[4]{(t^4+3t^2+8)^3}}$

86. $\dfrac{2(x+1)(-2x^2-3x+1)}{(x^2+1)^4}$

88. $\dfrac{2x^2+1}{\sqrt{1+x^2}}$

90. $\dfrac{1}{(1+u^2)^{3/2}}$ **92.** $\dfrac{1}{4\sqrt{x+x\sqrt{x}}}$

94. $\dfrac{2x+4x^3}{3(x^2+x^4)^{2/3}}$ **96.** [tw]

98. $(-1.4748, 9.48776)$,

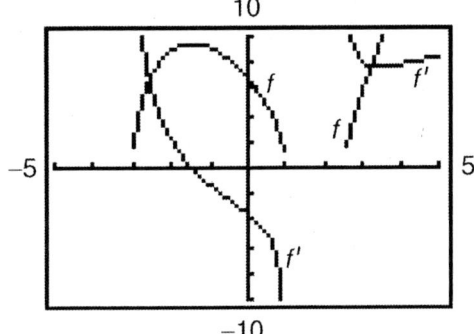

100.

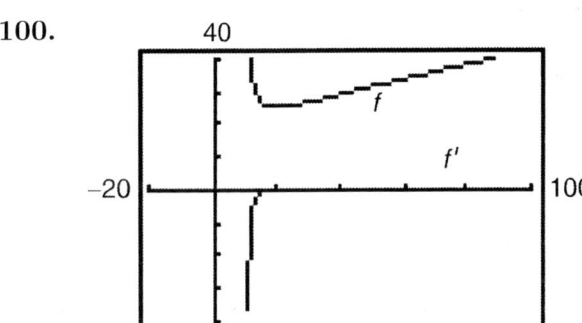

Exercise Set 2.9

2. 0   **4.** $-\dfrac{6}{x^3}$ **6.** $-\dfrac{1}{4x^{3/2}}$

8. $6x - \dfrac{6}{x^3}$ **10.** $\dfrac{20}{x^6}$

12. $n(n+1)\dfrac{1}{x^{n+2}}$ **14.** $12x^2 + 6x$

16. $\dfrac{-1}{4\sqrt{(x+1)^3}}$ **18.** $-\dfrac{b^2}{4}(a-bx)^{-3/2}$

20. $\dfrac{3}{2(x^2+1)^{1/4}} - \dfrac{3x^2}{4(x^2+1)^{5/4}}$

22. $60(x^3+15x)^{18}\left[59x^4 + 600x^2 + 1425\right]$

24. $12(x^2 - 3x + 1)^{10}(46x^2 - 138x + 101)$

26. $\dfrac{5}{16}x^{-3/4} + \dfrac{4}{25}x^{-6/5}$

28. $\dfrac{1}{2}x^3 - \dfrac{1}{6}x$

30. $120x^{-7} - \dfrac{140}{3}x^{-6}$

32. $120x$ **34.** $5040x$

36. $k(k-1)(k-2)(k-3)(k-4)x^{k-5}$

38. $12t^2 + 2$ **40.** $200{,}000$

42. $y' = \dfrac{1}{(1-x)^2}$, $y'' = \dfrac{2}{(1-x)^3}$, $y''' = \dfrac{6}{(1-x)^4}$

44. $y' = 15x^4 + \dfrac{4}{\sqrt{x}}$, $y'' = 60x^3 - \dfrac{2}{\sqrt{x^3}}$, $y''' = 180x^2 + \dfrac{3}{\sqrt{x^5}}$

46. $y' = -\dfrac{1}{2}(x-1)^{-3/2}$, $y'' = \dfrac{3}{4}(x-1)^{-5/2}$, $y''' = -\dfrac{15}{8}(x-1)^{-7/2}$

48. $y' = \dfrac{1}{\sqrt{x}\left(1+\sqrt{x}\right)^2}$, $y'' = \dfrac{-(1+3\sqrt{x})}{2x^{3/2}\left(1+\sqrt{x}\right)^3}$, $y''' = \dfrac{3\left(1+4\sqrt{x}+5x\right)}{4x^{5/2}\left(1+\sqrt{x}\right)^4}$

50. $\dfrac{6x^2-2}{(1+x^2)^3}$

52. $f'(x) = \dfrac{-5}{(x-2)^2}$, $f''(x) = \dfrac{10}{(x-2)^3}$, $f'''(x) = \dfrac{-30}{(x-2)^4}$, $f^{(4)}(x) = \dfrac{120}{(x-2)^5}$, $f^{(5)}(x) = \dfrac{-600}{(x-2)^6}$

54.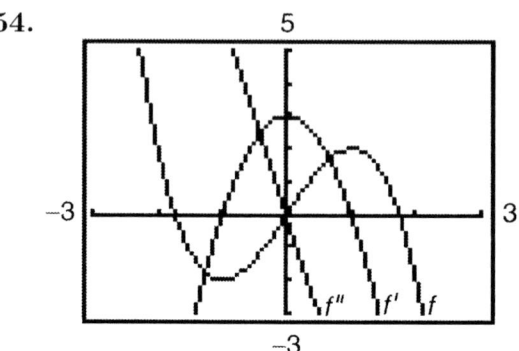

56.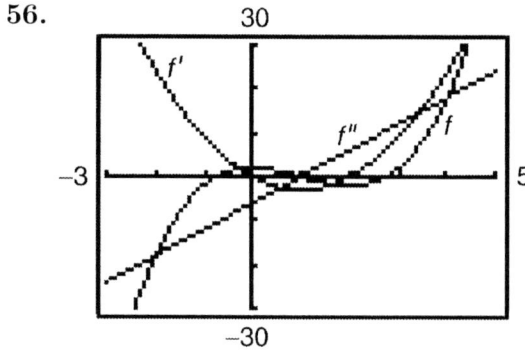

CHAPTER 3

Exercise Set 3.1

2. Relative minimum at $(3, -12)$.

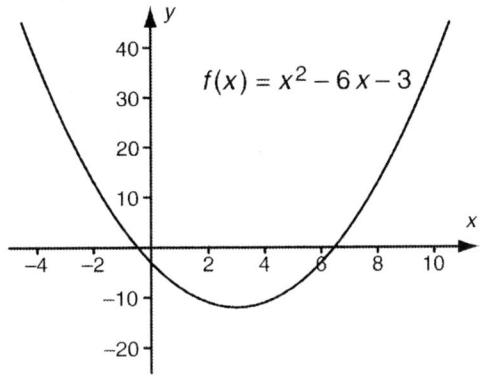

4. Relative maximum at $\left(-\frac{3}{4}, \frac{25}{8}\right)$.

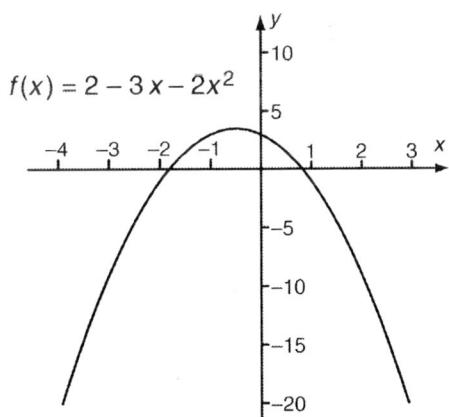

6. Relative minimum at $(2, -13)$.

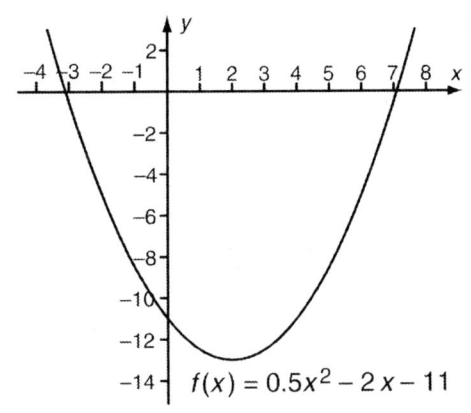

8. Relative maximum at $\left(-1, \frac{13}{2}\right)$; relative minimum at $\left(\frac{2}{3}, \frac{113}{27}\right)$.

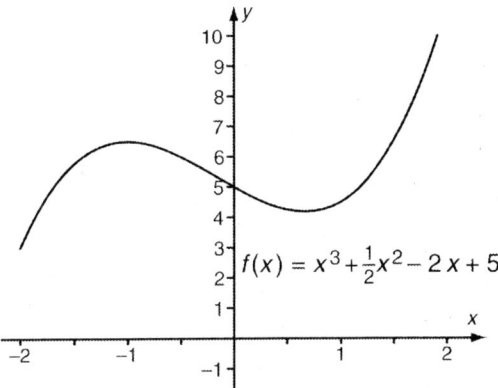

10. Relative maximum at $(0, 0)$; relative minimum at $(2, -4)$.

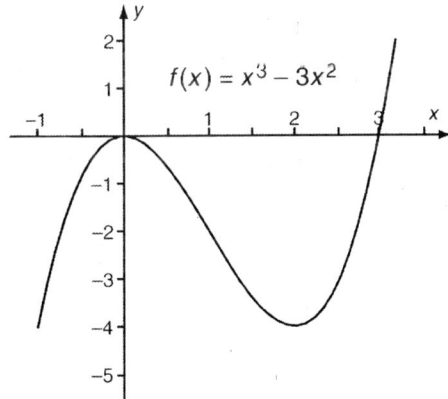

12. Relative maximum at $(-1, 2)$; relative minimum at $(1, -2)$.

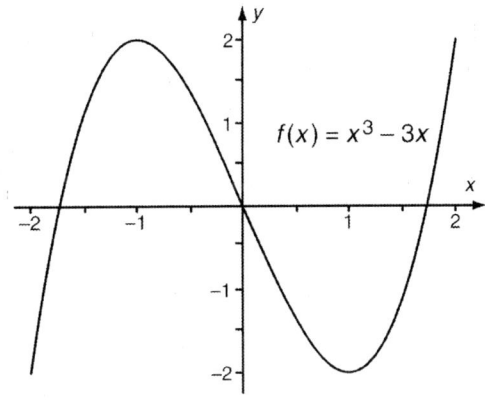

14. None

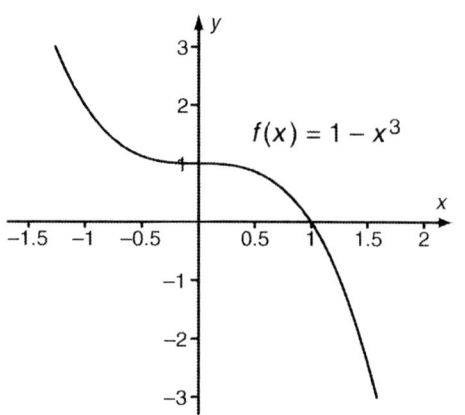

16. Relative maximum at $(1, 17)$; relative minimum at $(-3, -15)$.

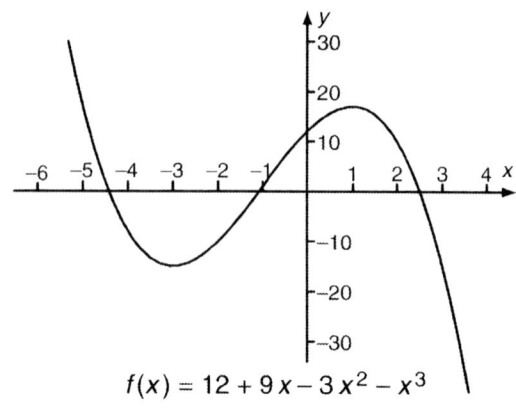

18. Relative minimum at $\left(\frac{3}{2}, -\frac{27}{16}\right)$.

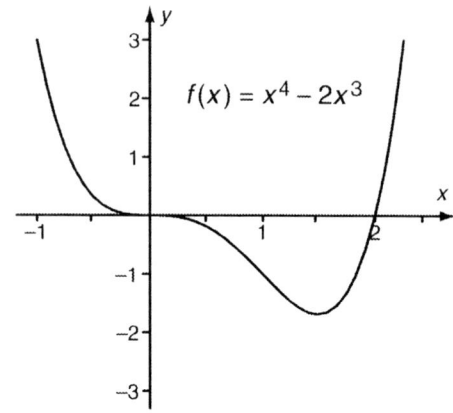

20. Relative minima at $(-1, 4)$ and $(1, 4)$; relative maximum at $(0, 5)$.

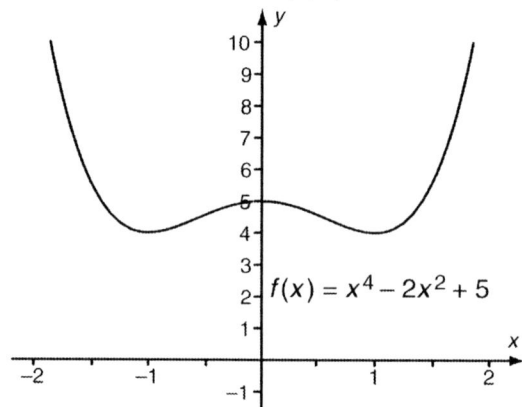

22. Relative minimum at $(-3, -5)$.

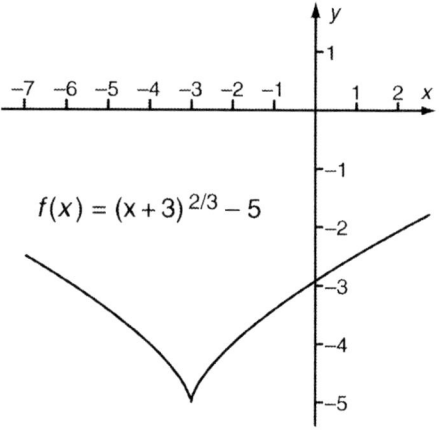

24. Relative maximum at $(0, 5)$.

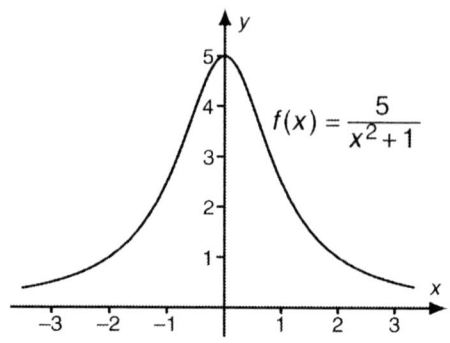

ANSWERS TO EVEN-NUMBERED EXERCISES

26. Relative minimum at $(0,0)$.

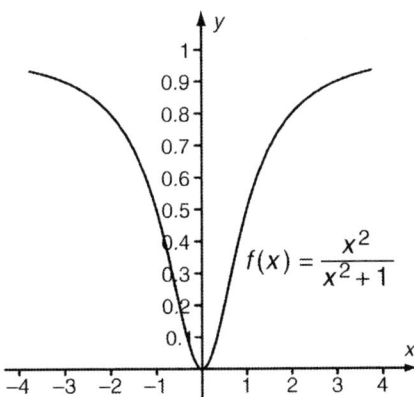

28. None

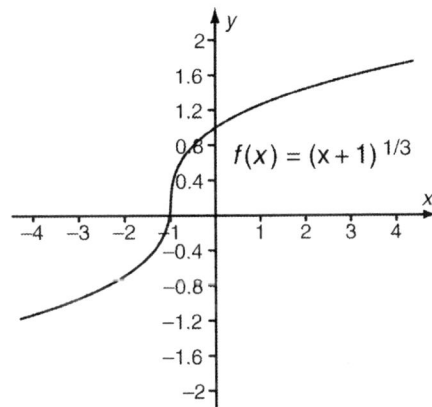

30. Relative maximum at $(0,1)$.

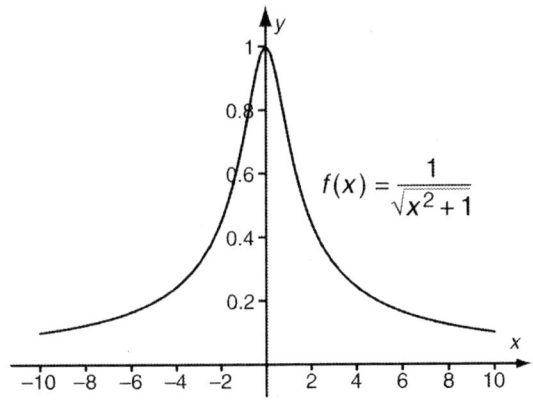

32.–60. Left to the student.

62. Relative maximum at $(6, 102.2°)$.

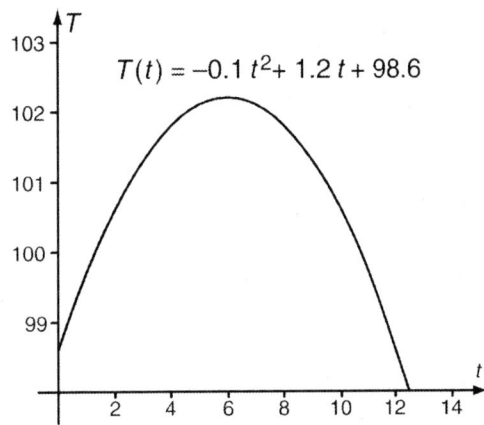

64. tw

66. Relative maxima at $(-6.2617, 3212.8)$, $(-0.5595, 1440.06)$ and $(5.0544, 6674.12)$; relative minima at $(-3.6829, -2288.03)$ and $(2.1162, -1083.08)$.

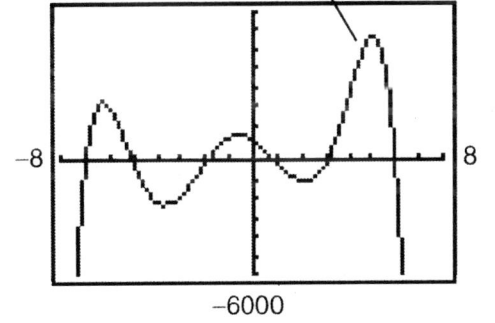

68. Relative minima at $(-2, 1)$ and $(2, 1)$; relative maximum at $(0, 2.5874)$.

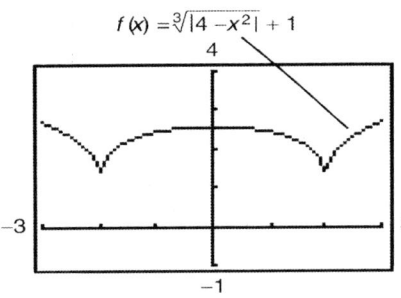

70. tw

Exercise Set 3.2

2. Relative maximum at $(0, 3)$.

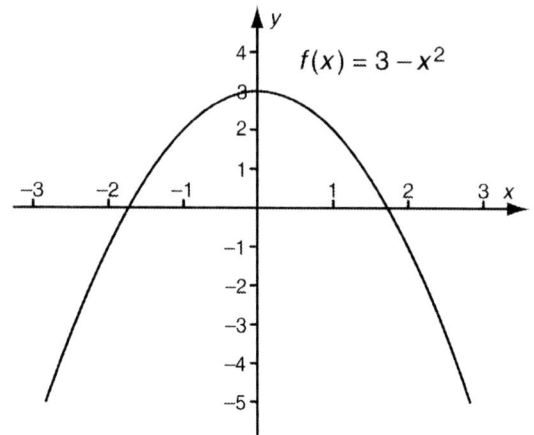

4. Relative minimum at $\left(\frac{1}{2}, -\frac{1}{4}\right)$.

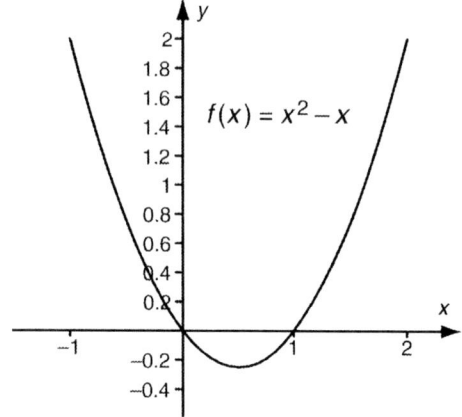

6. Relative minimum at $\left(\frac{4}{5}, \frac{19}{5}\right)$.

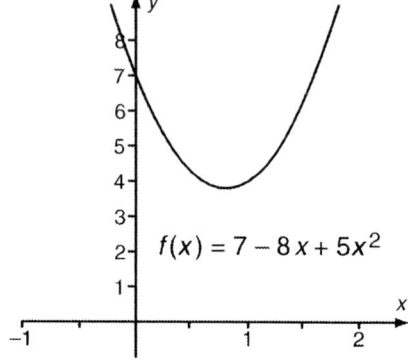

8. Relative maximum at $(-2, 45)$; relative minimum at $(2, -51)$.

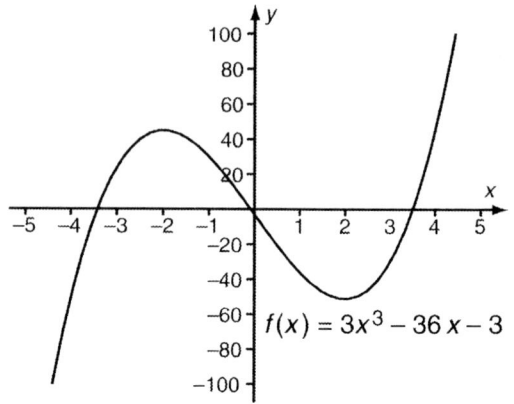

10. Relative minimum at $(-6, -28)$; relative maximum at $(0, 80)$.

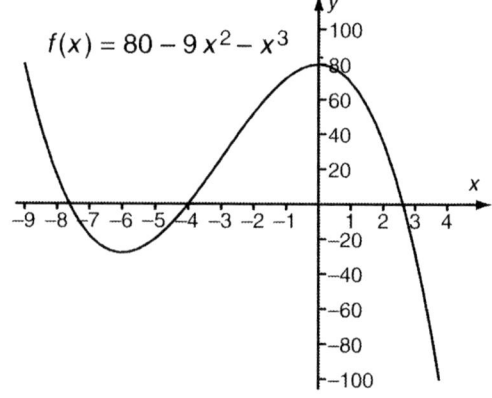

12. Relative minimum at $(-1, -4)$; relative maximum at $(1, 0)$.

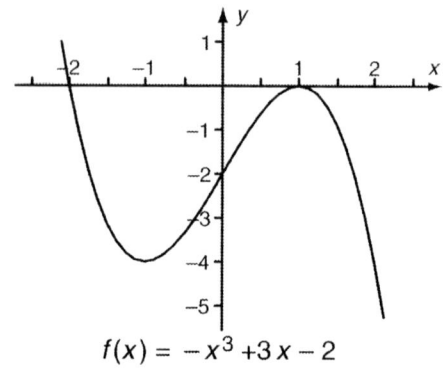

14. Relative maximum at $(0,5)$; relative minima at $(-2,-27)$ and $(1,0)$.

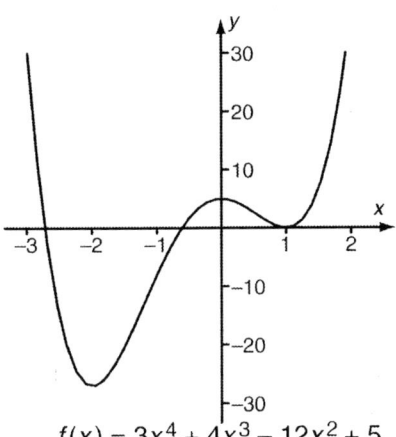

$f(x) = 3x^4 + 4x^3 - 12x^2 + 5$

16. Relative minimum at $(1,0)$.

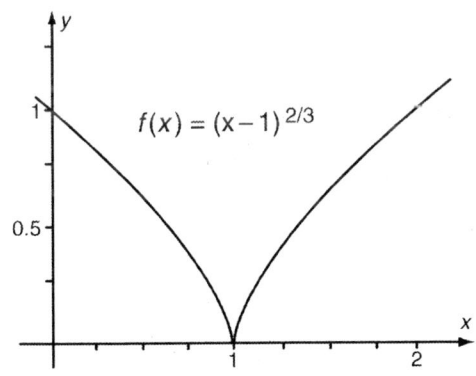

$f(x) = (x-1)^{2/3}$

18. Relative maxima at $(-1,1)$ and $(1,1)$; relative minimum at $(0,0)$.

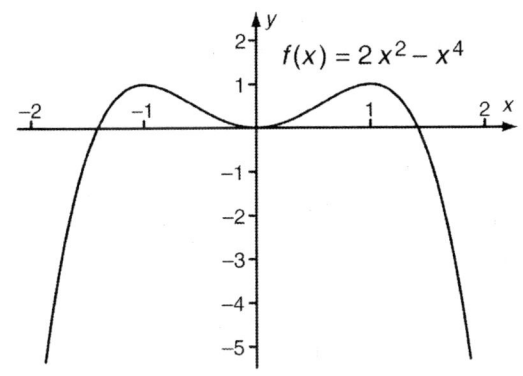

$f(x) = 2x^2 - x^4$

20. Relative minimum at $(3,1)$; relative maximum at $(1,5)$.

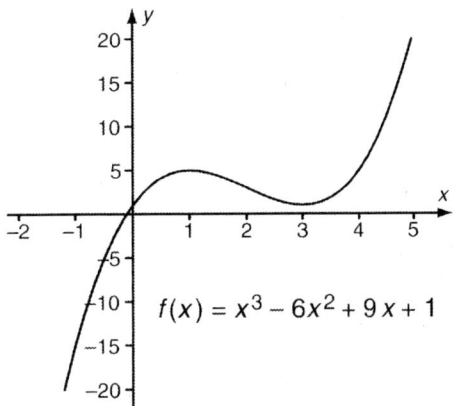

$f(x) = x^3 - 6x^2 + 9x + 1$

22. Relative minimum at $\left(\frac{3}{2}, -\frac{27}{16}\right)$.

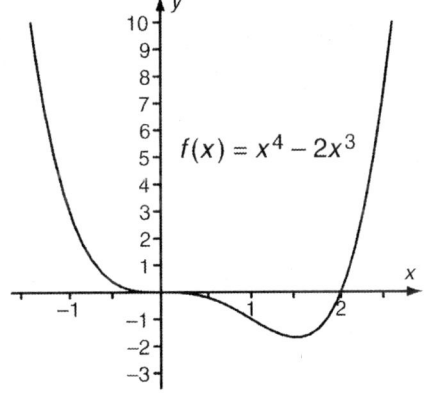

$f(x) = x^4 - 2x^3$

24. Relative maximum at $(-6, 400)$; relative minimum at $(8, -972)$.

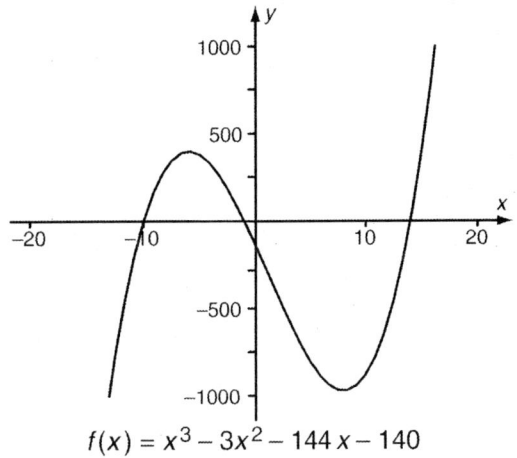

$f(x) = x^3 - 3x^2 - 144x - 140$

26. Relative minimum at $(-1, -4)$; relative maximum at $(1, 4)$.

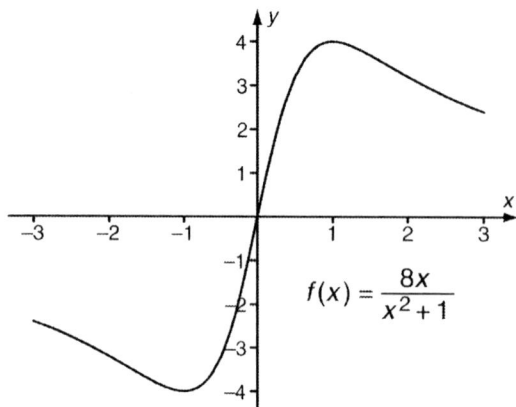

28. Relative minimum at $(0, -4)$.

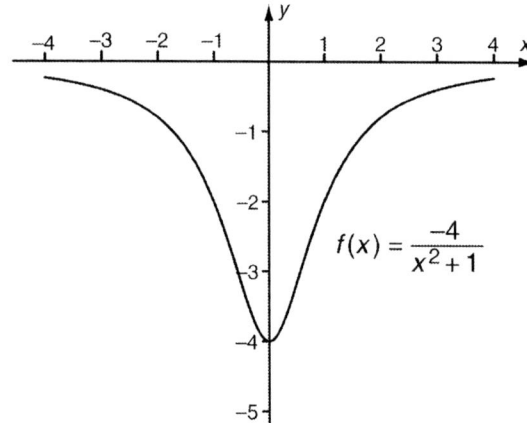

30. None

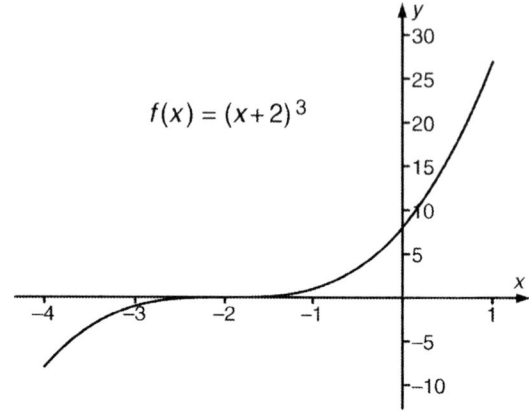

32. Relative minima at $(0, 0)$ and $(3, 0)$; relative maximum at $\left(\frac{3}{2}, \frac{81}{16}\right)$.

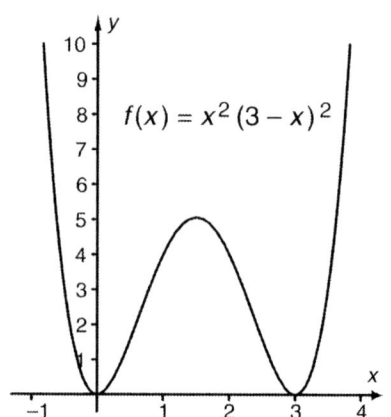

34. Relative minimum at $(-1, -2)$; relative maximum at $(1, 2)$.

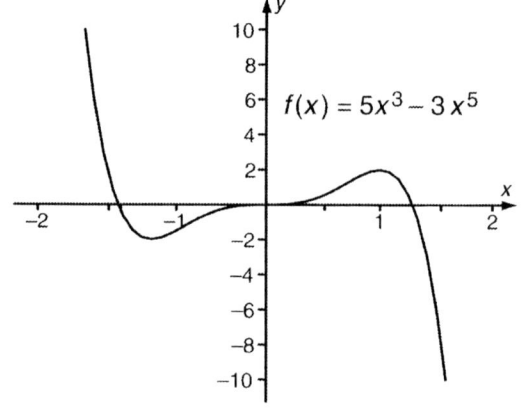

36. Relative maximum at $\left(-\frac{\sqrt{2}}{2}, \frac{1}{2}\right)$; relative minimum at $\left(\frac{\sqrt{2}}{2}, -\frac{1}{2}\right)$.

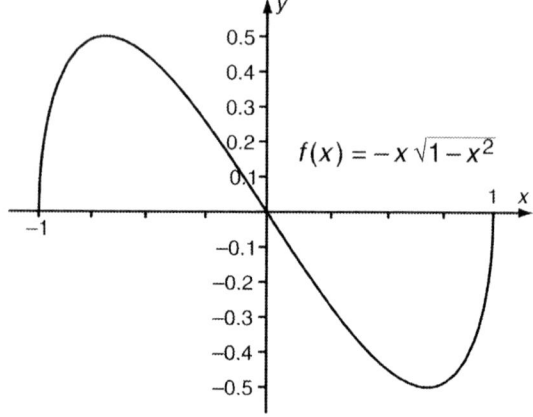

38. None

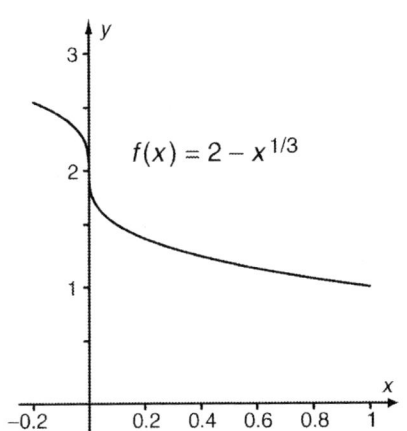

40. $(2,2)$

42. $(0,10)$, $(2,-6)$

44.–84. Left to the student.

86.

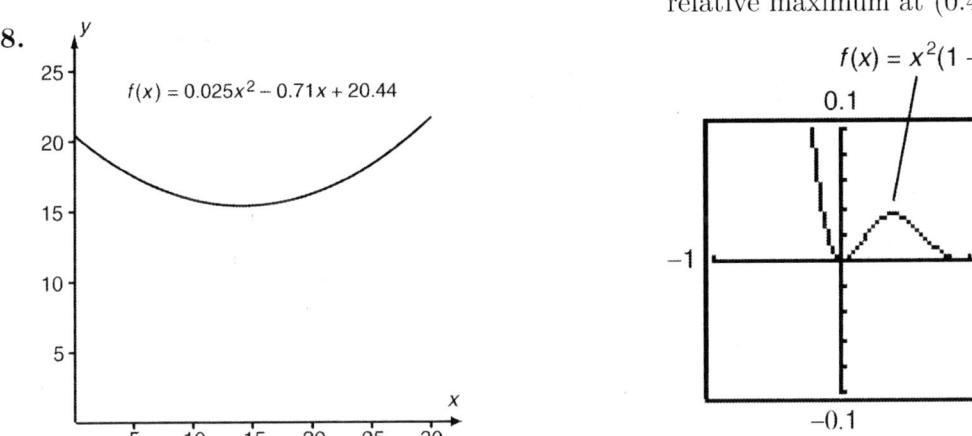

88.

90. (a) Around the end of July;
(b) around the end of January;
(c) tw

92. tw

94. tw

96. Relative maximum at $(0,0)$; relative minimum at $(1,-2)$.

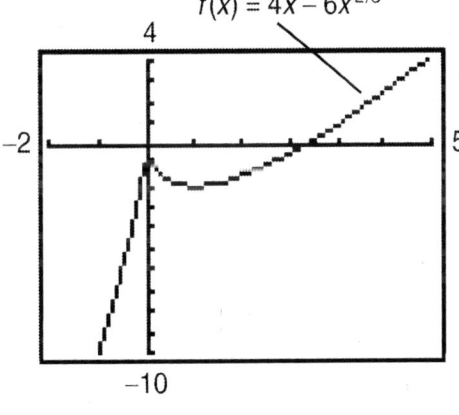

98. Relative minimum at $(0,0)$; relative maximum at $(0.4, 0.03456)$.

100. Relative maximum at $(-1, 1.5874)$; relative minimum at $(1, -1.5874)$.

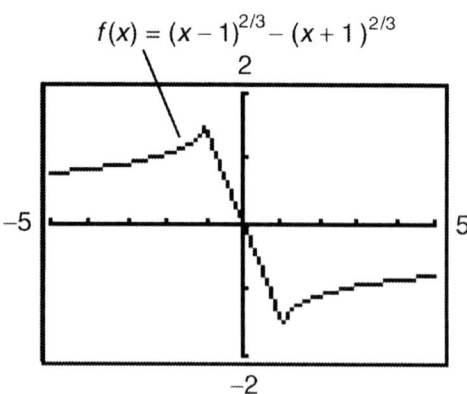

Exercise Set 3.3

2. $\dfrac{-5}{x^2}$ **4.** $\dfrac{3}{4}$ **6.** 7 **8.** $\dfrac{6}{5}$ **10.** -4 **12.** 0 **14.** $-\infty$ **16.** ∞

18.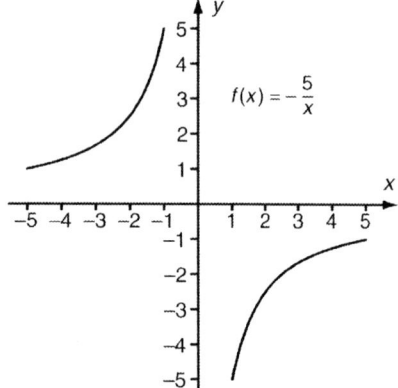

22.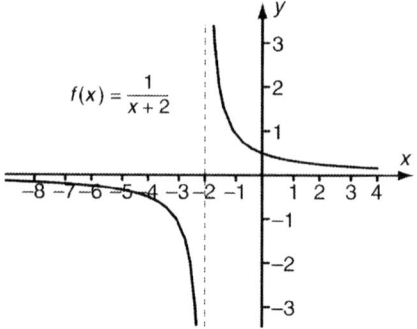

20.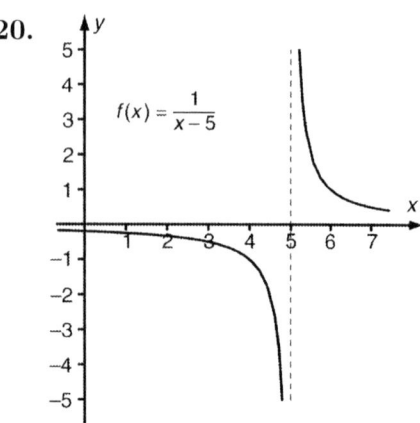

24.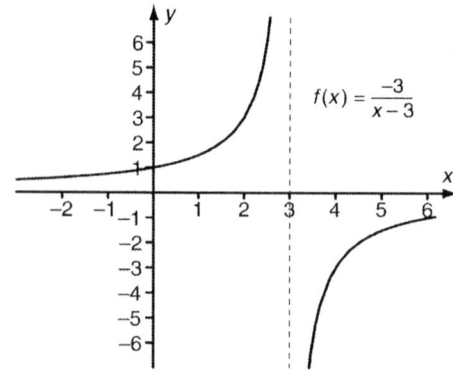

ANSWERS TO EVEN-NUMBERED EXERCISES

26.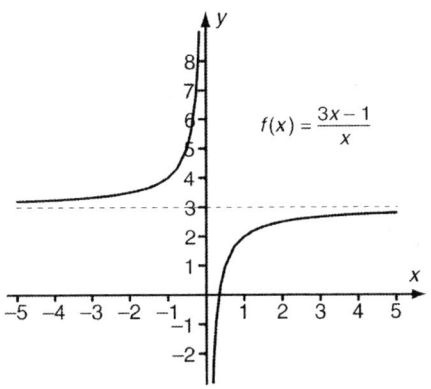

28.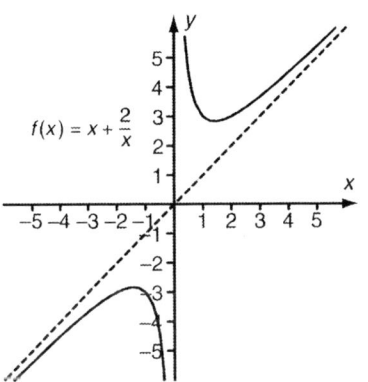

30.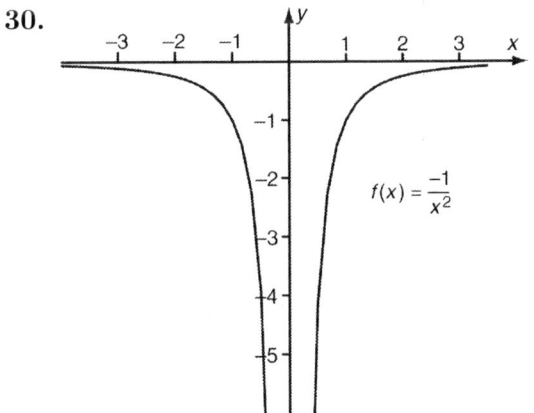

32.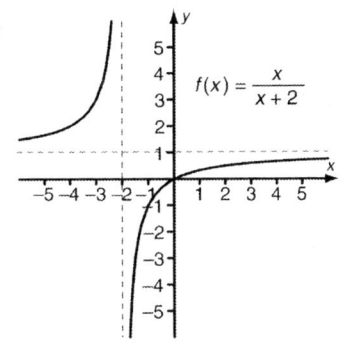

34.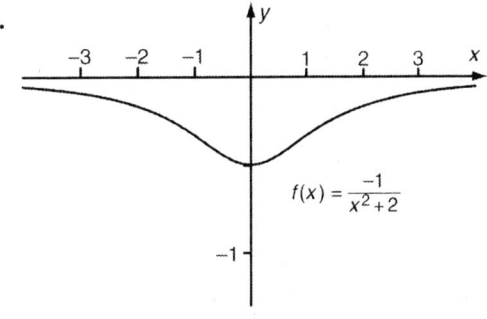

36.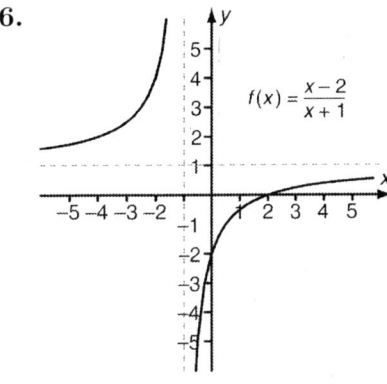

38.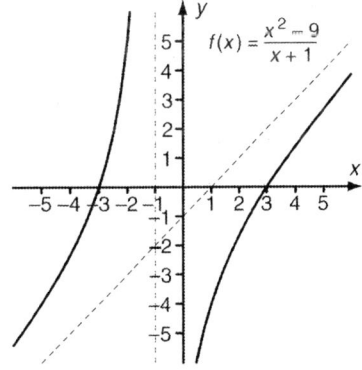

40.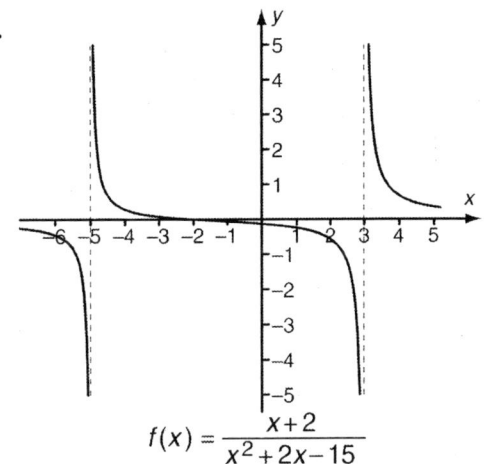

42.

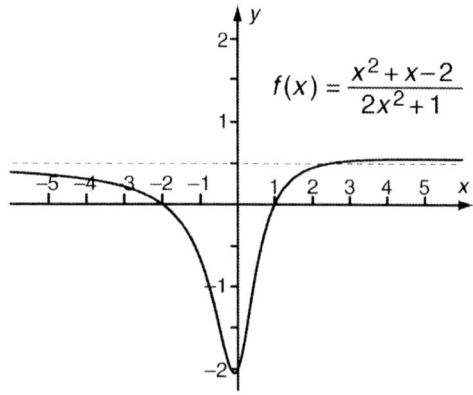

44.

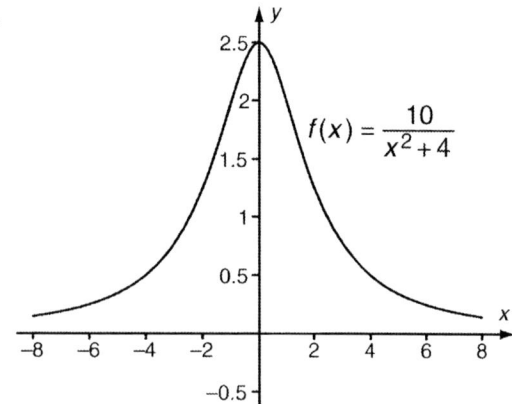

46.

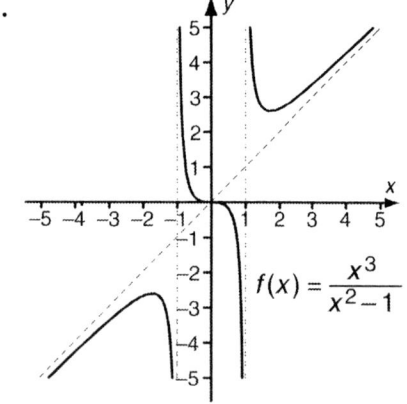

48. (a) $A(x) = 3x + \dfrac{80}{x}$;

(b)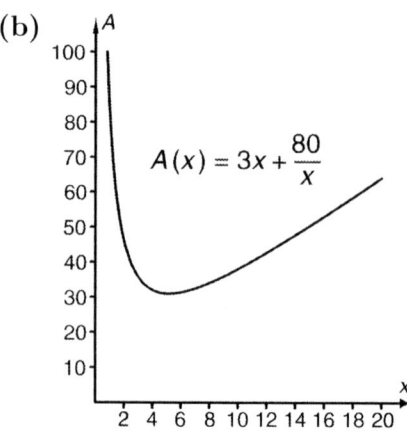

(c) $y = 3x$. When a large number of units are produced, the average cost can be found by multiplying the number of units produced by 3.

50. (a) $P(x) = -\dfrac{1}{2}x^2 + 400x - 5000$;

(b) $A(x) = -\dfrac{1}{2}x + 400 - \dfrac{5000}{x}$;

(c)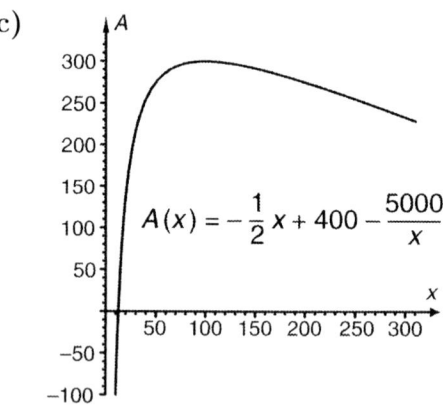

(d) $y = -\dfrac{1}{2}x + 400$; This is the long-term average profit per item.

52. (a) 100 cc, 50 cc, 20 cc, 2 cc, 0.99 cc;
(b) 0; (c) 100 cc at $t = 0$;
(d)

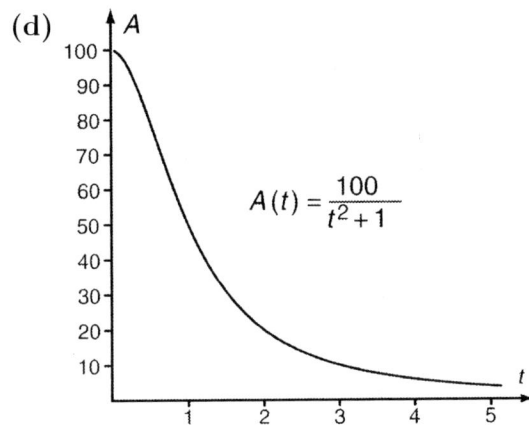

(e) tw

54. tw

56. $-\infty$ **58.** -3

60. ∞ **62.** ∞

64.

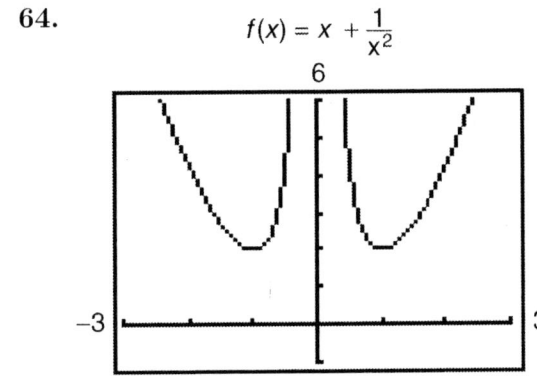

66. $f(x) = \dfrac{x^3 + 4x^2 + x - 6}{x^2 - x - 2}$

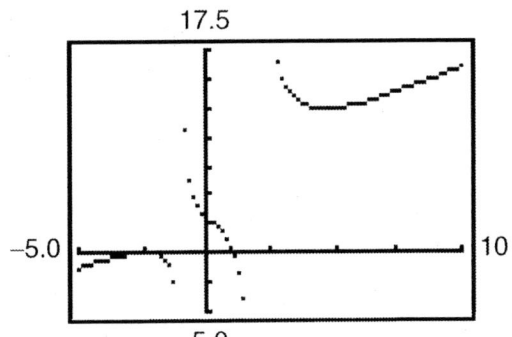

68. $f(x) = \dfrac{x^3 + 2x^2 - 3x}{x^2 - 25}$

70. $f(x) = \dfrac{x^2 - 3}{2x - 4}$

(a) $(-1.7, 0)$, $(1.7, 0)$;
(b) $(0, 0.75)$;
(c) $x = 2$, $y = 0.5x + 1$

Exercise Set 3.4

2. (a) 32 mph; (b) 80 mph; (c) 17.5 mpg; (d) 21 mpg; (e) 20%

4. Maximum $= 4\frac{1}{4}$ at $x = \frac{1}{2}$; minimum $= 2$ at $x = 2$

6. Maximum $= 5$ at $x = 0$; minimum $= \frac{113}{27}$ at $x = \frac{2}{3}$

8. Maximum $= 6\frac{1}{2}$ at $x = -1$; minimum $= 3$ at $x = -2$

10. Maximum $= 6$ at $x = 1$; minimum $= 2$ at $x = -1$

12. Maximum $= 28$ at $x = -10$; minimum $= -32$ at $x = 10$

14. Maximum $=$ minimum $= 24$ for all x in $[4, 13]$.

16. Maximum $= 10$ at $x = -1$; minimum $= 1$ at $x = 2$

18. Maximum $= 4$ at $x = 1$; minimum $= -23$ at $x = 4$

20. Minimum $= 4$ at $x = 1$; maximum $= 24$ at $x = 3$

22. Minimum $= 0$ at $x = 0$; maximum $= 325$ at $x = -5$

24. Maximum $= 2000$ at $x = 10$; minimum $= -2000$ at $x = -10$

26. Maximum $= 10$ at $x = 0$; minimum $= -22$ at $x = 4$

28. Minimum $= -2$ at $x = -1$; maximum $= \frac{27}{256}$ at $x = \frac{3}{4}$

30. Maximum $= 12$ at $x = -3$ and $x = 3$; minimum $= -13$ at $x = -2$ and $x = 2$

32. Minimum $= -3$ at $x = -8$ and $x = 8$; maximum $= 1$ at $x = 0$

34. Minimum $= -8\frac{1}{2}$ at $x = -8$; maximum $= -4$ at $x = -2$

36. Maximum $= 2$ at $x = 1$; minimum $= -2$ at $x = -1$

38. Minimum $= 2$ at $x = 8$; maximum $= 4$ at $x = 64$

40.–48. Left to the student.

50. Maximum $= 625$ at $x = 25$

52. Minimum $= 50$ at $x = 5$

54. Maximum $= \frac{64}{3}$ at $x = 2$

56. Maximum $= \frac{729}{4}$ at $x = \frac{27}{2}$

58. Minimum $= -\frac{10}{3}\sqrt{5}$ at $x = \sqrt{5}$; maximum $= \frac{10}{3}\sqrt{5}$ at $x = -\sqrt{5}$

60. Maximum $= 19$ at $x = 70$

62. Maximum $= \frac{148}{27}$ at $x = \frac{5}{3}$

64. Maximum $= 37\frac{25}{27}$ at $x = \frac{16}{3}$; minimum $= 0$ at $x = 0$ and $x = 8$

66. Minimum $= 120$ at $x = 60$

68. Minimum $= 75$ at $x = 5$

70. Maximum $= 3$ at $x = 1$; minimum $= -\frac{3}{8}$ at $x = -\frac{1}{2}$

72. Maximum $= 2$ at $x = 4$; minimum $= 0$ at $x = 0$

74. None

76. Maximum $= 59$ at $x = -10$; minimum $= -41$ at $x = 10$

78. None

ANSWERS TO EVEN-NUMBERED EXERCISES

80. Minimum $= 0$ at $x = 0$

82. None

84. Maximum $= \frac{10}{3} + 2\sqrt{3}$ at $x = -2 - \sqrt{3}$; minimum $= \frac{10}{3} - 2\sqrt{3}$ at $x = -2 + \sqrt{3}$

86. Minimum $= 0$ at $x = -1$ and $x = 1$

88.–96. Left to the student.

98. 22,506 units; $150,000

100. 1986

102. (a) $P(x) = -\frac{1}{2}x^2 + 400x - 5000$;
(b) 400 items

104. Maximum $= \frac{1}{4860}$ or about 0.000206 at $x = \frac{1}{9}$ cc

106. Maximum $= \frac{2}{3}\sqrt{\frac{1}{3}}$, or $\frac{2}{3\sqrt{3}}$, at $x = \frac{2}{3}$; minimum $= 0$ at $x = 0$ and $x = 1$

108. $x = \dfrac{a+b}{2}$ **110.** $\boxed{\text{tw}}$

112.

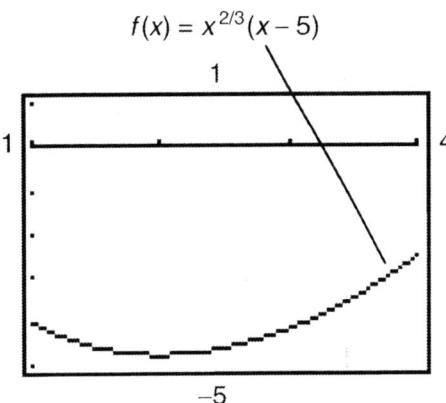

Minimum $= -4.76$ at $x = 2$; maximum $= -2.5$ at $x = 4$

114.

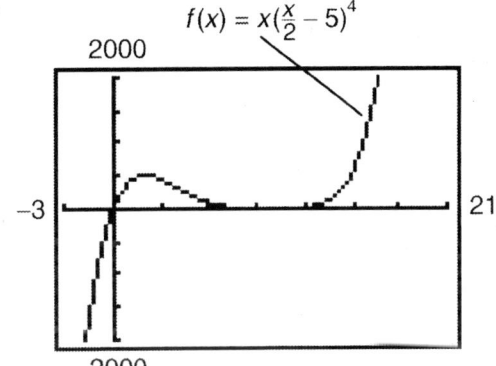

There are no absolute extrema.

Exercise Set 3.5

2. 35 and 35; maximum product $= 1225$

4. No; $Q = x(70 - x)$ has no minimum

6. 3 and -3; minimum product $= -9$

8. 2 and $\sqrt{2}$; maximum $= 4$

10. 2 and 1; minimum $= 6$

12. $\frac{2}{3}$ and $\frac{1}{2}$; maximum $= \frac{1}{3}$

14. 4800 yd^2

16. 8.5 ft by 8.5 ft; 72.25 ft^2

18. $13\frac{1}{3}$ in. by $13\frac{1}{3}$ in. by $3\frac{1}{3}$ in; $592\frac{16}{27}$ in^3

20. 4 ft by 4 ft by 2 ft; 48 ft^2

22. 40 units; $797

24. 1900 units; $3550

26. 11,000 units; $182,333

28. (a) $R(x) = 280x - 0.4x^2$;
(b) $P(x) = -x^2 + 280x - 5000$, $0 \le x < \infty$;
(c) 140 refrigerators; (d) $14,600; (e) $224

30. $201

32. (a) $D(p) = -0.293p + 1.293$;
(b) 2.21 times the January 2001 price.

34. $11.50 **36.** [tw]

38. $\sqrt{109.6875}$ in. by $\dfrac{73.125}{\sqrt{109.6875}}$ in., or about 10.47 in. by 6.98 in.

40. Reorder 20 times per year; lot size = 10

42. Reorder 12 times per year; lot size = 60

44. Reorder about 13 times per year; lot size = 54

46. 6 yd by 8 yd

48. $x = \dfrac{48}{8 + 3\pi}$ ft, $y = \dfrac{48 + 12\pi}{8 + 3\pi}$

50. $\dfrac{1}{2}$

52. Reorder 25 times per year; lot size = 100

54. S should be $3\frac{1}{4}$ miles down shore from A

56. The bridge should be located such that the distance x is $\dfrac{bp}{a+b}$ units.

58. (a) $A'(x) = \dfrac{xC'(x) - C(x)}{x^2}$;

(b) $A'(x)$ exists for all x in $(0, \infty)$. $A'(x) = 0$ when $xC'(x) - C(x) = 0$, or $C'(x) = \dfrac{C(x)}{x} = A(x)$.

60. Minimum $= -3\sqrt{2}$ at $x = -\sqrt{2}$ and $y = 0$

Exercise Set 3.6

2. 0.120601; 0.12

4. -0.1004, -0.1

6. -0.167, -0.2

8. 1, 1

10. $3.21, $3.20

12. $3, $3

14. (a) $P(x) = -0.01x^2 + 1.4x - 100$;
(b) $\Delta P = -\$0.21$, $P'(80) = -\$0.20$

16. 3.167

18. 10.15

20. 3.037

22. $dy = \dfrac{3}{2\sqrt{3x - 2}}dx$

24. $dy = 5x^2(2x + 5)(2x + 3)\,dx$

26. $dy = \left[\dfrac{x^4 + 8x^2 - 4x + 3}{(x^2 + 3)^2}\right]dx$

28. $dy = -8(7 - x)^7\,dx$

30. $-22{,}394.88$

32. 32 units

34. $2.51

36. About 1.256 cm^2

38. About 1.5422 ppm

40. [tw]

Exercise Set 3.7

2. $\dfrac{2-y}{x+2y}; -\dfrac{4}{3}$

4. $\dfrac{x}{y}; \dfrac{\sqrt{3}}{\sqrt{2}}$

6. $\dfrac{5+12x^2}{3+4y^3}; -\dfrac{17}{29}$

8. $\dfrac{x+y}{3y-x}$

10. $-\dfrac{x}{y}$

12. $\dfrac{5x^4}{3y^2}$

14. $\dfrac{5x^2y^2-3y}{2x-3x^3y}$

16. $\dfrac{dp}{dx} = \dfrac{2-p}{x-2}$

18. $\dfrac{dp}{dx} = \dfrac{-p}{3x}$

20. $\dfrac{dp}{dx} = \dfrac{1}{50p-300}$

22. $0, 2\sqrt{6}, -\dfrac{9}{4}$

24. $200 per day, $50 per day, $150 per day

26. $36,000 per day, $72,000 per day, $-$$36,000 per day

28. $R'(t) = -1.39$

30. -157 mm^2/day

32. (a) $\dfrac{dV}{dt} = 1000R \cdot \dfrac{dR}{dt}$; (b) 0.05 mm^3/min

34. $-2\dfrac{1}{12}$ ft/sec

36. $-\dfrac{y^3}{x^3}$

38. $\dfrac{2x}{y(x^2+1)^2}$

40. $\dfrac{5x^4-6x^2-6y^2}{12xy-5y^4}$

42. $\dfrac{dy}{dx} = \dfrac{y-2x}{2y-x}; \dfrac{d^2y}{dx^2} = \dfrac{-6(y^2-xy+x^2)}{(2y-x)^3}$

44. $\dfrac{dy}{dx} = \dfrac{x^2}{y^2}, \dfrac{d^2y}{dx^2} = \dfrac{2xy^3-2x^4}{y^5}$

46. [tw]

48.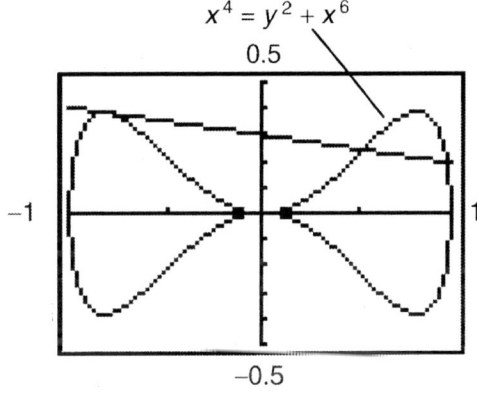

50.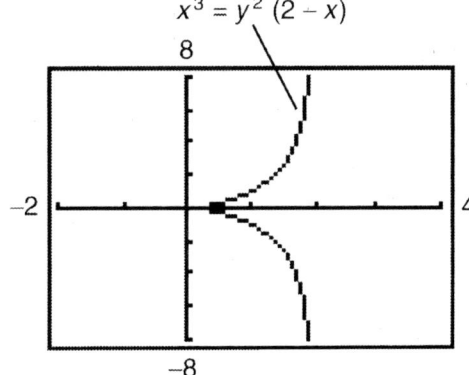

CHAPTER 4

Exercise Set 4.1

2.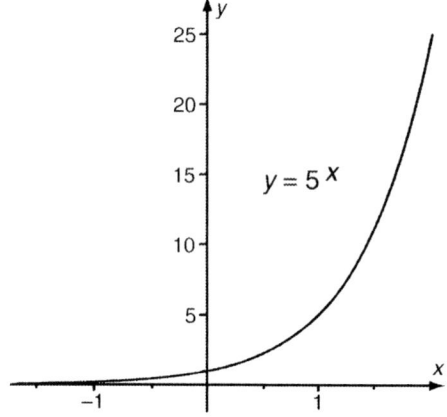

4.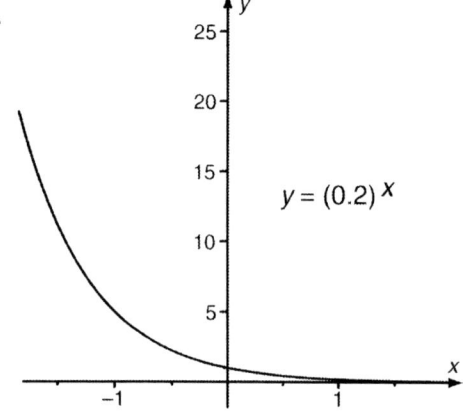

6.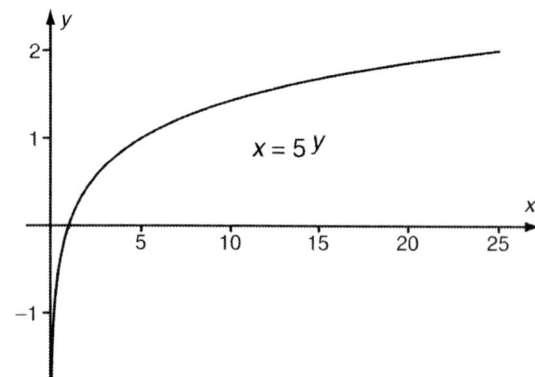

8. $2e^{2x}$ **10.** $-12e^{-3x}$

12. e^{-x} **14.** $-4e^x$ **16.** e^{4x}

18. $x^4 e^x (x+5)$ **20.** $x^2 e^x$

22. $\dfrac{e^x (x-5)}{x^6}$

24. $(-2x+8)e^{-x^2+8x}$

26. $xe^{x^2/2}$ **28.** $\dfrac{e^{\sqrt{x-4}}}{2\sqrt{x-4}}$

30. $\dfrac{e^x}{2\sqrt{e^x+1}}$ **32.** $x(3x - e^x)$

34. $3e^{-3x}$ **36.** me^{-mx}

38.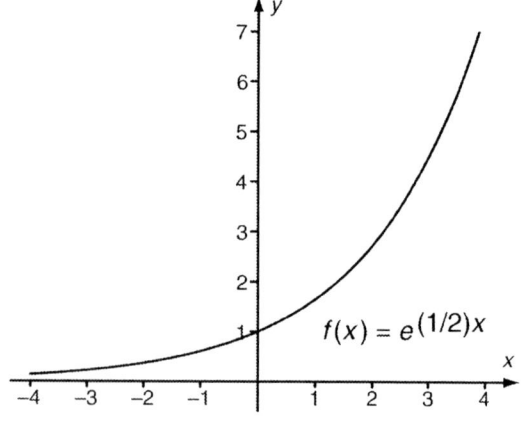

40.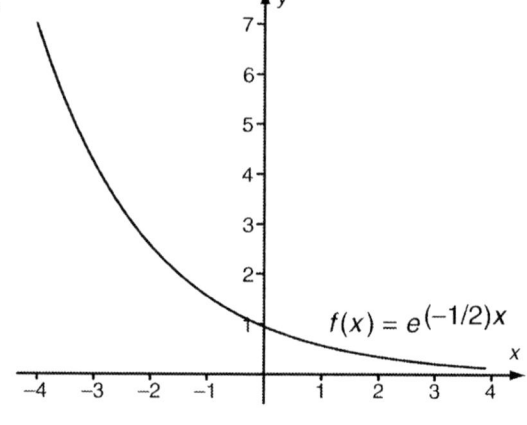

42.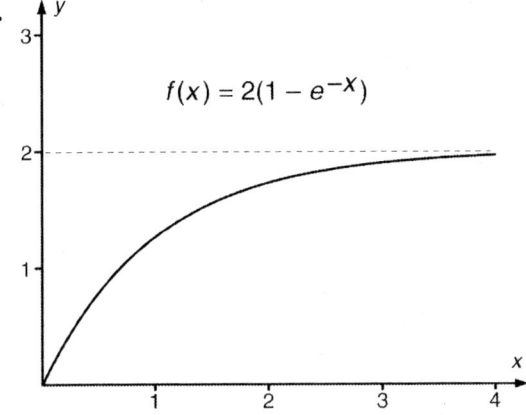

44.–48. Left to the student.

50. $y = -6x + 2$

52. Left to the student.

54. (a) $T'(x) = \dfrac{59{,}332 e^{-0.7x}}{(1 + 16.3 e^{-0.7x})^2}$, the rate of change of the total number of telephones with respect to the number of years after 1994; (b) 26,475

56. (a) $C'(t) = 40 e^{-t}$; (b) \$40 million; (c) \$0.270 million; (d) [tw]

58. (a) Approximately 242, 308, 584; (b)

(c) $S'(p) = 0.6 e^{0.004p}$; (d) [tw]

60. (a) 4.24 deaths per thousand people, 3.94 deaths per thousand people; (b) 3.92 deaths per thousand people; over the long haul, this is the annual death rate.

62. (a) 100%, 69.8%, 54.8%, 40.9%, 40.1%; (b) 40% (c)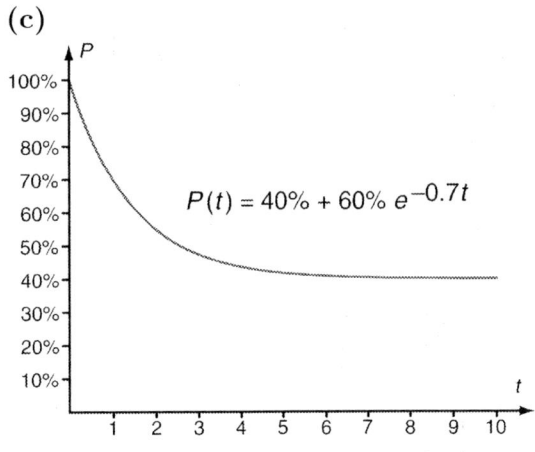

(d) $P' = -42\% e^{-0.7t}$; (e) [tw]

64. $8 x e^{x^2} \left(e^{x^2} - 2 \right)^3$

66. $\dfrac{1}{3} \left(e^{3t} + t \right)^{-2/3} \left(3 e^{3t} + 1 \right)$

68. $\dfrac{e^x}{(1 - e^x)^2}$

70. $-e^{-x} - e^{1/x} \cdot x^{-2}$

72. $\dfrac{e^{-x} \left(-x^3 - x^2 - x + 1 \right)}{(1 + x^2)^2}$

74. $e^{e^x + x}$

76. 4, 2.86797, 2.73200, 2.71964, 2.71855

78. Minimum $= -e^{-1} \approx -0.3679$ at $x = -1$

80. [tw]

82. Relative maximum at $(0, 1)$.

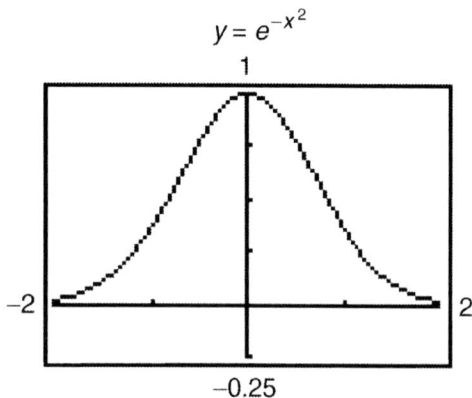

84. The graphs of f and f'' are identical.

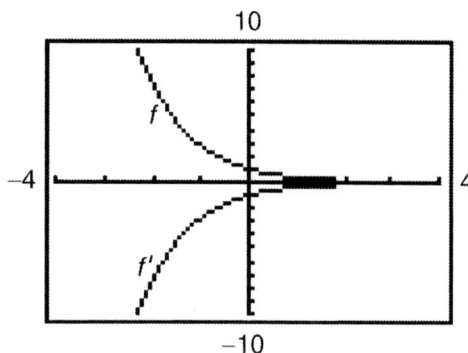

86. It is difficult to graph all three functions in the same window. We show f and f' in the window $[-20, 40, -1000, 1500]$, Xscl $= 10$, Yscl $= 500$. Then we show f'' in the window $[-50, 50, -2, 10]$, Xscl $= 10$.

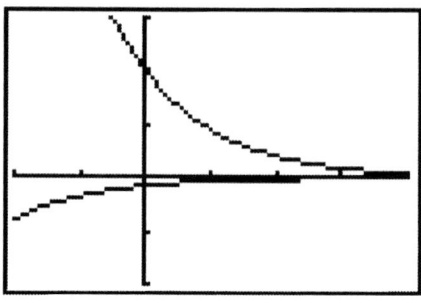

Exercise Set 4.2

2. $3^4 = 81$ **4.** $27^{1/3} = 3$

6. $a^K = J$ **8.** $b^{-w} = V$

10. $\log_e p = t$ or $\ln p = t$

12. $\log_{10} 1000 = 3$

14. $\log_{10} 0.01 = -2$ **16.** $\log_Q T = n$

18. -0.51 **20.** $\dfrac{3}{2}$ **22.** 4.317

24. 0.2231 **26.** 2.3863

28. 4.6051 **30.** 11.512915

32. -7.047017 **34.** -4.509860

36. $t \approx 6.9$ **38.** $t \approx 4.5$

40. $t \approx 4.6$ **42.** $t \approx 9.9$

44. $-\dfrac{4}{x}$ **46.** $x^4 (1 + 5 \ln x) - x^3$

48. $\dfrac{1 - 5 \ln x}{x^6}$ **50.** $\dfrac{1}{x}$

52. $\dfrac{14x + 5}{7x^2 + 5x + 2}$ **54.** $\dfrac{1}{x \ln 3x}$

56. $\dfrac{x^2 - 5}{x(x^2 + 5)}$ **58.** $e^{2x} \left(\dfrac{1}{x} + 2 \ln x \right)$

60. $\dfrac{e^x}{e^x - 2}$ **62.** $\dfrac{3 (\ln x)^2}{x}$

64. (a) 2000 units; (b) $N'(a) = \dfrac{500}{a}$, $N'(10) = 50$; (c) minimum $= 2000$ at $a = 1$; (d) $\boxed{\text{tw}}$

ANSWERS TO EVEN-NUMBERED EXERCISES

66. 40 days

68. (a) $R(p) = 800pe^{-0.125p}$;
(b) $R'(p) = e^{-0.125p}(800 - 100p)$;
(c) \$8/unit

70. $A \ln A - A \ln B - A$

72. (a) 78%; (b) 54%; (c) 30%;
(d) about 38.5%; (e) $S'(t) = -\dfrac{15}{t+1}$;
(f) maximum = 78% at $t = 0$;
(g) tw

74. (a) 26 words per min, 91 words per min;
(b) $W'(t) = 30e^{-0.3t}$; (c) 10 weeks;
(d) tw

76. $\dfrac{n(\ln x)^{n-1}}{x}$ **78.** $\dfrac{3(2t+1)}{t^2+t}$

80. $\dfrac{1}{x \cdot \ln(\ln 3x) \cdot \ln 3x}$ **82.** $\dfrac{-2}{1-t^2}$

84. $\dfrac{x}{5+x^2}$ **86.** $x^4 \ln x$

88. $\dfrac{\ln x - x^2 \ln x + 2x^2}{(x^2+1)^2}$

90. $\dfrac{\sqrt{x}}{x - x^2}$ **92.** $\dfrac{1 - (\ln x)^2}{x\left[1 + (\ln x)^2\right]^2}$

94. $t = \dfrac{\ln P_0 - \ln P}{k}$

96. Let $a = \log x$; then $10^a = x$.
$$\ln x = \ln 10^a$$
$$\ln x = a \ln 10$$
$$\dfrac{\ln x}{\ln 10} = a$$
$$\dfrac{\ln x}{\ln 10} = \log x \quad \text{Substituting } \log x \text{ for } a$$
$$\dfrac{\ln x}{2.3026} \approx \log x$$
$$0.4343 \ln x \approx \log x$$

98. tw   **100.** e^π **102.** 0

104. $y_1 = f(x) = \ln x$, $y_2 = f'(x) = \dfrac{1}{x}$

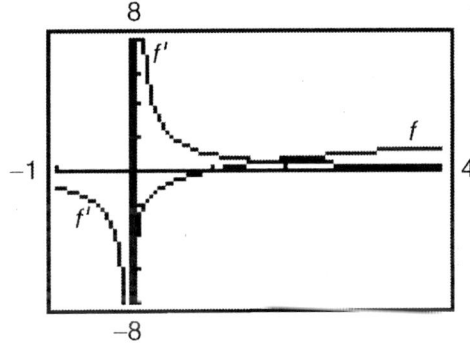

106. $y_1 = f(x) = x^2 \ln x$,
$y_2 = f'(x) = x + 2x \ln x$

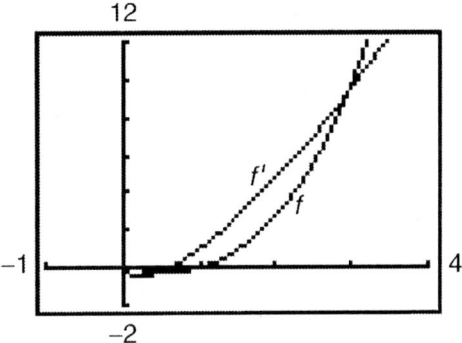

108. $-e^{-1} \approx -0.368$

Exercise Set 4.3

2. $R(t) = R_0 e^{kt}$

4. (a) $N(t) = 40e^{0.15t}$; (b) 803; (c) 4.6 yr

6. (a) $P(t) = P_0 e^{0.08t}$; (b) \$21,665.74, \$23,470.20; (c) 8.7 yr

8. 5.8%   **10.** 17.3 yr after 2004

12. 7.1%; 9.8 yr   **14.** 6.3%; \$12,800

16. (a) $k \approx 0.0788$, $P(t) = 39.4e^{0.0788t}$; (b) \$4454.9 billion; (c) 2020

18. (a) $k = 0.061248$, $P(t) = \$100e^{0.061248t}$; (b) \$1025.15; (c) 11.3 yr

20. (a) $k = 0.396$, $P(t) = 0.1e^{0.396t}$; (b) 25.6 million, 275.2 million; (c) 19.2 yr after 2000; (d) 1.8 yr; (e) $\boxed{\text{tw}}$

22. $k = 0.276137$, $R(t) = 1265e^{0.276137t}$; \$316,665 million

24. $k = 5.7033\%$ per year, 44¢, 62¢

26. (a) \$3.57, \$8.54; (b) $\boxed{\text{tw}}$

28. 1%   **30.** 4%   **32.** 3.5%

34. (a) $P(t) = 60e^{0.0037t}$; (b) 62.0 million; (c) 187 yr

36. 0.21%

38. (a) 1000, 1375, 1836, 3510, 5315, 5771; (b) $P'(t) = \dfrac{11{,}051.36e^{-0.4t}}{(1+4.78e^{-0.4t})^2}$
(c)
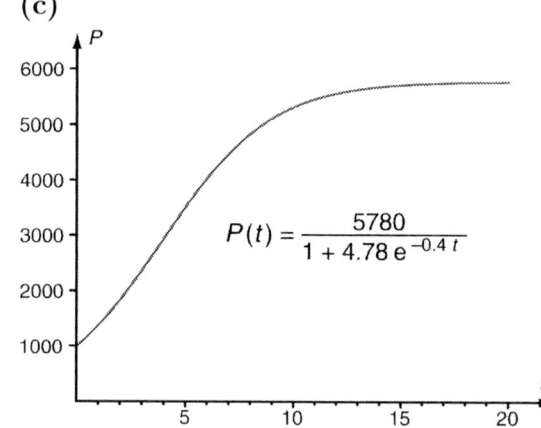

40. (a) 0.244, 0.429, 0.753, 0.954, 0.989, 0.996; (b) $p'(t) = 0.28e^{-0.28t}$;

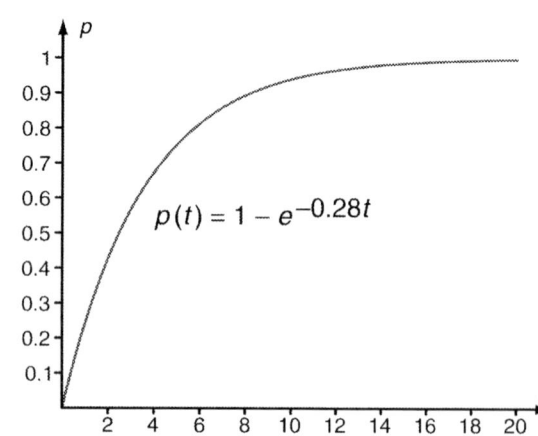

42. (a) $N(t) = \dfrac{29.47232081}{1 + 79.56767122 e^{-0.809743969t}}$;

(b) about 30 people

(c)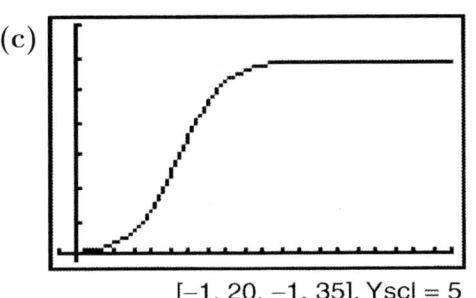

[−1, 20, −1, 35], Yscl = 5

(d) $N'(t) = \dfrac{1898.885181 e^{-0.809743969t}}{(1 + 79.56767122 e^{-0.809743969t})^2}$

(e) tw

44. 8.33%   **46.** 6.40%

48. $T_4 = \dfrac{\ln 4}{k}$ **50.** 2 yr

52. $y_1 = P_0 e^{kt_1}$ and $y_2 = P_0 e^{kt_2}$, so
$y_2/y_1 = e^{kt_2}/e^{kt_1}$,
$\ln(y_2/y_1) = \ln e^{k(t_2 - t_1)} = k(t_2 - t_1)$.
Then $k = \dfrac{\ln(y_2/y_1)}{t_2 - t_1}$.

54. tw

Exercise Set 4.4

2. $9477.11   **4.** $13,464.05

6. (a) $40,000; (b) $5413; (c) tw

8. (a) $k \approx 0.03$, $P(t) = \$100 e^{-0.03t}$, where t = years that precede 1967; (b) $13.40

10. (d)   **12.** (a)   **14.** (e)

16. 23.1% per min   **18.** 7.2 days

20. 0.015% per year   **22.** 40.8 g

24. 4239 yr   **26.** 24.9 days

28. (a) $A = A_0 e^{-kt}$; (b) 11 hr

30. (a) 0.9%; (b) 30 days

32. 37%I_0

34. (a) 60°; (b) $k \approx 0.04$; (c) 80.2°; (d) 102.4 min; (e) tw

36. 8 PM

38. (a) $k = 0.005832$, $P(t) = 150 e^{-0.005832t}$, where P is in millions and t is the number of years since 1995; (b) 137 million; (c) 2064

40. (a) 14.0 lb/in^2; (b) 5.4 lb/in^2; (c) 46,052 ft; (d) tw

42. 38%

44. (a) $k \approx 0.016$; $B(t) = 80 e^{-0.016t}$, where t is the number of years after 1985; (b) 62.9 lbs per person; (c) 2072

46. (a)–(e) Answers will vary; (f) yes; $\lim\limits_{t \to \infty} T = \lim\limits_{t \to \infty} \left(a e^{-kt} + C\right) = C$, for $k > 0$.
(g) tw

Exercise Set 4.5

2. $e^{2.0794}$ 4. $e^{0.941}$ 6. $e^{R \cdot \ln 5}$

8. $e^{kR \cdot \ln 10}$ 10. $(\ln 7) 7^x$

12. $(\ln 100) 100^x$

14. $(5.4)^x [x \ln 5.4 + 1]$

16. $5^x x^3 [x \ln 5 + 4]$

18. $\dfrac{1}{\ln 5} \cdot \dfrac{1}{x}$ 20. $\dfrac{5}{\ln 10} \cdot \dfrac{1}{x}$

22. $\dfrac{1}{\ln 10} \cdot \dfrac{1}{x}$ 24. $\dfrac{1}{\ln 6} + \log_6 x$

26. $2e^{2x}$ 28. $\dfrac{e^x}{x \ln 4} + e^x \log_4 x$

30. (a) $V'(t) = \$5200 (\ln 0.80)(0.80)^t$;
    (b) tw

32. (a) 20,487 thousand; (b) 240 thousand;
    (c) number of new businesses in 2005

34. 6.9

36. (a) $I = 10^{10} \cdot I_0$; (b) $I = 10 \cdot I_0$;
    (c) $10^{10} \cdot I_0 = 10^9 \cdot 10 \cdot I_0$
    (d) $\dfrac{dI}{dL} = (0.1 I_0 \cdot \ln 10) \cdot 10^{0.1L}$; (e) tw

38. (a) $\dfrac{dL}{dI} = \dfrac{10}{\ln 10} \cdot \dfrac{1}{I}$; (b) tw

40. $\ln 3$ 42. $2^{x^4} (4x^3 \ln 2)$

44. $\dfrac{2x}{(\ln 3)(x^2 + 1)}$

46. $f'(x)(\ln a) \left[a^{f(x)} \right]$

48. $\left[g'(x) \cdot \ln f(x) + g(x) \cdot \dfrac{f'(x)}{f(x)} \right] [f(x)]^{g(x)}$

50. tw

Exercise Set 4.6

2. (a) $E(p) = \dfrac{p}{500 - p}$; (b) $E(38) = \dfrac{19}{231}$,
   inelastic; (c) $p = \$250$

4. (a) $E(p) = \dfrac{p}{250 - p}$; (b) $E(57) = \dfrac{57}{193}$,
   inelastic; (c) $p = \$125$

6. (a) $E(p) = 1$; (b) $E(60) = 1$, unit elasticity; (c) Total revenue $= R(p) = 3000$, for all $p > 0$. It has 3000 as a maximum for all $p > 0$.

8. (a) $E(p) = \dfrac{p}{600 - 2p}$; (b) $E(250) = \dfrac{5}{2}$,
   elastic; (c) $p = \$200$

10. (a) $E(p) = \dfrac{p}{20}$; (b) $E(80) = 4$, elastic;
    (c) $p = \$20$

12. (a) $E(p) = \dfrac{2p}{p + 6}$; (b) $E(8) = \dfrac{8}{7}$, elastic;
    (c) $p = \$6$

14. (a) $E(p) = \dfrac{50p^2 - 50p}{63{,}000 + 50p - 25p^2}$;
    (b) $E(10) \approx 0.07377$, inelastic;
    (c) $E(20) \approx 0.35185$, inelastic;
    (d) $E(30) \approx 1.03571$, elastic; (e) $\$29.66$;
    (f) About 42,490 million barrels per day;
    (g) decrease

16. (a) $E(p) = \dfrac{1489p}{(10p + 11)(p + 150)}$;
    (b) $E(3) \approx 0.712$; (c) increases it

18. (a) $E(p) = kp$; (b) yes; (c) yes, at $p = \$\dfrac{1}{k}$

20. tw

CHAPTER 5

Exercise Set 5.1

2. $\dfrac{x^8}{8} + C$ **4.** $4x + C$

6. $\dfrac{3}{4}x^{4/3} + C$

8. $\dfrac{x^3}{3} - \dfrac{x^2}{2} + 2x + C$

10. $t^3 - 2t^2 + 7t + C$

12. $\dfrac{3}{5}e^{5x} + C$

14. $\dfrac{x^5}{5} - \dfrac{5}{11}x^{11/5} + C$

16. $500 \cdot \ln x + C$

18. $-\dfrac{1}{2}x^{-2} + C$, or $-\dfrac{1}{2x^2} + C$

20. $\dfrac{3}{5}x^{5/3} + C$ **22.** $100x^{1/5} + C$

24. $-28e^{-0.25x} + C$

26. $\dfrac{x^5}{5} + \dfrac{1}{4}x^{1/2} - \dfrac{4}{3}x^{3/5} + C$

28. $\dfrac{1}{3}x^3 + 4x^2 + 16x + C$

30. $f(x) = \dfrac{x^2}{2} - 5x + \dfrac{21}{2}$

32. $f(x) = \dfrac{x^3}{3} + x + 8$

34. $27,510.8 billion

36. $C(x) = \dfrac{x^4}{4} - \dfrac{x^2}{2} + 200$

38. (a) $R(x) = \dfrac{x^3}{3} - x$; (b) $\boxed{\text{tw}}$

40. $S(p) = 0.08p^3 + 2p^2 + 10p + 11$

42. (a) $E(t) = 40t - 5t^2 + 12$; (b) 92%, 12%

44. $s(t) = t^2 + 10$

46. $v(t) = 3t^2 + 30$

48. $s(t) = -t^3 + \dfrac{7}{2}t^2 + 10t + 20$

50. 5.1 sec

52. (a) $I(t) = 32.31e^{0.1049t} - 32.31$; (b) 549; (c) 1111; (d) 595

54. $f(t) = \dfrac{2}{3}t^{3/2} + 2\sqrt{t} - \dfrac{28}{3}$

56. $\dfrac{25}{3}t^3 + 20t^2 + 16t + C$

58. $\dfrac{2}{3}t^{3/2} - \dfrac{2}{5}t^{5/2} + C$

60. $\dfrac{x^2}{2} - 6\ln x + \dfrac{7}{2}x^{-2} + C$

62. $\dfrac{\ln x}{\ln 10} + C$, or $\log x + C$

64. $2x^3 - \dfrac{7}{2}x^2 - 5x + C$

66. $\dfrac{x^2}{2} - x + C$ **68.** $\boxed{\text{tw}}$

Exercise Set 5.2

2. 10   **4.** 9   **6.** 4

8. $\dfrac{4}{3}$, or $1\dfrac{1}{3}$ **10.** $e^2 - 1 \approx 6.3891$

12. $2\ln 4 \approx 2.773$ **14.** $\dfrac{128}{3}$, or $42\dfrac{2}{3}$

16. $\dfrac{2}{3}$ **18.** $-\dfrac{2}{3}e^{3b} + \dfrac{2}{3}$, or $-\dfrac{2}{3}\left(e^{3b} - 1\right)$

20.–36. Left to the student.

38. $\dfrac{a^3}{6}$ **40.** -609 **42.** $\dfrac{e^2}{2} - \dfrac{3}{2}$

44. 162 **46.** $\frac{4}{13}$ **48.** $\frac{1}{5}$ **66.** $5\frac{1}{3}$, or $\frac{16}{3}$ **68.** $\frac{44}{3}$ **70.** $\frac{5}{6}$

50. $4\frac{1}{2}$ **52.** $e^3 - e^{-2}$, or $e^3 - \frac{1}{e^2}$ **72.** tw **74.** -13.75

54. $15\frac{3}{4}$ **56.** tw **58.** 155.52 **76.** 4068.79 **78.** 0.43

60. $16\frac{1}{4}$ **62.** $3\frac{1}{3}$ **64.** 30 **80.** 1250 **82.** 9.52 **84.** 10.99

Exercise Set 5.3

2. (a) 35; (b) 40.625; (c) $\frac{140}{3}$, or $46\frac{2}{3}$

4. An antiderivative, distance

6. An antiderivative, total number of divorces in time t

8. An antiderivative, total cost

10. An antiderivative, total sales

12. An antiderivative, number of pages typed in t hours

14. $\frac{8}{3}$ **16.** $-e^{-1} + 1$, or $1 - \frac{1}{e}$

18. $\frac{16}{3}$ **20.** $2a + 5$ **22.** $\frac{2^{n+1} - 1}{n+1}$

24. (a) \$1474.13; (b) \$1456.95; (c) the 9th day

26. \$3600 **28.** \$56,000

30. \$7173.2 billion **32.** 148

34. (a) 114,688 lb; (b) 11.5 months

36. 56°F

38. 7

40. (a) $\frac{a100^{b+1}}{b+1}$; (b) 3348 hr

42. (a) 4.9 million; (b) 3 million; (c) 5.5 million

44. 1.0016 **46.** 8.4 **48.** 4

Exercise Set 5.4

2. $\frac{3}{10}$ **4.** $\frac{9}{2}$ **6.** $\frac{125}{6}$

8. $\frac{1}{3}$ **10.** $\frac{9}{2}$ **12.** $\frac{1}{12}$

14. 2 **16.** $\frac{32}{3}$ **18.** $\frac{3}{2}$

20. $\frac{625}{10}$, or $62\frac{1}{2}$ **22.** 5

24. \$2,201,557 **26.** $\frac{608}{15}$

28. $\frac{2276}{15}$ **30.** 6 **32.** 4, -1, -1

34. 16 **36.** 5.8855 **38.** 0.2375

Exercise Set 5.5

2. $\ln(1+x^3) + C$ **4.** $\frac{1}{3}e^{3x} + C$

6. $3e^{x/3} + C$ **8.** $\frac{1}{5}e^{x^5} + C$

10. $-\frac{1}{2}e^{-t^2} + C$ **12.** $\frac{(\ln 5x)^2}{2} + C$

14. $\ln(12+x) + C$ **16.** $-\ln(1-x) + C$

18. $\frac{1}{12}(t^2-1)^6 + C$

20. $\frac{1}{10}(x^3 - x^2 - x)^{10} + C$

22. $\ln(3+e^t) + C$ **24.** $\frac{(\ln x)^3}{3} + C$

26. $\frac{1}{2}\ln(\ln x^2) + C$

28. $\frac{1}{3a}(ax^2+b)^{3/2} + C$

30. $\frac{P_0}{k}e^{kt} + C$

32. $\frac{1}{24(2-x^4)^6} + C$

34. $\frac{5}{6}(1+6x^2)^{6/5} + C$

36. $e - 1$ **38.** $\frac{6561}{16}$

40. $\frac{e^8-1}{4}$ **42.** $\ln\frac{9}{2}$

44. $-e^{-2b} + 1$, or $1 - \frac{1}{e^{2b}}$

46. $-e^{-kb} + 1$, or $1 - \frac{1}{e^{kb}}$

48. 39   **50.** $-\frac{21}{512}$ **52.** 5

54. $D(p) = 2000\sqrt{25-p^2} + 5000$

56. $P(x) = \frac{1500}{x^2 - 6x + 10}$

58. $\frac{16}{3}$ **60.** $\frac{1}{a}\ln(ax+b) + C$

62. $2e^{\sqrt{t}} + C$ **64.** $\frac{(\ln x)^{100}}{100} + C$

66. $\frac{(e^t+2)^2}{2} + C$, or $\frac{1}{2}e^{2t} + 2e^t + k$, where $k = 2 + C$

68. $\frac{4}{9}(2+t^3)^{3/4} + C$

70. $\frac{(\ln x)^3}{3} + \frac{3}{2}(\ln x)^2 + 4\ln x + C$

72. $\frac{1}{8}[\ln(t^4+8)]^2 + C$

74. $x + \frac{9}{x+3} + C$, where $C = 3 + k$

76. $t - \ln(t-4) + C$, where $C = -4 + k$

78. $-\ln(1+e^{-x}) + C$

80. $\frac{1}{n+1}(\ln x)^{n+1} + C$

82. $-\frac{1}{am}\ln(1+ae^{-mx}) + C$

84. $\frac{5(2x^3-7)^{n+1}}{6(n+1)} + C$

Exercise Set 5.6

2. $xe^{2x} - \dfrac{1}{2}e^{2x} + C$

4. $\dfrac{1}{2}x^4 + C$

6. $\dfrac{2}{5}xe^{5x} - \dfrac{2}{25}e^{5x} + C$

8. $-xe^{-x} - e^{-x} + C$

10. $\dfrac{x^4}{4}\ln x - \dfrac{x^4}{16} + C$

12. $x^3 \ln x - \dfrac{x^3}{3} + C$

14. $(x+1)\ln(x+1) - x + C$

16. $\left(\dfrac{x^2}{2} + x\right)\ln x - \dfrac{x^2}{4} - x + C$

18. $\left(\dfrac{x^2}{2} - 2x\right)\ln x - \dfrac{x^2}{4} + 2x + C$

20. $\dfrac{2}{3}x(x+5)^{3/2} - \dfrac{4}{15}(x+5)^{5/2} + C$

22. $\dfrac{x^3}{3}\ln 5x - \dfrac{x^3}{9} + C$

24. $x(\ln x)^2 - 2x\ln x + 2x + k$, where $k = -2C$

26. $-\dfrac{1}{4}x^{-4}\ln x - \dfrac{1}{16}x^{-4} + C$

28. $e^{4x}\left(\dfrac{1}{4}x^5 - \dfrac{5}{16}x^4 + \dfrac{5}{16}x^3 - \dfrac{15}{64}x^2 + \dfrac{15}{128}x - \dfrac{15}{512}\right) + C$

30. $-e^{-x}(x^3 + 3x^2 + 5x + 6) + C$ **32.** $4\ln 2 - \dfrac{15}{16}$ **34.** $6\cdot\ln 6 - 5$ **36.** $\dfrac{-51e^{-2} + 19}{8}$

38. $P(x) = e^{-0.2x}(-5000x^2 - 50{,}000x - 250{,}000) + 248{,}000$

40. (a) $\dfrac{1}{k^2} - e^{-kT}\left(\dfrac{T}{k} + \dfrac{1}{k^2}\right)$; (b) 14.850 mg

42. $\dfrac{x^{n+1}}{n+1}\ln x - \dfrac{x^{n+1}}{(n+1)^2} + C$ **44.** $\dfrac{1}{3}x^3(\ln x)^2 - \dfrac{2}{9}x^3 \ln x + \dfrac{2}{27}x^3 + C$, where $C = -k$

46. $\dfrac{x^{n+1}(\ln x)^2}{n+1} - \dfrac{2x^{n+1}\ln x}{(n+1)^2} + \dfrac{2x^{n+1}}{(n+1)^3} + C$, where $C = -\dfrac{2k}{n+1}$

48. $\dfrac{2}{7}(27x^3 + 83x - 2)(3x+8)^{7/6} - \dfrac{4}{91}(81x^2 + 83)(3x+8)^{13/6} + \dfrac{1296}{1729}x(3x+8)^{19/6} - \dfrac{2592}{43{,}225}(3x+8)^{25/6} + C$

50. Let $u = (\ln x)^n$ and $dv = dx$. Then $du = n(\ln x)^{n-1} \cdot \dfrac{1}{x}dx$ and $v = x$.
$$\int (\ln x)^n \, dx = x(\ln x)^n - \int x \cdot \dfrac{n}{x}(\ln x)^{n-1}\, dx$$
$$= x(\ln x)^n - n\int (\ln x)^{n-1}\, dx$$

52. tw

Exercise Set 5.7

2. $\dfrac{2}{9}e^{3x}(3x-1) + C$

4. $\ln\left(x + \sqrt{x^2 - 9}\right) + C$

6. $-\dfrac{1}{2}\ln\left(\dfrac{2 + \sqrt{4 + x^2}}{2}\right) + C$

8. $\dfrac{1}{1-x} + \ln(1-x) + C$

10. $\dfrac{1}{2}\left[x\sqrt{x^2+9} + 9\ln\left(x + \sqrt{x^2+9}\right)\right] + C$

12. $x\ln\dfrac{4}{5} + x\ln x - x + C$, or $x\ln\left(\dfrac{4}{5}x\right) - x + C$

14. $-\dfrac{1}{2}x^3 e^{-2x} - \dfrac{3}{4}x^2 e^{-2x} - \dfrac{3}{4}x e^{-2x} - \dfrac{3}{8}e^{-2x} + C$

16. $x^5 \ln x - \dfrac{x^5}{5} + C$

18. $-3\ln\left(\dfrac{1 + \sqrt{1 - x^2}}{x}\right) + C$

20. $\dfrac{1}{5}\ln\left(\dfrac{x}{7x + 2}\right) + C$

22. $\dfrac{3}{2}\left[t\sqrt{t^2 - \dfrac{1}{9}} - \dfrac{1}{9}\ln\left(t + \sqrt{t^2 - \dfrac{1}{9}}\right)\right] + C$

24. $\dfrac{3}{x}(-\ln x - 1) + C$

26. $x(\ln x)^4 - 4x(\ln x)^3 + 12x(\ln x)^2 - 24x\ln x + 24x + C$

28. $\dfrac{3}{2}\ln\left(x + \sqrt{x^2 + 25}\right) + C$

30. $\dfrac{2}{135}(9x - 4)(2 + 3x)^{3/2} + C$

32. $p(t) = \dfrac{1}{2(2+t)} + \dfrac{1}{4}\ln\dfrac{t}{2+t} + 0.8750$

34. $\dfrac{-3}{4(2x - 3)} + \dfrac{1}{4}\ln(2x - 3) + C$

36. $\dfrac{1}{2}\left[e^x\sqrt{e^{2x} + 1} + \ln\left(e^x + \sqrt{e^{2x} + 1}\right)\right] + C$

38. $\dfrac{1}{4}\left[(\ln x)\sqrt{(\ln x)^2 + 49} + 49\ln\left(\ln x + \sqrt{(\ln x)^2 + 49}\right)\right] + C$

CHAPTER 6

Exercise Set 6.1

2. (a) $(1, \$3)$; (b) $\$1.50$; (c) $\$1$

4. (a) $(1, \$4)$; (b) $\$2.33$; (c) $\$1.67$

6. (a) $(4, \$16)$; (b) $\$85.33$; (c) $\$42.67$

8. (a) $(40, \$7600)$; (b) $\$24,000$; (c) $\$12,000$

10. (a) $(3, \$4)$; (b) $\$4.50$; (c) $\$2.67$

12. (a) $(899, \$60)$; (b) $\$50,400$; (c) $\$17,941.33$

14. (a) $(7, \$7)$; (b) $\$1.71$; (c) $\$24.50$

16. [tw]

18. (a) $(2, \$3)$;
(b)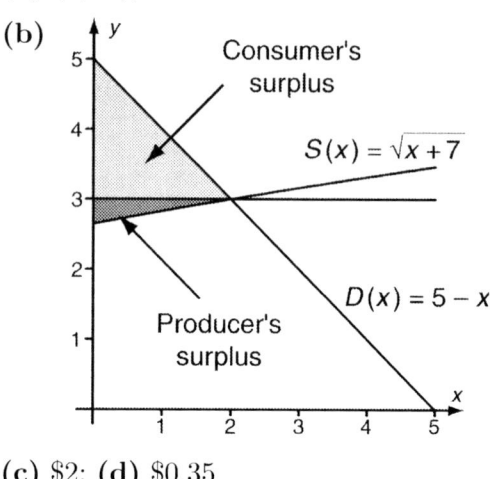
(c) $\$2$; (d) $\$0.35$

Exercise Set 6.2

2. $\$149.18$ 4. $\$6389.06$

6. $\$254,473.83$ 8. $\$1077.06$

10. $\$209.58$ 12. $\$8120.12$

14. $\$15,302.60$ 16. $\$17,802.90$

18. $\$456,177.97$ 20. $\$535,847$

22. $\$732,121$

24. 2239 trillion cubic feet

26. After 38 yrs (2038)   28. 16.031 lb

30. $\$50,872.58$ 32. $\$20,585.02$

34. [tw]

36. The area and the number of tons of aluminum ore are the same. Both are 7302.6 million tons.

Exercise Set 6.3

2. Convergent, $\dfrac{1}{4}$ 4. Divergent

6. Convergent, 1   8. Convergent, $\dfrac{1}{3}$

10. Divergent   12. Convergent, 7

14. Divergent   16. Divergent

18. Divergent   20. Convergent, $\dfrac{Q}{k}$

22. Divergent   24. $\dfrac{1}{2}$

26. Infinite   28. $\dfrac{1}{6}$

30. $\$50,000$ 32. $\$6250$

ANSWERS TO EVEN-NUMBERED EXERCISES

34. 4762 tires   **36.** $25,000

38. $1,300,000   **40.** 43.5 lb

42. (a) $k = 1489.1\%$ per year;
(b) 0.47739 rems; (c) 0.67155 rems

44. Divergent   **46.** Convergent, $\dfrac{1}{2}$

48. Convergent, 0   **50.** $k = 0.1$ mg/hr

52. $\boxed{\text{tw}}$ **54.** 3.14

Exercise Set 6.4

2. $\displaystyle\int_1^3 \dfrac{1}{4}\,dx = \left[\dfrac{x^2}{8}\right]_1^3 = \dfrac{3^2}{8} - \dfrac{1^2}{8} = \dfrac{9}{8} - \dfrac{1}{8} = 1$ **4.** $\displaystyle\int_0^{1/3} 3\,dx = [3x]_0^{1/3} = 3\left(\dfrac{1}{3} - 0\right) = 1$

6. $\displaystyle\int_0^4 \dfrac{3}{64}x^2\,dx = \left[\dfrac{3}{64} \cdot \dfrac{x^3}{3}\right]_0^4 = \dfrac{1}{64}(4^3 - 0^3) = 1$

8. $\displaystyle\int_0^1 \dfrac{1}{e-1}e^x\,dx = \left[\dfrac{1}{e-1}e^x\right]_0^1 = \dfrac{1}{e-1}e^1 - \dfrac{1}{e-1}e^0 = \dfrac{1}{e-1}(e-1) = 1$

10. $\displaystyle\int_{-2}^1 \dfrac{1}{3}x^2\,dx = \left[\dfrac{x^3}{9}\right]_{-2}^1 = \dfrac{1}{9}\left(1^3 - (-2)^3\right) = 1$

12. $\displaystyle\int_0^\infty 4e^{-4x}\,dx = \lim_{b\to\infty}\int_0^b 4e^{-4x}\,dx$

$= \lim_{b\to\infty}\left[\dfrac{4}{-4}e^{-4x}\right]_0^b$

$= \lim_{b\to\infty}\left[-e^{-4b} - \left(-e^{-4\cdot 0}\right)\right]$

$= \lim_{b\to\infty}\left(-\dfrac{1}{e^{4b}} + e^0\right)$

$= 1$

14. $\dfrac{2}{15}$; $f(x) = \dfrac{2}{15}x$

16. $\dfrac{3}{16}$; $f(x) = \dfrac{3}{16}x^2$

18. $\dfrac{1}{6}$; $f(x) = \dfrac{1}{6}$

20. $\dfrac{1}{8}$; $f(x) = \dfrac{1}{8}(4-x)$, or $\dfrac{4-x}{8}$

22. $\dfrac{1}{\ln 2}$; $f(x) = \dfrac{1}{x\ln 2}$

24. $\dfrac{1}{e^2-1}$; $f(x) = \dfrac{e^x}{e^2-1}$

26. (a) $\dfrac{63}{125}$; (b) $\boxed{\text{tw}}$

28. $\dfrac{2}{3}$ **30.** 0.0488   **32.** 0.9817

34. $R(T) = e^{-0.01T}$

36. (a) $k = 1.01 \times 10^{-6}$; (b) 0.0348931

38. 0.6321   **40.** $\dfrac{1}{2}$ **42.** $\boxed{\text{tw}}$

Exercise Set 6.5

2. $\mu = E(x) = 5$, $E(x^2) = \dfrac{79}{3}$,

$\sigma^2 = \dfrac{4}{3}$, $\sigma = \dfrac{2}{\sqrt{3}}$

4. $\mu = E(x) = \dfrac{8}{3}$, $E(x^2) = 8$,

$\sigma^2 = \dfrac{8}{9}$, $\sigma = \dfrac{2}{3}\sqrt{2}$

6. $\mu = E(x) = \dfrac{13}{6}$, $E(x^2) = 5$,

$\sigma^2 = \dfrac{11}{36}$, $\sigma = \dfrac{1}{6}\sqrt{11}$

8. $\mu = E(x) = 0$, $E(x^2) = \dfrac{3}{5}$,

$\sigma^2 = \dfrac{3}{5}$, $\sigma = \sqrt{\dfrac{3}{5}}$

10. $\mu = E(x) = \dfrac{1}{\ln 2}$, $E(x^2) = \dfrac{3}{2\ln 2}$,

$\sigma^2 = \dfrac{3\ln 2 - 2}{2(\ln 2)^2}$, $\sigma = \dfrac{1}{\ln 2}\sqrt{\dfrac{3\ln 2 - 2}{2}}$

12. 0.0160   **14.** 0.4955

16. 0.9756   **18.** 0.1501

20. 0.0384   **22.** 0.1562

24. (a) 0.9544; (b) 95.44%

26. 0.3413   **28.** 0.5762

30.–46. Left to the student.

48. 2.28%   **50.** 0.6797

52. (a) −0.52; (b) 0; (c) 1.645   **54.** tw

56. $E(x) = \mu = \dfrac{3a}{2}$, $E(x^2) = 3a^2$,

$\sigma^2 = \dfrac{3a^2}{4}$, $\sigma = \dfrac{a}{2}\sqrt{3}$

58. 0   **60.** $\mu = 60.164$

62. tw   **64.** 1.772

Exercise Set 6.6

2. 2π **4.** 84π **6.** $\dfrac{\pi}{2}(e^4 - e^{-6})$

8. $\dfrac{3\pi}{4}$ **10.** $\pi \ln 4$ **12.** 50π

14. 9π **16.** 6π **18.** $\dfrac{4}{3}\pi r^3$

20. $\dfrac{\pi(2e - 3)}{e^2}$

22. Using the fnInt feature on a grapher with successively larger values for the upper limit, we find that the integral diverges.

Exercise Set 6.7

2. $y = x^6 + C$; $y = x^6 - 19$, $y = x^6$, $y = x^6 + 0.05$; answers may vary

4. $y = \dfrac{1}{4}e^{4x} - \dfrac{1}{2}x^2 + x + C$; $y = \dfrac{1}{4}e^{4x} - \dfrac{1}{2}x^2 + x + 9$; $y = \dfrac{1}{4}e^{4x} - \dfrac{1}{2}x^2 + x$,

$y = \dfrac{1}{4}e^{4x} - \dfrac{1}{2}x^2 + x + 2.67$; answers may vary

6. $y = 5\ln x + \dfrac{1}{3}x^3 - \dfrac{1}{5}x^5 + C$; $y = 5\ln x + \dfrac{1}{3}x^3 - \dfrac{1}{5}x^5$, $y = 5\ln x + \dfrac{1}{3}x^3 - \dfrac{1}{5}x^5 - 1011$,

$y = 5\ln x + \dfrac{1}{3}x^3 - \dfrac{1}{5}x^5 + 4$; answers may vary

8. $y = x^3 - \dfrac{1}{2}x^2 + 5x + 6$

10. $f(x) = \dfrac{5}{7}x^{7/5} + \dfrac{1}{2}x^2 - \dfrac{115}{14}$

12. $y'' = \dfrac{1}{x}$. Then

$$y'' - \dfrac{1}{x} = 0$$

$$\begin{array}{c|c} \dfrac{1}{x} - \dfrac{1}{x} & 0 \\ 0 & \end{array}$$

14. $y' = xe^x - e^x$, $y'' = xe^x$. Then

$$y'' - 2y' + y = 0$$

$$\begin{array}{c|c} (xe^x) - 2(xe^x - e^x) + (-2e^x + xe^x) & 0 \\ xe^x - 2xe^x + 2e^x - 2e^x + xe^x & \\ 0 & \end{array}$$

16. $y = C_1 e^{x^5}$, where $C_1 = e^C$

18. $y = \sqrt[3]{\dfrac{7}{2}x^2 + C}$

20. $y = \sqrt{\dfrac{x^2}{2} + C}$, $y = -\sqrt{\dfrac{x^2}{2} + C}$

22. $y = \sqrt[3]{3x + C_1}$, where $C_1 = 3C$

24. $y = 2 + 7e^{-x^2/2}$ **26.** $y = \sqrt[3]{21x + 6}$

28. $y = C_1 e^{4x}$, where $C_1 = e^C$

30. $P = C_1 e^{4t}$, where $C_1 = e^C$

32. $R(x) = 300x - x^2$

34. (a) $I = C_1 e^{hkt}$, where $C_1 = e^C$; (b) $I = I_0 e^{hkt}$

36. $V(t) = \$24.81 - \$4.81 e^{-kt}$

38. $x = e^{4/p}$ **40.** $x = \dfrac{C_1}{p^2}$

42. (a) $P = C_1 e^{kt}$, where $C_1 = e^C$; (b) $P = P_0 e^{kt}$

44. ⬚ tw

CHAPTER 7

Exercise Set 7.1

2. $f(-2, 0) = 0$, $f(3, 2) = 10{,}648$, $f(-5, 10) = -125{,}000$

4. $f(3, 7) = 28$, $f(1, 99) = 5$, $f(2, -1) = -12$

6. $f(0, 0) = 0$, $f(1, 1) = -1$, $f(2, 2) = -5$

8. $f(0, 1, -3) = -14$, $f(1, 0, -3) = 1$

10. 18.94 **12.** 0.025 **14.** 1.91485 m^2

16. $Q(21, 20) = 105$, $Q(19, 20) = 95$

18. $\boxed{\text{tw}}$ **20.** $-10°$ **22.** $-64°$

24.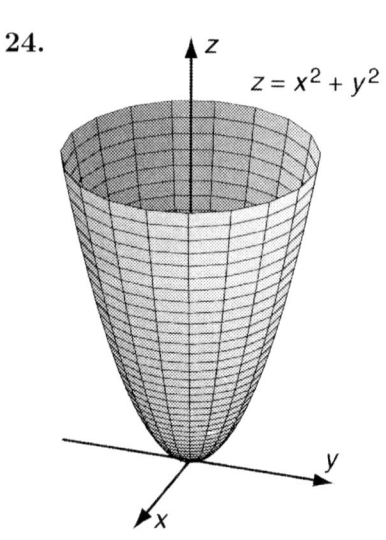

26.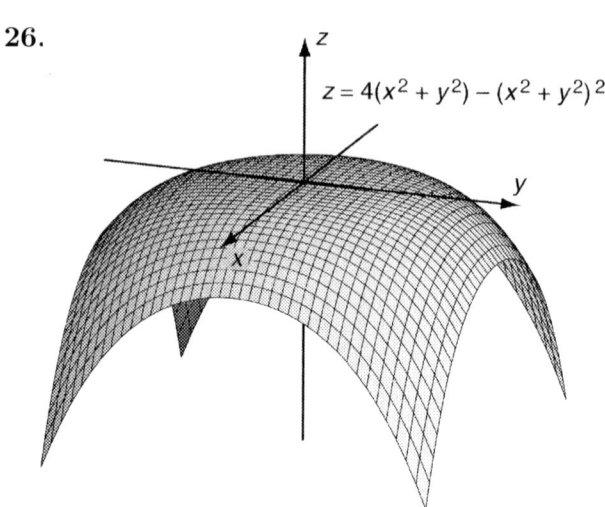

28.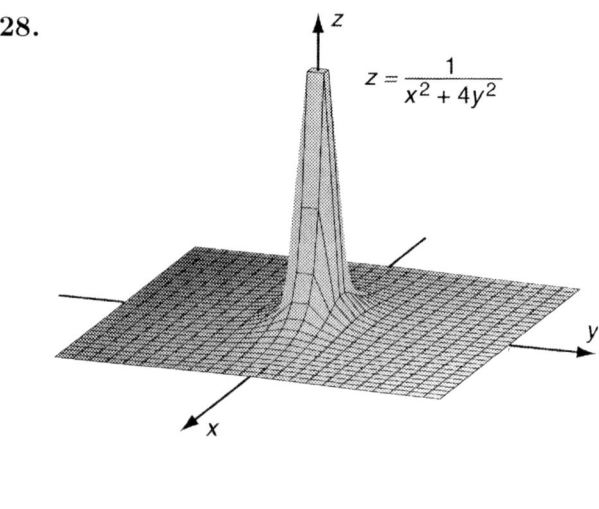

Exercise Set 7.2

2. $\dfrac{\partial z}{\partial x} = 3(x - y)^2$, $\dfrac{\partial z}{\partial y} = -3(x - y)^2$,

$\left.\dfrac{\partial z}{\partial x}\right|_{(-2,-3)} = 3$, $\left.\dfrac{\partial z}{\partial y}\right|_{(0,-5)} = -75$

4. $\dfrac{\partial z}{\partial x} = 6x^2 + 3y - 1$, $\dfrac{\partial z}{\partial y} = 3x$,

$\left.\dfrac{\partial z}{\partial x}\right|_{(-2,-3)} = 14$, $\left.\dfrac{\partial z}{\partial y}\right|_{(0,-5)} = 0$

6. $f_x = 5$, $f_y = 7$, $f_x(-2, 4) = 5$, $f_y(4, -3) = 7$

8. $f_x = \dfrac{x}{\sqrt{x^2 - y^2}}$, $f_y = \dfrac{-y}{\sqrt{x^2 - y^2}}$,

$f_x(-2, 1) = \dfrac{-2}{\sqrt{3}}$, $f_y(-3, -2) = \dfrac{2}{\sqrt{5}}$

10. $f_x = 3e^{3x-2y}$, $f_y = -2e^{3x-2y}$

ANSWERS TO EVEN-NUMBERED EXERCISES

12. $f_x = 2ye^{2xy}$, $f_y = 2xe^{2xy}$

14. $f_x = \dfrac{x}{x-y} + \ln(x-y)$,

$f_y = \dfrac{-x}{x-y}$, or $\dfrac{x}{y-x}$

16. $f_x = \dfrac{y}{x}$, $f_y = 1 + \ln xy$

18. $f_x = \dfrac{1}{y} - \dfrac{y}{x^2}$, $f_y = -\dfrac{x}{y^2} + \dfrac{1}{x}$

20. $f_x = 24(3x + y - 8)$,
$f_y = 8(3x + y - 8)$

22. $\dfrac{\partial f}{\partial b} = 12m + 6b - 46$,

$\dfrac{\partial f}{\partial m} = 28m + 12b - 98$

24. $f_x = 4y - 3\lambda$, $f_y = 4x + \lambda$,
$f_\lambda = -(3x - y + 7)$

26. $f_x = 2x - 4\lambda$, $f_y = -2y + 7\lambda$,
$f_\lambda = -(4x - 7y - 10)$

28. **(a)** 614,400 units;

(b) $\dfrac{\partial p}{\partial x} = 960\left(\dfrac{y}{x}\right)^{3/5}$, $\dfrac{\partial p}{\partial y} = 1440\left(\dfrac{x}{y}\right)^{2/5}$

(c) $\boxed{\text{tw}}$; **(d)** 7680, 360

30. **(a)** $\dfrac{\partial p}{\partial w} = -0.005075w^{-1.638}r^{1.038}s^{0.873}t^{2.468}$,

$\dfrac{\partial p}{\partial r} = 0.008257w^{-0.638}r^{0.038}s^{0.873}t^{2.468}$,

$\dfrac{\partial p}{\partial s} = 0.006945w^{-0.638}r^{1.038}s^{-0.127}t^{2.468}$,

$\dfrac{\partial p}{\partial t} = 0.019633w^{-0.638}r^{1.038}s^{0.873}t^{1.468}$;

(b) $\boxed{\text{tw}}$; **(c)** $0.25\dfrac{2p}{\partial w} = -\$10,161$.

The profit will decrease by about $10,200

32. 117.8° **34.** 97.5° **36.** $\boxed{\text{tw}}$

38. **(a)** $\dfrac{\partial s}{\partial h} = 0.00961865h^{-0.6036}w^{0.5378}$;

(b) $\dfrac{\partial s}{\partial w} = 0.0130497h^{0.3964}w^{-0.4622}$;

(c) -0.0251 m^2

40. 48.465 **42.** -1.015

44. $f_x = \dfrac{-x^4 + 3x^2t + 2xt}{(x^3 + t)^2}$, $f_t = \dfrac{-x^3 - x^2}{(x^3 + t)^2}$

46. $f_x = \dfrac{3}{4}\sqrt[4]{\dfrac{t^5}{x}}$, $f_t = \dfrac{5}{4}\sqrt[4]{x^3 t}$

48. $f_x = -\dfrac{20xt^2(x^2 + t^2)^4}{(x^2 - t^2)^6}$,

$f_t = \dfrac{20x^2t(x^2 + t^2)^4}{(x^2 - t^2)^6}$

50. $\boxed{\text{tw}}$

Exercise Set 7.3

2. $\dfrac{\partial^2 f}{\partial x^2} = 2$, $\dfrac{\partial^2 f}{\partial y \partial x} = \dfrac{\partial^2 f}{\partial x \partial y} = -10$, $\dfrac{\partial^2 f}{\partial y^2} = 6$

4. $f_{xx} = 0$, $f_{xy} = f_{yx} = 4$, $f_{yy} = 0$

6. $f_{xx} = 12x^2 y^3 - 2y^3$, $f_{xy} = f_{yx} = 12x^3 y^2 - 6xy^2$, $f_{yy} = 6x^4 - 6x^2 y$

8. $f_{xx} = 0$, $f_{yx} = f_{xy} = 0$, $f_{yy} = 0$

10. $f_{xx} = y^2 e^{xy}$, $f_{yx} = f_{xy} = xy e^{xy} + e^{xy}$, $f_{yy} = x^2 e^{xy}$

12. $f_{xx} = -e^x$, $f_{yx} = f_{xy} = 0$, $f_{yy} = 0$

14. $f_{xx} = 0$, $f_{yx} = f_{xy} = \dfrac{1}{y}$, $f_{yy} = -\dfrac{x}{y^2}$

16. $f_{xx} = \dfrac{2y^2}{(x-y)^3}$, $f_{yx} = f_{xy} = \dfrac{-2xy}{(x-y)^3}$, $f_{yy} = \dfrac{2x^2}{(x-y)^3}$

18. $f_{xy} = -10y$, $f_y = -10xy$, so
$xf_{xy} - f_y = x(-10y) - (-10xy)$
$= -10xy + 10xy$
$= 0$

20. $f_{xx} = 20 \left[\ln(x^3 + e^y)\right]^3 \dfrac{9x^4}{(x^3 + e^y)^2} + 5 \left[\ln(x^3 + e^y)\right]^4 \dfrac{6xe^y - 3x^4}{(x^3 + e^y)^2}$,

$f_{xy} = f_{yx} = 5 \left[\ln(x^3 + e^y)\right]^4 \dfrac{-3x^2 e^y}{(x^3 + e^y)^2} + 20 \left[\ln(x^3 + e^y)\right]^3 \dfrac{3x^2 e^y}{(x^3 + e^y)^2}$,

$f_{yy} = 20 \left[\ln(x^3 + e^y)\right]^3 \dfrac{e^{2y}}{(x^3 + e^y)^2} + 5 \left[\ln(x^3 + e^y)\right]^4 \dfrac{x^3 e^y}{(x^3 + e^y)^2}$

Exercise Set 7.4

2. Relative minimum $= -\dfrac{25}{3}$ at $\left(-\dfrac{5}{3}, \dfrac{10}{3}\right)$

4. Relative maximum $= \dfrac{256}{27}$ at $\left(\dfrac{8}{3}, \dfrac{16}{3}\right)$

6. Relative minimum $= -8$ at $(2, 2)$

8. Relative minimum $= -7$ at $(-3, 3)$

10. Relative maximum $= 13$ at $(3, 2)$

12. None

14. Relative minimum $= e^{-3}$ at $(1, 2)$

16. 2 thousand of the $18 gloves and 3 thousand of the $25 gloves

18. 418 million dollars when $a = 5$ million dollars and $n = 1$ thousand

20. (a) $R = 78p_1 - 6p_1^2 - 6p_1 p_2 + 66p_2 - 6p_2^2$; (b) $p_1 = \$50$, $p_2 = \$30$; (c) 39 hundred units of q_1, 33 hundred units of q_2; (d) $294,000

22. $-18°$ at $(4, -1)$; there is no maximum.

24. Relative minimum $= 6$ at $(1, 2)$

26. Relative minimum $= \dfrac{1}{6}$ at $\left(\dfrac{211}{3}, \dfrac{3}{2}\right)$

28. tw

30. Relative minimum $= 1$ at $(-1, -1)$

32. Relative minimum $= -6$ at $(-0.5, 4)$

ANSWERS TO EVEN-NUMBERED EXERCISES

Exercise Set 7.5

2. (a) $y = 1.74x + 132.4$;
(b) $158.5 million, $167.2 million

4. (a) $y = 0.174286x - 275.014$;
(b) 75.3 yr, 76.2 yr

6. (a) $y = 0.2272315x - 357.555$;
(b) 98.180 in., 108.270 in.; (c) tw

8. tw

Exercise Set 7.6

2. Maximum $= 32$ at $(2, 8)$

4. Maximum $= -34$ at $(1, 6)$

6. Minimum $= 17$ at $(1, 4)$

8. Minimum $= 28$ at $(3, 5)$

10. Minimum $= \dfrac{1}{3}$ at $\left(\dfrac{1}{3}, \dfrac{1}{3}, \dfrac{1}{3}\right)$

12. 25 and 25

14. 2 and -2

16. Minimum $= 29$ at $(4, 8, 4)$

18. 20 ft by 20 ft; 400 ft^2

20. $r = 2.5$ in.; $h = 5.0$ in.; 117.8 in^2

22. Maximum $= 1200$ at $L = 20$, $M = 40$

24. 2.13 ft by 2.13 ft by 2.66 ft

26. Minimum $= -2$ at $\left(\sqrt{2}, -\sqrt{2}\right)$ and $\left(-\sqrt{2}, \sqrt{2}\right)$

28. Maximum $= \sqrt{3}$ at $\left(\dfrac{1}{\sqrt{3}}, \dfrac{1}{\sqrt{3}}, \dfrac{1}{\sqrt{3}}\right)$

30. Maximum $= 6$ at $\left(\dfrac{2}{3}, \dfrac{4}{3}, -\dfrac{4}{3}\right)$

32. Minimum $= \dfrac{1}{30}$ at $\left(\dfrac{1}{30}, -\dfrac{1}{15}, \dfrac{1}{6}\right)$

34. $750,000 for labor, $250,000 for capital

36. tw

38.–44. Left to the student.

Exercise Set 7.7

2. 1   **4.** 3   **6.** 1

8. $\dfrac{1}{2}e^4 - e^2 + \dfrac{1}{2}$

10. 1   **12.** $e^9 - e$

14. $\dfrac{1}{3}$ **16.** $\dfrac{5}{48}$

18. 0   **20.** $\dfrac{26}{3}$

22. Left to the student.

EXTENDED TECHNOLOGY ANSWERS

CHAPTER 1

1. (a) $y = 0.1051383399x + 58.57956522$, where x is the number of years after 1975;

   (b)
   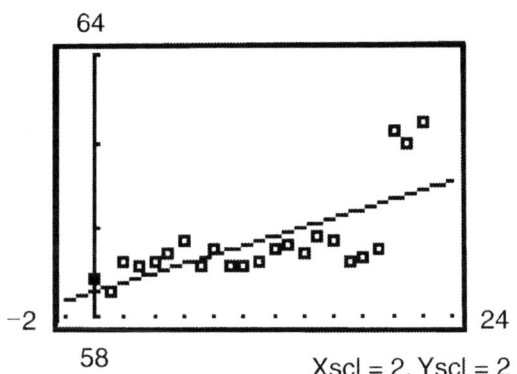

   (c) $61.10°$, $61.21°$, $66.46°$
   (d) 2084, 2179

2. (a) $y = 0.0017997365x^3 - 0.048595144x^2 + 0.3787479767x + 58.5793913$, where x is the number of years after 1975;

   (b)
   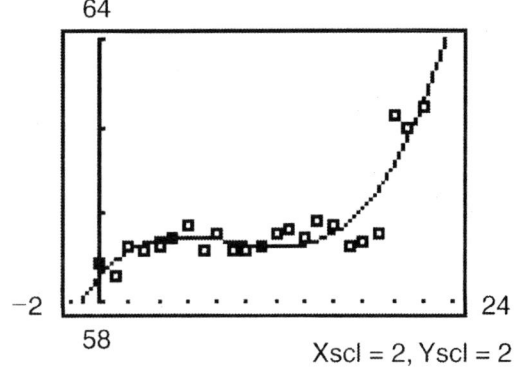

   (c) $64.56°$, $65.80°$, $572.90°$
   (d) 2003, 2007

3. (a) $y = 0.000099987732x^4 - 0.0025997237x^3 + 0.0127544713x^2 + 0.0937258058x + 58.83016054$, where x is the number of years after 1975;

   (b)
   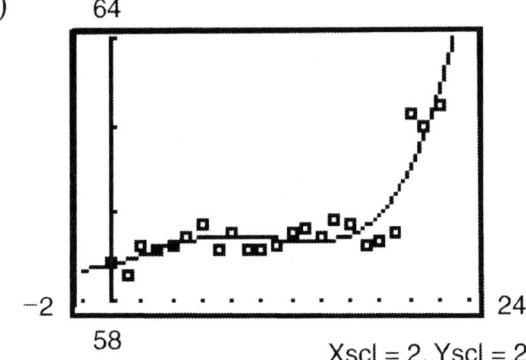

   (c) $65.66°$, $67.58°$, $2204.52°$; (d) 2001, 2004

4. The discussion of the merits of the functions is left to the student.

5. Answers will vary.

CHAPTER 2

1. Yes

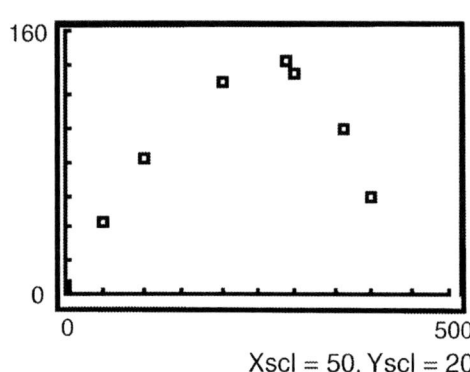

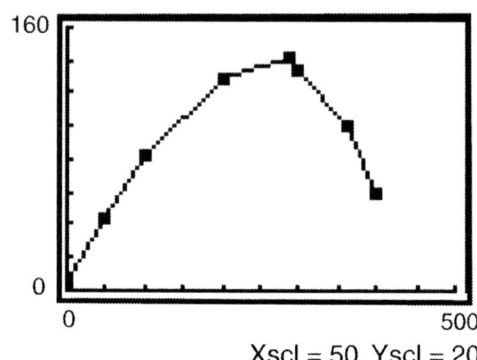

2. (a) $y = -0.0000045x^3 + 0.000204x^2 + 0.7806x + 4.605;$

(b)

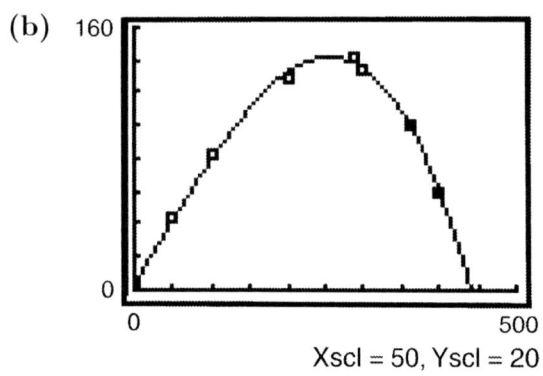

(c) Acceptable; **(d)** About 441 ft;

(e) $\dfrac{dy}{dx} = -0.0000135x^2 + 0.000408x + 0.7806;$

(f) Max $\approx$ 142 ft at a horizontal distance of about 256 ft.

3. (a) $y = -0.0000000024x^4 - 0.0000026x^3 - 0.00026x^2 + 0.82x + 4.3;$

(b)

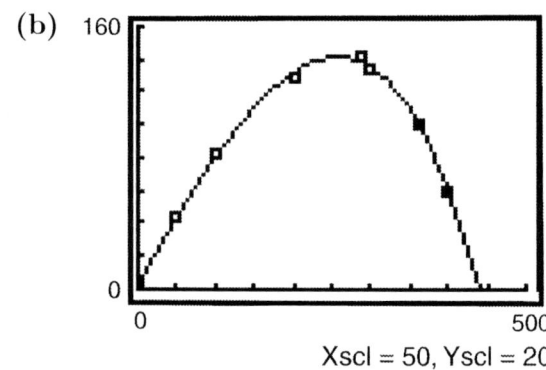

(c) Acceptable; **(d)** About 440 ft;

(e) $\dfrac{dy}{dx} = -0.0000000096x^3 - 0.0000078x^2 - 0.00053x + 0.815;$

(f) Max $\approx$ 142 ft at a horizontal distance of about 257 ft.

EXTENDED TECHNOLOGY ANSWERS

4. (a)

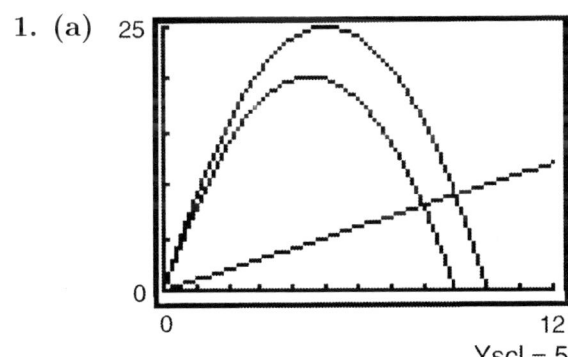

(b) 450 ft; (c) $\dfrac{dy}{dx} = \dfrac{303.75 - 0.003x^2}{\sqrt{202{,}500 - x^2}}$;

(d) Max ≈ 152 ft at a horizontal distance of about 318 ft.

5. The discussion is somewhat open-ended. The models are very similar. The main difference seems to be that the maximum height is reached further from home plate in the model in Exercise 4.

6. 466 ft, 442 ft, 430 ft

7. The prediction is way off. The student is not expected to know this, but the wind was a factor that day. Nevertheless, the ball would at best have gone about 511 ft.

CHAPTER 3

1. (a)

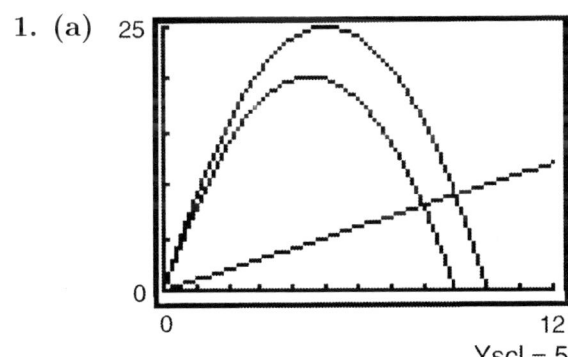

(b) 4500; (c) 20,250

2. (a)

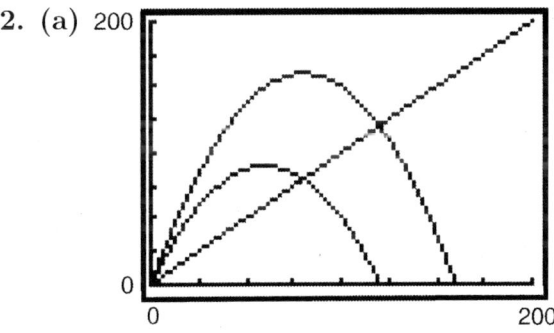

(b) 60,000; (c) 90,000

3. (a)

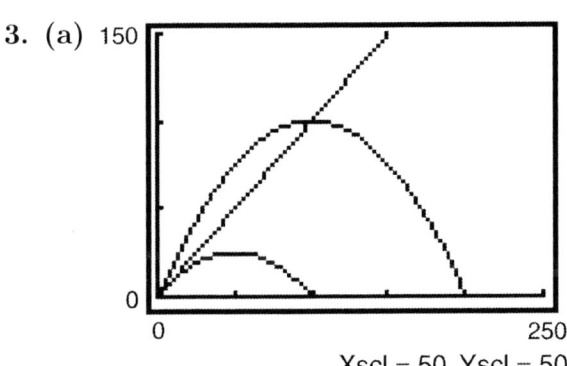

(b) 50,000; (c) 25,000

4. (a)

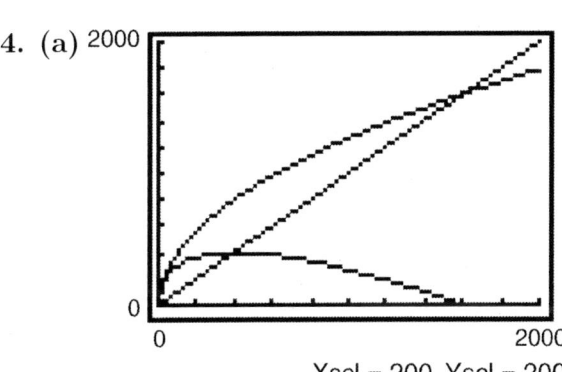

(b) 400,000; (c) 400,000

5. (a)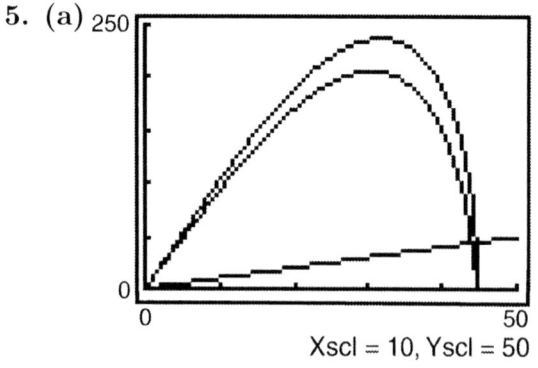
(b) 30,513; (c) 205,923

6. (a) $y = -0.0011P^3 + 0.0715P^2 - 0.0338P + 4$;
(b)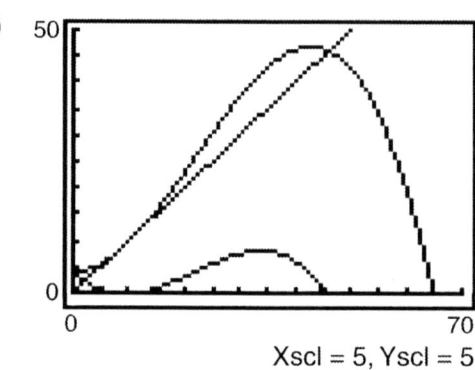
(c) 33,841

CHAPTER 4

1.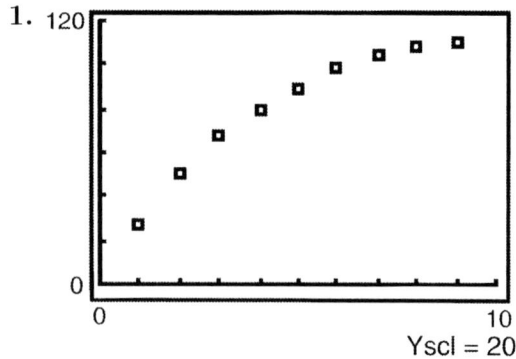

A logistic equation seems to fit the data, because there appears to be a limiting value for R.

2. $R(t) = \dfrac{110.6770351}{1 + 4.762662697e^{-0.6354791317t}}$;

3. $R'(t) = \dfrac{334.972099e^{-0.6354791317t}}{\left(1 + 4.762662697e^{-0.6354791317t}\right)^2}$;

$R'(t)$ represents the rate of change of the total revenue.
$\lim\limits_{t \to \infty} R'(t) = 0$; this means that eventually the total revenue does not change.

4.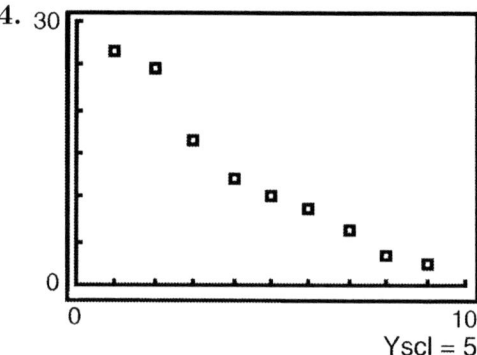

Answers will vary.

5. Answer will depend on the answer to Exercise 4.

6.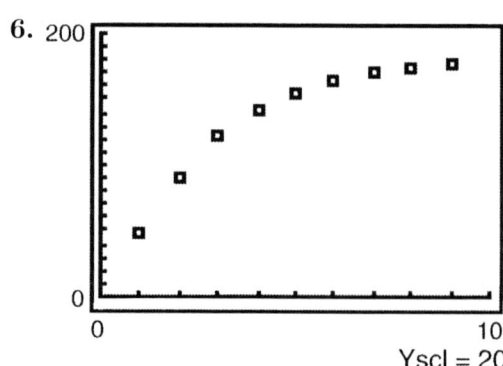

A logistic equation seems to fit the data, because there appears to be a limiting value for R.

EXTENDED TECHNOLOGY ANSWERS

7. $R(t) = \dfrac{173.2679399}{1 + 5.082147351 e^{-0.8045386337 t}}$

8. $R'(t) = \dfrac{708.4551606 e^{-0.8045386337 t}}{(1 + 5.082147351 e^{-0.8045386337 t})^2}$;

   $R'(t)$ represents the rate of change of the total revenue.

   $\lim\limits_{t \to \infty} R(t) = 0$; this means that eventually the total revenue does not change.

9.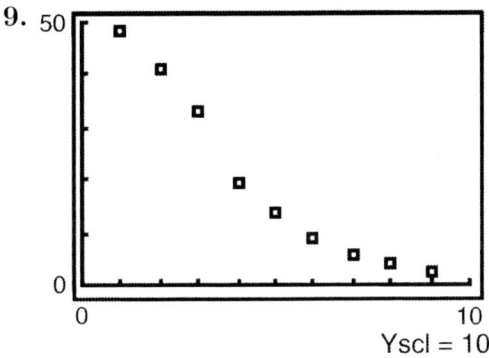

   Answers will vary.

10. Answer will depend on the answer to Exercise 9.

11.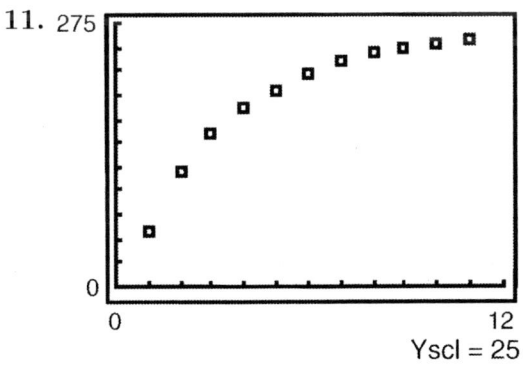

    A logistic equation seems to fit the data, because there appears to be a limiting value for R.

12. $R(t) = \dfrac{251.726825}{1 + 4.627020527 e^{-0.6358366492 t}}$

13. $R'(t) = \dfrac{740.5876766 e^{-0.6358366492 t}}{(1 + 4.627020527 e^{-0.6358366492 t})^2}$;

    $R'(t)$ represents the rate of change of the total revenue.

    $\lim\limits_{t \to \infty} R(t) = 0$; this means that eventually the total revenue does not change.

14.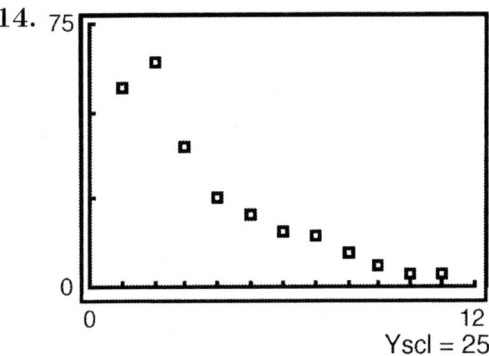

    Answers will vary.

15. Answers will depend on the answer to Exercise 14.

16.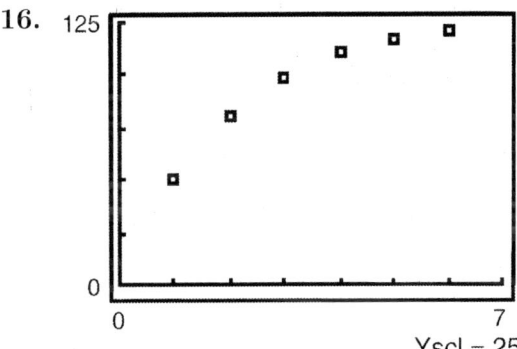

    A logistic equation seems to fit the data, because there appears to be a limiting value for R.

17. $R(t) = \dfrac{120.8336781}{1 + 3.397600606 e^{-0.9163360234 t}}$

18. $R'(t) = \dfrac{376.1967861 e^{-0.9163360234 t}}{(1 + 3.397600606 e^{-0.9163360234 t})^2}$;

    $R'(t)$ represents the rate of change of the total revenue.

    $\lim\limits_{t \to \infty} R(t) = 0$; this means that eventually the total revenue does not change.

19.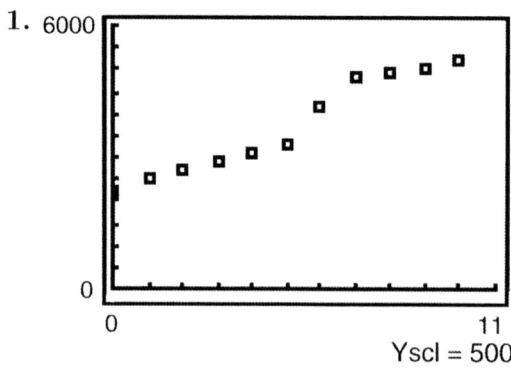

Answers will vary.

20. Answers will depend on the answer to Exercise 19.

21. Yes; eventually a movie will no longer be shown and its total revenue will approach a limiting value.

22. Answers will vary.

CHAPTER 5

1.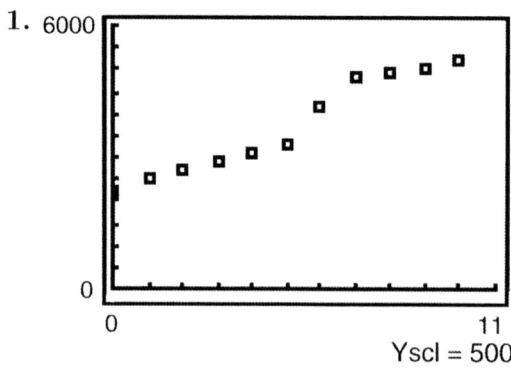

Answers will vary.

2.–4. Answers will depend on the answer to Exercise 1.

5.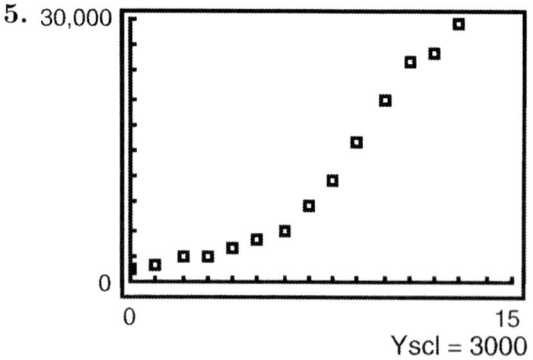

Answers will vary.

6.–8. Answers will depend on the answer to Exercise 5.

9.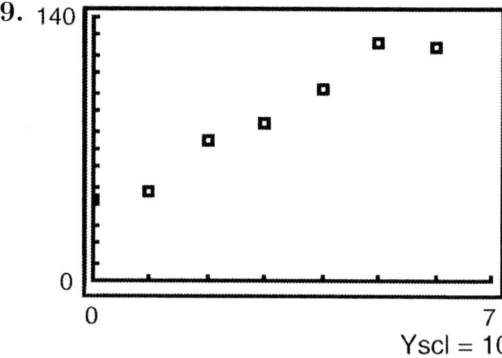

Answers will vary.

10.–12. Answers will depend on the answer to Exercise 9.

13.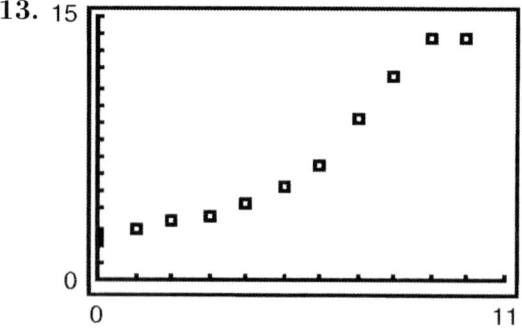

Answers will vary.

14.–16. Answers will depend on the answer to Exercise 13.

EXTENDED TECHNOLOGY ANSWERS

CHAPTER 6

1. $y = 0.1x^3 - 11x^2 + 326x - 820$

2. 344,343,322

3. $y = 0.0010900071x^4 - 0.0221927289x^3 + 0.1194408838x^2 - 0.1158504244x + 1.046614361$

4. 35.1705 in³, 19.493 oz

5. It seems good since our estimate was 19.493 oz.

6. A cubic: $y = 0.0014972917x^3 - 0.0519369633x^2 + 0.3476322163x + 0.6880524071$. We get 35.4635 in³ or 19.655 oz, which is a better estimate.

CHAPTER 7

1.

Case	Building	n	k	A	h	$t(h,k)$
1	B1	2	40	3200	30	23
	B2	3	32.66	3200	45	20.83
2	B1	2	60	7200	30	33
	B2	3	48.99	7200.06	45	28.995
3	B1	4	40	6400	60	26
	B2	5	35.777	6399.969	75	25.389
4	B1	5	60	18000	75	37.5
	B2	10	42.426	17999.65	150	36.213
5	B1	5	150	112500	75	82.5
	B2	10	106.066	112499.96	150	68.033
6	B1	10	40	16000	150	35
	B2	17	30.679	16000.42	255	40.84
7	B1	10	80	64000	150	55
	B2	17	61.357	63999.58	255	56.179
8	B1	17	40	27200	255	45.5
	B2	26	32.344	27199.49	390	55.172
9	B1	17	50	42500	255	50.5
	B2	26	40.43	42499.21	390	59.215
10	B1	26	77	154154	390	77.5
	B2	50	55.525	154151.3	750	102.76

2. Yes

3. Answer is open-ended. Expanded table is left for student exploration. The use of the analytic method of LaGrange Multipliers.

4. $t(h,k) = \dfrac{h}{10} + \dfrac{k}{2}$

5. $\dfrac{h}{8}k^2 = 40{,}000$ or $hk^2 = 320{,}000$

6. The dimensions of the building are about 50.4 ft by 50.4 ft by 134 ft.

7. $h = \sqrt[3]{\dfrac{Aca^2}{b^2}}$, $k = \sqrt[3]{\dfrac{Abc}{a}}$; dimensions are k by k by $h + c$.